Biomanufacturing

Chander Prakash · Sunpreet Singh ·
Rupinder Singh · Seeram Ramakrishna ·
B. S. Pabla · Sanjeev Puri · M. S. Uddin
Editors

Biomanufacturing

Springer

Editors
Chander Prakash
School of Mechanical Engineering
Lovely Professional University
Jalandhar, Punjab, India

Sunpreet Singh
School of Mechanical Engineering
Lovely Professional University
Jalandhar, Punjab, India

Rupinder Singh
Department of Production Engineering
Guru Nanak Dev Engineering College
Ludhiana, Punjab, India

Seeram Ramakrishna
Centre for Nanofibers and Nanotechnology
National University of Singapore
Singapore, Singapore

B. S. Pabla
Department of Mechanical Engineering
National Institute of Technical Teachers
Training and Research
Chandigarh, India

Sanjeev Puri
Center for Stem Cell and Tissue Engineering
Panjab University
Chandigarh, India

M. S. Uddin
School of Engineering
University of South Australia
Mawson Lakes, SA, Australia

ISBN 978-3-030-13950-6 ISBN 978-3-030-13951-3 (eBook)
https://doi.org/10.1007/978-3-030-13951-3

Library of Congress Control Number: 2019932676

This Springer imprint is published by the registered company Springer Nature Switzerland AG
The registered company address is: Gewerbestrasse 11, 6330 Cham, Switzerland

Preface

The book entitled *Biomanufacturing*, by Springer, aims to present a comprehensive and broad-spectrum picture of the state-of-the-art research, development, and commercial prospective of various discoveries conducted in the real-world biomedical applications. This book is prompted by universal applications of several biomaterials, their processing, and evaluations through the conduct of standard procedures. It has been seen by us that although myriads of book publications are readily available, yet only a very few of them have compiled the different aspects of biomedical science and engineering, all together. We believe that the major reason behind such a scenario is the limited collaborations among the specialized research communities that otherwise can play a vital role to refurbish the current situation of the biomedical industry. Therefore, the primary focus of ours was to invite the experts of materials science, design, manufacturing, testing, optimization, and of course medical practitioners to come up with their recent practices over a unique platform in order to provide the readers a database which can help them to understand the challenges involved and to channelize their energies to resolve the same. Moreover, this edited book features a different variety of chapters which include exhaustive literature review, case studies, technical papers, and experimental investigations. The few book chapters include the review of different types of biomaterials and biomanufacturing processes used in the present scenario. After that, few chapters illustrate the implications of 3D printing technologies by highlighting their variety of applications in the subject matter. Next few chapters involve great efforts to develop hybrid technologies, surface integrity, and effective techniques and tools to improve the surface bioperformance of the medical devices. Finally, last chapter focuses on the review and case studies of bone remodeling and biomechanics by providing the full-proof conceptualization of the theories and practices. The anticipated audience of this book will mainly consist of researchers, research students, and practitioners in biomedical science and engineering. This edited book is also of interest to the commercial businesses, service providers, decision support systems, and research and development laboratories. We would like to convey our appreciation to all contributors including the accepted chapters' authors and many other participants who submitted their chapters that cannot be included in this book due to space limits. Moreover, the real success of this book is dedicated to the eminent reviewers who provided their valuable time to conduct the critical reviews on assigned chapters. With this, we really hope our best that this contribution will benefit all the readers in different ways.

Jalandhar, India — Chander Prakash
Jalandhar, India — Sunpreet Singh
Ludhiana, India — Rupinder Singh
Singapore — Seeram Ramakrishna
Chandigarh, India — B. S. Pabla
Chandigarh, India — Sanjeev Puri
Mawson Lakes, Australia — M. S. Uddin

Contents

Editors and Contributors

About the Editors

Dr. Chander Prakash is Associate Professor in the School of Mechanical Engineering, Lovely Professional University, Jalandhar, India. He has received his Ph.D. in mechanical engineering from Panjab University, Chandigarh, India. His areas of research are biomaterials, rapid prototyping and 3-D printing, advanced manufacturing, modeling, simulation, and optimization. He has more than 11 years of teaching experience and 6 years of research experience. He authored 50 research papers and 10 chapters. He is also Editor of three books and three journals.

Dr. Sunpreet Singh is Assistant Professor in the School of Mechanical Engineering, Lovely Professional University, Jalandhar, India. He has received his Ph.D. in mechanical engineering from Guru Nanak Dev Engineering College, Ludhiana, India. His areas of research are additive manufacturing and application of 3D printing for the development of new biomaterials for clinical applications. He authored 50 research papers and 10 chapters.

Prof. Rupinder Singh is Professor in the Department of Production Engineering, Guru Nanak Dev Engineering College, Ludhiana, India. His areas of research are non-traditional machining, additive manufacturing, and development of porous biomaterials using 3D printing and rapid prototyping techniques. He has more than 18 years of teaching and research experience. He authored 17 books and ~3 chapters.

Prof. Seeram Ramakrishna is Co-Director, NUS Nanoscience and Nanotechnology Initiative (NUSNNI). He has received his Ph.D. from University of Cambridge. He is Global Leader in electrospinning and nanostructured materials. His research resulted in ~1000 peer-reviewed articles with ~70,000 citations and ~120 H-index. He authored 5 books and ~25 chapters. He has been recognized as Highly Cited Researcher in materials science.

Prof. B. S. Pabla is Dean in Extension Services and Consultancy and Professor in mechanical engineering, National Institute of Technical Teachers Training and Research, Chandigarh, India. His areas of research are non-traditional machining, CAD/CAM, biomaterials, and biomanufacturing. He has more than 37 years of teaching and research experience. He has granted two patents, published five books, and completed more than ten research projects from various funding agencies.

Prof. Sanjeev Puri is Head of Center for Stem Cell and Tissue Engineering and Department of Biotechnology, Panjab University, Chandigarh, India. His areas of research are biomaterials, cell culture, and tissue engineering. He has more than 22 years of teaching and research experience. He authored 50 research articles, 6 books, and ~3 chapters. He is also Editorial Member of various journals. He has received many research grants from various funding agencies such as DBT-SERB. He has granted three patents and completed more than five research projects.

Dr. M. S. Uddin is Program Director and Lecturer in the School of Engineering, University of South Australia (UniSA). He has more than 15 years of teaching and research experience. His area of research is surface engineering of biomaterials. He authored more than 50 research articles.

Contributors

Ahmad Majdi Abdul-Rani Mechanical Engineering Department, Universiti Teknologi PETRONAS, Bandar Seri Iskandar, Perak, Malaysia

IPS Ahuja Department of Mechanical Engineering, Punjabi University, Patiala, India

Abdul' Azeez Abdu Aliyu Mechanical Engineering Department, Universiti Teknologi PETRONAS, Bandar Seri Iskandar, Perak, Malaysia

Yunus Ziya Arslan Department of Mechanical Engineering, Faculty of Engineering, Istanbul University-Cerrahpasa, Istanbul, Turkey

Atul Babbar Mechanical Engineering Department, Thapar Institute of Engineering and Technology, Patiala, India

Jagat Bhushan Dr. Harvansh Singh Judge Institute of Dental Sciences, Panjab University, Chandigarh, India

Erol Cansiz Department of Oral and Maxillofacial Surgery, Faculty of Dentistry, Istanbul University, Istanbul, Turkey

Nikit Deoray Department of Plastic and Polymer Engineering, Maharashtra Institute of Technology, Aurangabad, Maharashtra, India

Swaroop Gharde Rapid Prototyping Laboratory, Department of Metallurgical and Materials Engineering, Defence Institute of Advanced Technology (DU), Ministry of Defence, Girinagar, Pune, India

Rajendra Goud Rapid Prototyping Laboratory, Department of Metallurgical and Materials Engineering, Defence Institute of Advanced Technology (DU), Ministry of Defence, Girinagar, Pune, India

Vishakha Grover Dr. Harvansh Singh Judge Institute of Dental Sciences, Panjab University, Chandigarh, India

Dheeraj Gupta Mechanical Engineering Department, Thapar Institute of Engineering and Technology, Patiala, India

Munish Kumar Gupta Mechanical Engineering, National Institute of Technology, Hamirpur, India;
Mechanical Engineering, Ludhiana College of Engineering and Technology, Ludhiana, India

Vivek Jain Mechanical Engineering Department, Thapar Institute of Engineering and Technology, Patiala, India

Balasubramanian Kandasubramanian Rapid Prototyping Laboratory, Department of Metallurgical and Materials Engineering, Defence Institute of Advanced Technology (DU), Ministry of Defence, Girinagar, Pune, India

Jay M. Korde Department of Plastic and Polymer Engineering, Maharashtra Institute of Technology, Aurangabad, Maharashtra, India

Kalipada Maity Mechanical Engineering Department, National Institute of Technology, Rourkela, Odisha, India

Anup Malik Department of Mechanical Engineering, Malaviya National Institute of Technology, Jaipur, Rajasthan, India

Guravtar Singh Mann School of Mechanical Engineering, Lovely Professional University, Phagwara, Punjab, India

Alakesh Manna Department of Mechanical Engineering, Punjab Engineering College, Chandigarh, India

Sunil Nimje Department of Mechanical Engineering, Defence Institute of Advanced Technology (DU), Ministry of Defence, Pune, India

Faruk Ortes Department of Mechanical Engineering, Faculty of Engineering, Istanbul University-Cerrahpasa, Istanbul, Turkey

Swastik Pradhan School of Mechanical Engineering, Lovely Profesional University, Phagwara, Punjab, India

Chander Prakash School of Mechanical Engineering, Lovely Professional University, Phagwara, Punjab, India

Jitendra Prasad Department of Mechanical Engineering, Indian Institute of Technology Ropar, Rupnagar, Punjab, India

Seeram Ramakrishna Mechanical Engineering, National University of Singapore, Singapore, Singapore

Nishant Ranjan Department of Production Engineering, Guru Nanak Dev Engineering College, Ludhiana, India;
Department of Mechanical Engineering, Punjabi University, Patiala, India

Muhammad Al'Hapis Abdul Razak Mechanical Engineering Department, Universiti Teknologi PETRONAS, Bandar Seri Iskandar, Perak, Malaysia;
Manufacturing Section, Universiti Kuala Lumpur Malaysian Spanish Institute, Kulim, Kedah, Malaysia

Shubham Saini Dr. B. R. Ambedkar National Institute of Technology, Jalandhar, India

Harjit Singh School of Mechanical Engineering, Lovely Professional University, Phagwara, Punjab, India

Manjeet Singh School of Mechanical Engineering, Lovely Professional University, Phagwara, Punjab, India

Rupinder Singh Department of Production Engineering, Guru Nanak Dev Engineering College, Ludhiana, India

Sunpreet Singh School of Mechanical Engineering, Lovely Professional University, Phagwara, Punjab, India

Aarsha Surendren Centre for Biopolymer Science and Technology, Central Institute of Plastics Engineering & Technology, Udyogamandal, Eloor, Kochi, Kerala, India

Abhishek Kumar Tiwari Department of Applied Mechanics, Motilal Nehru National Institute of Technology Allahabad, Allahabad, Uttar Pradesh, India

Abdul Wasy Zia City University of Hong Kong, Kowloon, Hong Kong SAR

Current Trends in Biomaterials and Bio-manufacturing

Harjit Singh, Sunpreet Singh and Chander Prakash

Abstract The current research work presents the critical review of current trends on the synthesis/development of biomaterials and surface modification/processing/treatment of biomaterials for biomedical application. In the first phase, the significance of biomaterial is presented. The technique for the development of porous and solid biomedical implants was discussed in detail for their successful applications. Powder metallurgical, additive manufacturing, and 3-D printing technologies were reported good potential techniques for the development of porous mechanically tuned of metallic and ceramic-based implants for medical applications. In the second phase, an innovative engineering technique for surface modification, processing, and treatment of implants was discussed to enhance the bioactivity, mechanical properties, and corrosion and wear resistance properties. Electric discharge machining, electrochemical deposition, and plasma spar deposition were reported the best and potential innovative engineering technique to improve the mechanical properties and bioactivity. The chapter also presents the future scope for the development and surface modification of biomedical implants.

Keywords Biomaterials · Implants · Porous · Powder metallurgy · Additive manufacturing · Temporary space alloying · 3-D printing · Electrospinning · Coating · Nanofinishing

1 Introduction and Requirement of Biomaterials

From recent studies, it is confirmed that the demand for joint and bone replacement increases rapidly. It is estimated that maximum human beings having the age of more than 45 years are fighting with the problem of weakening the bone, inflammation of the human's bone joints and weakening of the cartilages resulted in the bone degradation under continuous cyclic loading conditions [1, 2]. This is the main cause of

H. Singh · S. Singh · C. Prakash (✉)
School of Mechanical Engineering, Lovely Professional University,
Phagwara 144411, Punjab, India
e-mail: chander.mechengg@gmail.com

C. Prakash et al. (eds.), *Biomanufacturing*,
https://doi.org/10.1007/978-3-030-13951-3_1

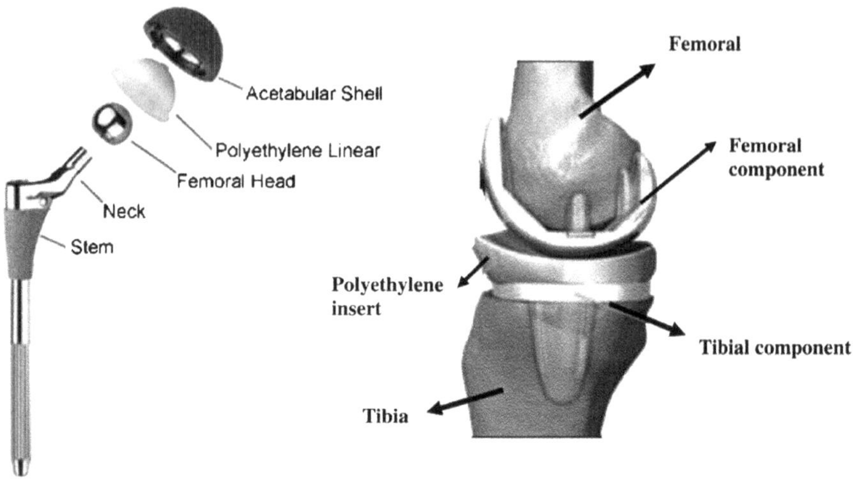

Fig. 1 Artificial implants used in the human body to restore the functionality of dysfunctional joints or parts [2]

the degradation of the mechanical and physical behavior of natural human's bone and bone joint loss in friction. Total joint replacement (TJR) is the best way to solve these problems, where load-bearing joints are replaced with artificial implants [3, 4]. Artificial implants comprise whole or part of a living structure which replaces a natural function to improve the quality of the human body made up of biomaterials. Figure 1 presents the possible implants used in the human body such as intraocular lenses, bone plates, dentistry joint replacements to more complex material like pacemakers, blood vessels, artificial heart, histopathology, prosthetic arterial grafts, etc.

The stability of implants depends upon the three factors: (i) properties of biomaterials such as mechanical, chemical, and morphology; (ii) biocompatibility; and nontoxic of biomaterials [5, 6]. According to the responses of implanted biomaterials, the major dominant metallic biomaterials are cobalt-based (Co) alloys, stainless steel (SS), magnesium-based alloys, and titanium (Ti). These biomaterials are superior metallic materials due to of their ability to carry significant stress and go through the plastic deformation prior unit to the fail, as this show their specific UTS and fracture toughness (Table 1).

It is essential to cite that SS, Co alloys, and Ti alloys have a much larger value of stiffness than to cortical bone [1]. They ought to be decent mechanical properties as well as chemical also. Generally, host reaction occurs after the implantation of biomaterial such as blood–material interaction, acute inflammation, chronic inflammation, and foreign body reaction. There are various types of reaction to a broad spectrum of implant biomaterials nearby implant site. Although these alloys are pleasant mechanical and chemical properties, they have a lifespan of 10–15 years only.

Table 1 Characteristic of biomaterials and bone

Materials	Young's modulus (GPa)	Ultimate tensile strength (MPa)	Fracture toughness (MPa $\sqrt{m}$)
Co	240	900–1540	~100
SS	200	540–1000	~100
Ti	105–125	900	~80
Cortical bone	10–30	130–150	2–12

In order to deal with such problems, researchers have taken so my initiatives to make implantation successful. These innovations are the development of new fabrication route for the synthesis of porous implants and surface engineering (coating/deposition and finishing of implants) [7, 8].

2 Fabrication Technique to Synthesize Porous Implants

In spite of the incredible advancement that has been accomplished in biomaterials, the fixation of the implant with the bone is still an issue because of the elastic modulus mismatch [1]. The elastic modulus of the metallic implant is much greater than bones which lead to creating stress shielding and results in immolation failed [9]. One thought to accomplish this has been the advancement of materials that exhibit porous structure [10]. The structural porosity reduces the stiffness of Ti alloy and therefore results in the low elastic modulus [11, 12]. Moreover, the porous structure not only solves the problem of stress shielding but also provides suitable conditions to promote tissue growth around the implant and improved the implant stability [13].

The use of porous-coated and fully or partially porous orthopedic implants' usages is increasing day by day due to better implant–bone ingrowth performance. Percentage of porosity, pore size, and the implant material has significant effects on the porous implant performance [14]. In order to achieve bone ingrowth (osseointegration) into porous implant surface and inner pores, the implant must have adequate porosity, which ranges between 35 and 70%. Cook et al. showed the bone ingrowth of 35–40% with porous implants implanted to mongrel dogs [15]. Cameron et al. investigated the rate of bone growth in 65% dense cobalt-based implants [16]. Hofmann showed more bone ingrowth around Ti-6Al-4V-coated implants as compared to Ti-coated implants [17]. Research has also been performed to determine optimum parameters for bone formation in porous implant surfaces. Experimental results based on the animal implantation using porous materials have shown that the pore sizes affected bone ingrowth. In an investigation performed by Bobyn et al., bone growth was observed into porous surface down to 50 μm pore size and porous cobalt-based porous implant with pore size ranging between 50 μm and 400 reached the maximum fixation strength after 8 weeks of implantation [18]. Hulbert et al. reported that the minimum pore size is between 75 and 100 μm to regenerate a significant level of mineralized natural bone [19]. Barth et al. reported an optimum pore size of 400 μm for bone ingrowth in porous fiber titanium implant. Melikyan and Itin implanted sintered porous titanium specimens of 100–300 μm pore size in rabbits and reported successful growth of bone tissue into the implant [20]. The optimum pore size for the growth of osteoblasts and vascularization was reported in the range of 300–400 μm by Tsuruga et al. This result was obtained by implanting 106–212, 212–300, 300–400, 400–500, and 500–600 μm pore size hydroxyapatite implants in rats [21]. Similar results for optimum size of pores, i.e., 200–500 μm in a porous-surface-coated Ti-6Al-4V specimens, implanted into the femoral modularly canal of dogs for six months, were reported by Clemow et al. [22]. Studies also showed that higher bone and implant interlocking strength occurred when porous tantalum implants with an average pore size of 430 μm are used [23]. The in vivo studies were performed in rabbit to evaluate bone ingrowth into titanium scaffolds with variable pore sizes, i.e., 60, 100, 200, and 600 μm [24]. However, when the size of the pore is increased to 1.0 mm, the formation of fibrous tissue was more than the growth of bone [25]. There are a number of techniques which have been used to synthesize porous structure in bulk material. The potential techniques are powder metallurgical process using temporary space holder techniques and foaming agent, and additive manufacturing/3-D printing [26].

2.1 Powder Metallurgy Route Using Temporary Space Alloying and Foaming Agent

The powder metallurgy (PM) route has been most widely used for the development of porous scaffolds, because in this process pore characteristics are easy to control and capability to generate very close to the final shape of the product [27, 28]. The fabrication of porous structure in PM route consists first alloying of removable temporary space holders with powder particles then pressure-less sintering [29]. The most commonly used removable temporary space holders (RTSH) are sugar pallet, sodium chloride (NaCl), carbamide, magnesium, tapioca, saccharose, dextrin, and woven wire mesh, which removed by evaporation or by dissolution followed by the sintering process [7, 10, 30]. Figure 2 presents the schematic representation of fabrication method to developed porous structure through RTSH technique.

High porosity (90%) NiTi alloys were synthesized by powder metallurgical route using NaCl as a RTSH. The finding presented that the shape and size of structural porosity depend upon the NaCl particles [31]. Zhang et al. used carbonate (NH_4HCO_3) as a temporary space holder to fabricate Ti-HA composite by using spark plasma sintering technique. The developed composite not only possessed near to human bone's elastic modulus (~ 8.2–15.8 GPa) but also high compressive strength (~ 86–388 MPa) [32]. Exhaustive research efforts have been made for the fabrication of Ti-based porous scaffold. But, the drawback of Ti-based porous scaffold is non-degradable nature, which was removed from the body after another surgical pro-

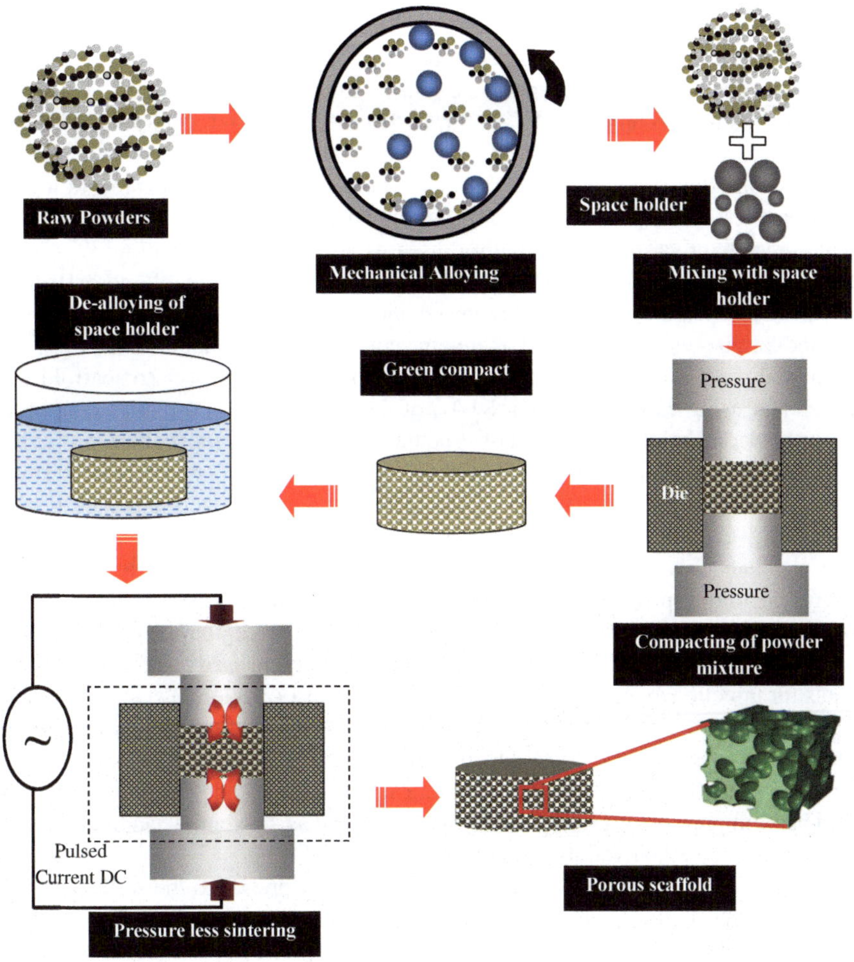

Fig. 2 Fabrication route of the porous scaffold by powder metallurgy process

cess [33, 34]. In concern of the metallic biomaterials issues, magnesium (Mg)-based alloys/composites have gained attentions as an alternative material for the developing scaffolds, because of their superior biodegradability, biocompatibility, and low elastic modulus [35]. Yazdimamaghani et al. critically reviewed and presented the studies carried out in the development of porous magnesium-based scaffolds for tissue engineering (TE) applications using temporary space holder technique and other methodologies [36]. It has been reported that temporary space holder technique was a potential methodology for the development of 3-D open porous structure [10]. The major techniques to control the pore shape and size of Mg scaffolds were 3-D-printed negative salt pattern, woven mesh wire, and removable space holders. Kirkland et al. (2011) used 3-D-printed negative salt pattern to fabricate porous Mg-based scaffold with controlled architecture. The findings indicate that the technique can successfully produce an ordered open-cell porous Mg scaffold [37]. Further, a highly porous (72–76%) biodegradable AZ91D alloy was developed using negative NaCl salt pattern technique with a controlled pore size of 10–1000 μm. The in vivo results indicate that the Mg scaffold was degraded after three months of implantation and has consistent biocompatibility with a required inflammatory host reaction [38].

Moreover, the major drawback of Mg is high degradation rate in the human body, which results in various inflammatory diseases after post-implantation [39, 40]. In this context, Hedayati et al. (2008) fabricated AZ-61 biodegradable porous alloys using NaCl as temporary space alloying technique and studied the degradation behavior and fatigue resistance. The element zinc controls the degradation rate of Mg and improved the corrosion properties [41]. Furthermore, porous Mg–Zn scaffold was developed using carbamide ($CO(NH_2)_2$) of the size of 200–400 μm as temporary space holder technique. The effect of structural porosity on the mechanical properties was studied and found that mechanical properties degraded with the increase in structural porosity. Bakhsheshi-Rad et al. (2018) fabricated doxycycline-loaded porous hybrid Mg–Ca–TiO_2 scaffold using spherical carbamide as a temporary space holder. The elements Ca and TiO_2 were used to enhance the corrosion properties and bioactivity. The developed scaffold possesses a interconnected porosity in the range of 65–67% of size 600–800 μm with upright compressive strength (5 ± 0.1 MPa). The antibacterial property was improved by the doxycycline and used as alternative applicants for infection avoidance for bone TE applications [42]. Ghomi et al. (2018) fabricated bioactive porous bredigite ($Ca_7MgSi_4O_{16}$) scaffold via space holder method using NaCl as temporary space holders. Highly interconnected porous scaffold with a porosity of 86% was developed with pore size 400–600 μm. The developed scaffold has compressive strength 1.1 MPa and proposed as a potential candidate for the bone repair and regeneration in bone TE application [43].

Singh and Batanagr (2017) critically reviewed and presented the studies carried out on the development of porous titanium and magnesium-based scaffolds for TE applications using various fabrication techniques [44]. The powder metallurgy (PM) technique has been best quoted due to being a economical method and more convenient to control the pore size, pore density, and morphology of the scaffolds. Further, they fabricated 3-D open Ti-woven wire mesh-based cellular structure of Mg10Zn alloy by using as a space holding. The developed scaffold has 60% porosity, and ultimate tensile strength was measured around 76 MPa [45]. The developed scaffold was proposed for the bone scaffolding material for TE. Dutta et al. (2017) developed porous biodegradable Mg-based scaffolds using spherical naphthalene (porogen) particle as temporary space holder. The developed scaffolds have a porosity ranging from 8–40% with pore size 30–60 μm. The compressive strength of the developed scaffold decreased with increase in the porogen content from 184 to 24 MPa [46].

2.2 *Additive Manufacturing and 3-D Printing*

These days, additive manufacturing technique in the form of 3-Dimension printing has been widely used to develop 3-D porous biomedical implants with controlled pore characteristics such as size, shape, and distribution [47]. In 3-D printing technique, metal-implant parts were fabricated by a layer-by-layer printing process in which parts are designed using CAD. In metal 3-D printing process, metallic parts were fabricated by layer-by-layer printing of metallic powders and together fusion/sintering of powder carried out by laser technique. Here, two types of most widely laser sintering technique were used such as Selective Laser Melting (SLM) and Direct Metal Laser Sintering (DMLS) [48]. Figure 3 shows the schematic line diagram of working principle of SLM/DMLS process.

In the early ages, the SLM technology was used to developed fully-dense components [49]. The utilization of SLM to develop porous component was reported by Wang et al. [50]. Highly porous patient-specific Ti-based implant was developed with 70% porous structure exhibiting pore size from 200–500 μm. Pattanayak et al. investigated the effect of SLM process parameters on the pore characteristics and optimized the process parameters to synthesize porous Ti-implant [51]. Further, the attainability of manufacturing grid structures in Ti-6Al-4V with high geometrical exactness utilizing SLM was explored by Weißmann et al. [52]. Moreover, it has been reported that SLM has potential to fabricate customize complicated structure. Recently, Shipley et al. critically reviewed the application of SLM to develop Ti-6Al-4V-based porous implant [53]. They presented and addressed the various challenges during optimization of SLM process parameters and their influence on the porous

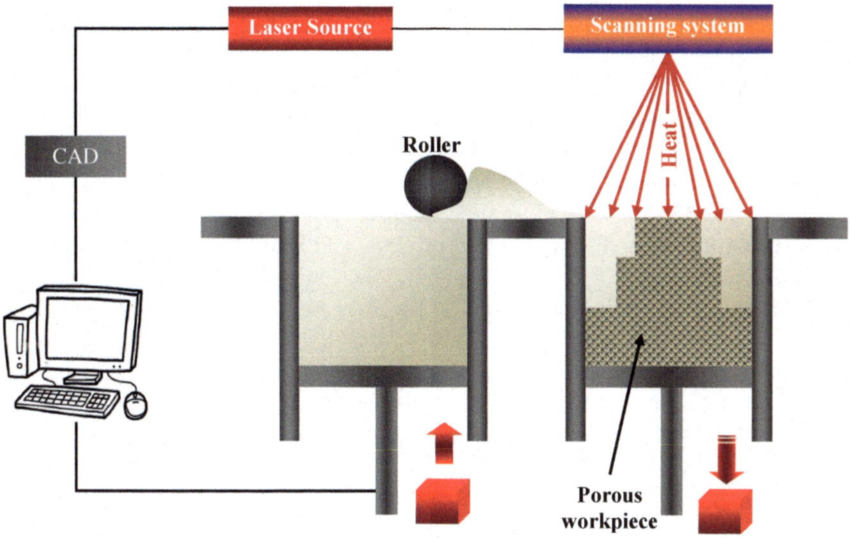

Fig. 3 Schematic representation of the SLM/DMLS process

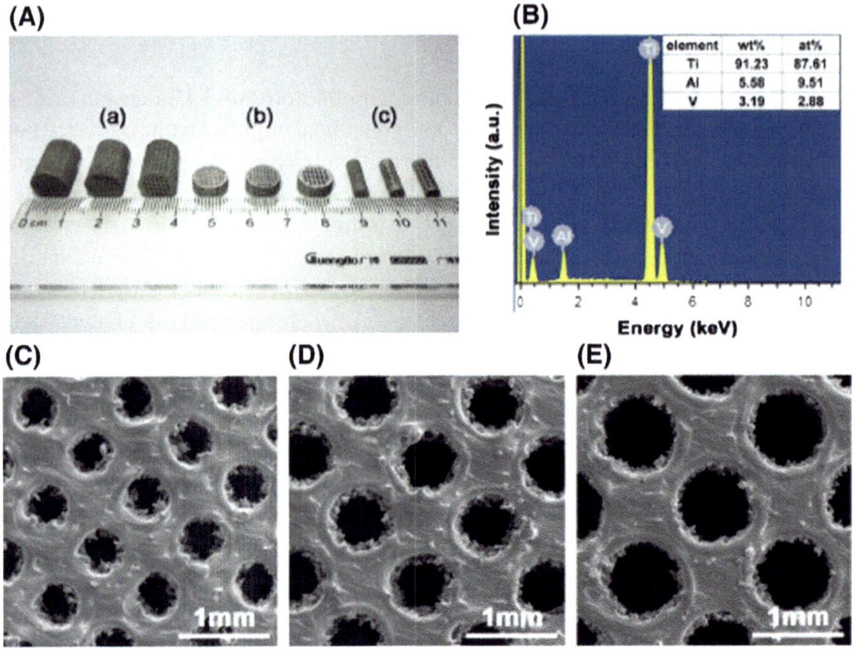

Fig. 4 **A** Macrographs of 3-D-printed porous Ti-6Al-4V specimens for mechanical test (a), *in vitro* (b), and *in vivo* (c) studies; **B** chemical compositions of the fabricated samples detected by EDS; and SEM images of **C** 401 μm, **D** 607 μm, and **E** 801 μm

structure. Khorasani et al. studied the effect of SLM process parameters (laser power, scan speed, hatch space, laser pattern angle coupling, along with heat treatment as a post-process) on the mechanical properties (density, hardness, tensile strength), and pore characteristics (Surface quality) have been studied [54]. Zao et al. fabricated the Ti-6Al-4V alloy by electron beam melting (EBM) and SLM technique and compared their in vitro corrosion and bioactivity performance. The SLM-synthesized implant possessed excellent bioactivity and EBM-synthesized exhibits good corrosion resistance [55]. Ran et al. studied the osteogenesis of 3-D-printed Ti-6Al-4V implant at different pore size by SLM process, and it has been found that implant with 600 μm pore size promotes excellent bone ingrowths [56]. Figure 4 presents the 3-D-printed Ti-6Al-4V implant, and Fig. 5 presents the bioactivity analysis of synthesized implants.

Yan et al. fabricated gyroid triply-periodic minimal surface (TPMS) lattice-based Ti-6Al-4V alloy using SLM. Figure 6 presented the CAD model of pore morphology. The effect of heat treatment on mechanical properties and effect of HA coating of bioactivity of implant were studied. The compressive strength of Ti-6Al-4V implant was decreased with heat treatment. On the other hand, the HA coating enhances the bioactivity of implant. Figure 7 presented the porous Ti-implant coated with HA layer in simulated body fluid [57].

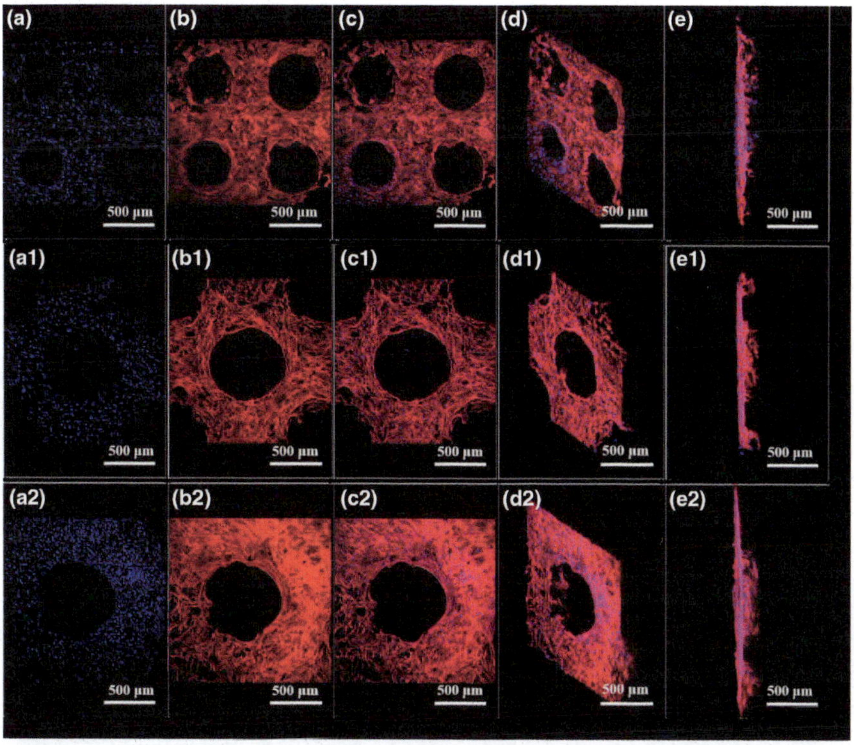

Fig. 5 Representative fluorescence images of osteoblasts adhered to porous Ti-6Al-4V implants after culture for 14 days. Cells were stained with actin filament (red), cell nuclei (blue). Three regions of each sample were randomly captured, and all experiments were performed in triplicate

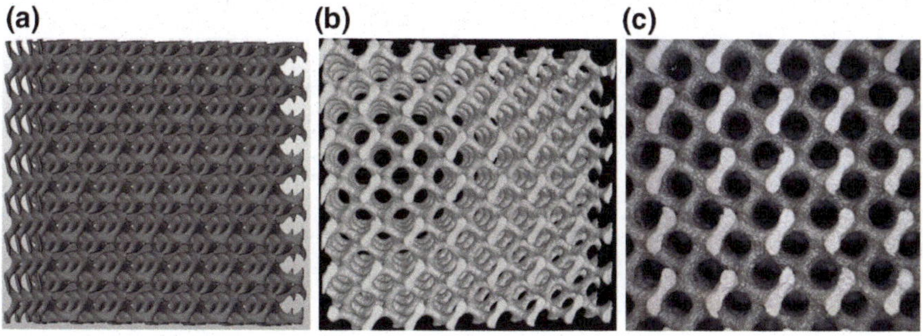

Fig. 6 **a** CAD model, **b** micro-CT reconstruction model, and **c** optical

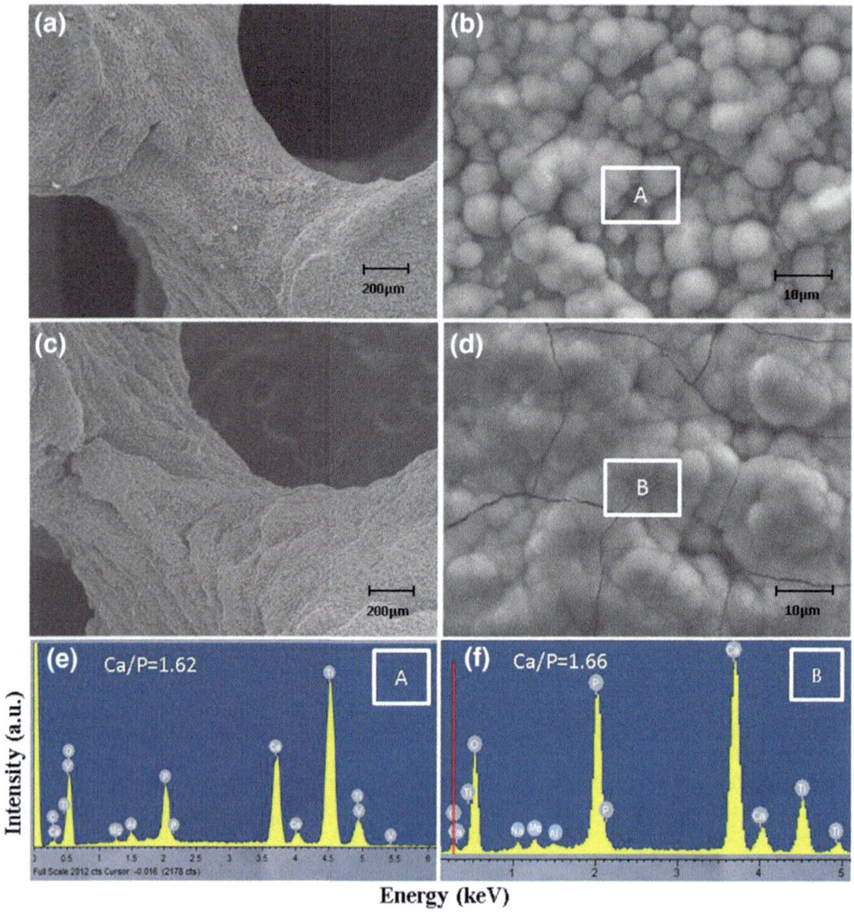

Fig. 7 Surface morphologies of the Ti-6Al-4V TPMS lattices after soaked in SBF for: (**a**, **b**) 2 weeks and (**c**, **d**) 3 weeks. EDX analysis on the Ti-6Al-4V TPMS lattices soaked in SBF for (**e**) 7 days and (**f**) 14 days, corresponding to area A and B in (**b**, **d**), respectively

2.3 *Electrospinning*

From the last decade, the electrospinning process gained great attention because of being a highly versatile method for spinning a wide range of polymeric fibers of nanosize [58, 59]. Electrospun nanofibers have been widely used biomedical scaffolds for TE application. In the electrospinning process, an electrostatic force is used to the spine polymer solution or melts to produce fine-fibers from nanometer to a micrometer to enhance surface area [60, 61]. Figure 8 presents the standard vertical and horizontal electrospinning processes. A high-voltage power supply, a spinner, and a collecting plate are the three main components of electrospinning process. A polymeric solution is transferred to a syringe and pumped through capillaries or tubes to the spinneret, which forms a suspended droplet at the tip. A DC voltage source of several tens of KVs is used to generate an electric field between syringe tip and a collector [62, 63]. Due to an electric field, electrostatic charge controls surface tension and a charged jet is produced and accumulates on the collector plate. The polymer fiber produced is called electrospun or nanofibers. A large number of natural and synthetic polymers including polylactic acid, polyurethanes, silk fibroin, collagen, hyaluronic, polycaprolactone, acid, cellulose, and chitosan collagen were used for the development of scaffolds for TE application [64, 65]. Ruckh et al. reported that polycaprolactone (PCL) was the potential and ultimate polymer for the scaffold TE application owing to its superior biocompatibility and processability [66]. The developed nanofibrous scaffold improved the osseointegration activities, and excellent cell adhesion and growth were observed.

Doustgani et al. fabricated a fibrous nanocomposite-scaffold of PCL along with PVA (average diameter ~ 123 ± 32 nm) and HA (average diameter ~123 ± 32 nm) nanoparticles for aligned and random nanofibers, respectively. The bioactivity of the fibrous nanocomposite was estimated by in vitro cell culture test using mesenchymal stem cells (MSCs) cell lines, and it was reported that the developed nanocomposite supported cell adhesion, growth, proliferation, and differentiation [67]. A critical downside of electrospinning process is the low-efficiency level (low production) due to a single needle. An innovative advancement has been accomplished to enhance the productivity of current electrospinning process by multiplication of the jets using multi-nozzle construction [68]. The production was increased but causes several problems low reliability, quality, and high maintenance cost. Petrík et al. developed an innovative nozzle-free electrospinning technology, which overcome the associated problem with conventional electrospinning and multi-nozzle electrospinning process [69]. In this process, a polymer wrapped over a rotating drum and nanofiber mats were formed when subjected to the high-voltage electric field. Esmaeilzadeh et al. synthesized cellulose-based nanofibers for TE applications using the nozzle-free electrospinning process [70]. Despite several advantages, solution electrospinning processes have few drawbacks. The fabricated scaffold exhibited too small pore characteristics and tightly-packed nonwoven networks which obstruct cell infiltration. Recently developed melt electrospinning was an innovative technique, which fabricates highly porous (with porosities of 98%) scaffold by direct writing technique [71]. In this procedure, the ideal polymer is dissolved and afterward pushed to a spinneret, at which high voltage is connected between spinneret and collector, bringing about the electrostatic shock and stacking filaments over one another [72]. Because of the viscosity of the liquid polymer, this procedure can create micro- and sub-micron scale fibers [73]. Additionally, it need not bother with the utilization of dangerous solvents and is able to do creating structures with adjustable pore sizes,

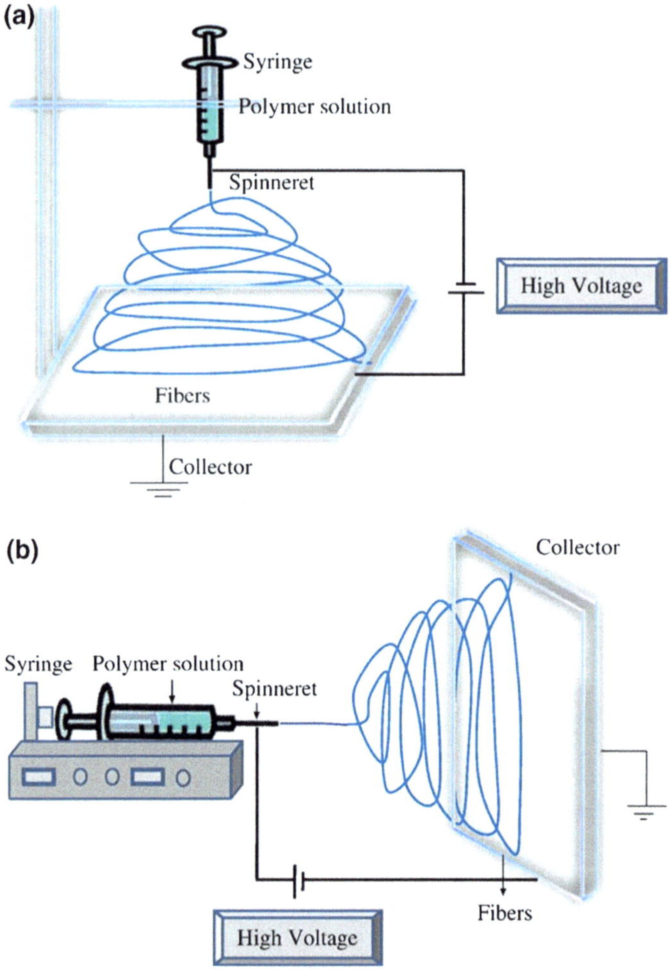

Fig. 8 Schematic diagram of setting up of electrospinning apparatus: **a** typical vertical setup and **b** horizontal setup of electrospinning apparatus

which are appropriate for cell entrance. Melt electrospinning has been used to process a variety of biodegradable polymers such as PCL [74, 75], PLA [76], PLGA [77].

3 Innovative Surface Engineering/Surface Modification Techniques

Surface engineering is processing/alteration of the surface characteristics of the object obtained through removing or adding the material. The surface engineering processes are majorly classified into two categories: surface modification (treatment, cladding, and machining) and surface finishing (polishing and grinding). To enhance the service life of biomaterials, by the special treatment or technique used on the surface to increase the productivity of biomaterial in the harsh environment of the human body. By this life expectancy has been increasing because of the improvements in the biomaterial surface by the surface alteration. Table 2 presents the various surface modification techniques, and everyone have their own advantages and drawbacks.

Among all the coating methods, thermal spray coating is potentially good technique to deposit a hard, corrosion, and wear resistance layer on implant surface [4].

3.1 Electrodeposition Technique

Electrodeposition is an electrochemical process to deposit a protective micro-, submicron, and nanoscale layer on a substrate of conducting materials [78]. The electrodeposition process is based on Faraday's law of electrolysis that expresses the delivery amount of the chemical charge by an electrical current is corresponding to the amount of electricity that passes [79]. Figure 9 presents the schematic illustration of electrodeposition setup which consists electrochemical cell for electrolyte handling, a working electrode, a counter-electrode, a reference electrode, and an external electrical circuit. The working electrode is connected to cathode terminal where electrodeposition takes place, and counter-electrode is connected to the anode to complete the electrical circuit. Various metal salts are dissolved to make an aqueous solution of electrolytes containing positive and negative ions. The electric current that flows between the two conductive electrodes in the presence of an external voltage is because of the motion of charged species, via migration and diffusion, towards the surfaces of the polarized electrodes. At the surface of the anodes, the conduction system must change from ionic to electronic, an interface procedure intervened by the event of electrochemical responses that advance the decrease or the oxidation (redox responses) of the ionic species. In addition to the above components, thermometer is used to measure temperature during the process and a magnetic/ultrasonic stirrer is used to shake the electrolyte in order to bring new supply of metallic salts/ions to the cathode, therefore, facilitates replacement of metal ions at the cathode allowing the use of advanced current density and better current scattering. It sweeps away the foams which may cause pitting at deposit surface [80].

There are two types of electrodeposition system, galvanostatic and potentiostatic. The current between the electrodes in the galvanostatic plating system is kept constant. In pontentiostatic plating system, the potential of the working electrode is kept constant. Reactions that occur at cathode and anode are called half-cell reactions.

Table 2 Surface modification techniques for metallic alloy as a biomedical implant [4]

Surface modification methods	Modified layer	Objective
Mechanical methods • Machining • Grinding • Polishing • Blasting	The rough or smooth surface formed by the subtraction process	Produce specific surface topographies; Clean and roughen surface; improve adhesion in bonding
Chemical methods • Acidic treatment • Alkaline treatment	<10 nm of the surface oxide layer ~1 μm of sodium titanate gel ~5 nm of dense inner oxide and porous outer layer	Remove oxide scales and contamination Improve biocompatibility, bioactivity, or bone conductivity
Sol-gel	~10 μm of the thin films, such as CaP, TiO_2, and silica	Improve biocompatibility, bioactivity, or bone conductivity
Anodic oxidation	~10 nm to 40 μm of TiO_2 layer, adsorption, and incorporation of electrolyte anions	Produce specific surface topographies; improve corrosion resistance and biocompatibility
Chemical vapor deposition	~1 μm of TiN, TiC, TiCN, diamond, and diamond-like carbon thin film	Improve wear resistance, corrosion resistance, and blood compatibility
Glow discharge plasma	~1 nm to ~100 nm of surface modified layer	Clean, sterilize, nitride surface; remove the native oxide layer
Biochemical methods	Modification through silanized titania, self-assembled monolayers, photochemistry, protein resistance, etc.	Induce specific cell and tissue response by means of surface immobilized peptides, proteins, or growth factors
Physical methods • Thermal spray • Flame spray • HVOF • DGON	~30 to 200 μm of coatings, such as titanium, HA, calcium silicate, Al_2O_3, ZrO_2, TiO_2	Improve wear resistance, corrosion resistance, and biological properties
PVD • Evaporation • Ion plating • Sputtering	~1 μm of TiN, TiC, TiCN, diamond, and diamond-like carbon thin film	Improve wear resistance, corrosion resistance, and blood compatibility
Ion implantation and deposition Beam-line ion implantation and P-III	~10 nm of the surface modified layer and/or ~μm of thin film	Modify surface composition; improve wear, corrosion resistance, and biocompatibility

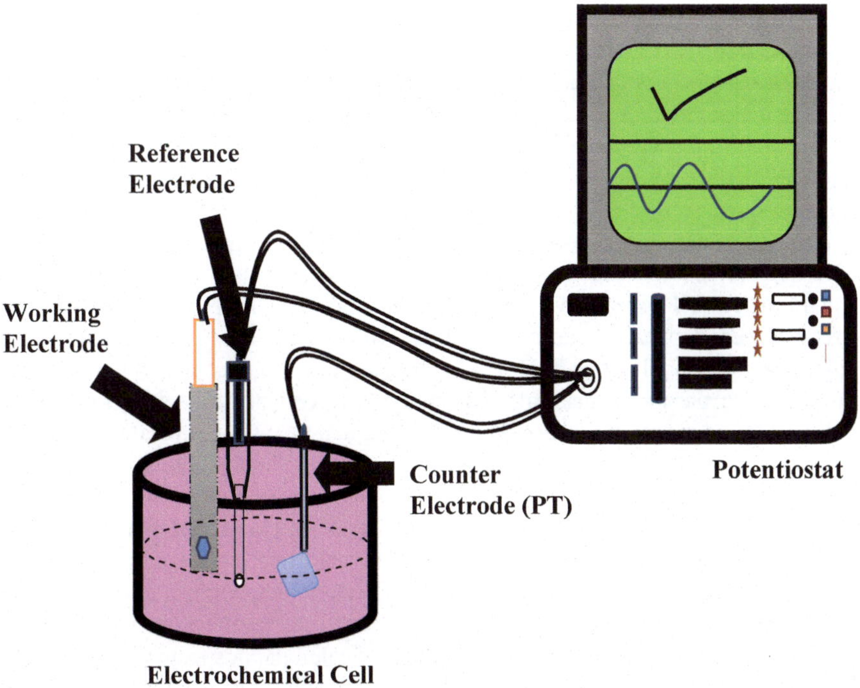

Fig. 9 Three electrode cells with potentiostat/galvanostat

$$\text{Cathode Reaction: } M^{n+} + ne^{-} \rightarrow M^{0}$$

$$\text{Anode Reaction: } M(s) \rightarrow Mz^{+}(aq) + ze^{-}$$

The various parameters affecting the process are current density, electrolyte concentration, potential, temperature of the electrolyte, pH of electrolyte, duty cycle, and RPM of the stirrer. With the increase in current density, the thickness of layer increased. Low current density produces gross grain structures due to little scope for creation of new nuclei. High concentrations of electrolytes are used to increase the conductivity of solution and cathode efficiency. Firm and adherent deposits can be obtained with higher electrolyte concentration and intermediate current density.

A number of research studies have been reported by various researchers about the electrodeposition of biomimetic layer to enhance the wear and corrosion resistance of implant. Qiu et al. [81] electrodeposited HAP-TiO_2 composite coatings on the surface of Ni–Ti alloy by the process. Addition of TiO_2 to the electrolyte results in needle-flower-like crystals of HAP. The adhesive strength of HAP-TiO_2 coating and HAP coating was measured to be 23.7 and 13.4 MPa, respectively. Thicknesses of coatings were 7 μm in both cases. Qiu et al. [82] also produced HAP-ZrO_2 composite coatings on Ni–Ti alloy by the same process. The needle-flower-like crystals of HAP were obtained due to **the** addition of ZrO_2 to the electrolyte solution. The HAP-

ZrO_2-coated surface exhibited higher corrosion resistance and potential as compared to uncoated surface. The thicknesses of the HAP and HAP-ZrO_2 coatings were about 7 and 8 μm, respectively. The adhesive strength of HAP-ZrO_2 coating and HAP coating was measured to be 24.2 and 13.4 MPa, respectively. Peng et al. [83] deposited optical semi-transparent and adherent thin CaP coatings on CP-Ti alloy surface in modified SBF using periodic pulsed potentials. The average thickness of coatings was measured around 200 nm with surface nanoporosities with a pore size in the range of a few nanometers to 1 μm. The deposited CaP coating layer possessed Ca/P ratio of 1.65. The root-mean-square (RMS) roughness values on Ti-CaP surfaces and untreated Ti substrates were about 300 nm and 110 nm, respectively. Adhesion strength, wear resistance, and corrosion resistance of coatings are enhanced due to pulsed voltage supply. Kar et al. [84] fabricated the surfaces of CP-Ti with HAP in order to make these materials bioactive and to enhance osseointegration with pulsed electrodeposition process. The effect of alkali treatment (annealing) at different temperature (450 to 600 °C) without treatment on the bond strength of coating was studied and it has been observed that the adhesion strength of the coatings was 16–21, 32–37 and 38–44 MPa. Saremi and Golshan [85] produced nano-HAP coatings by cathodic electrodeposition on Ti-6Al-4V substrate. The effects of the variation of the electrochemical process variable which include voltage, current density, and pulse conditions were analyzed to determine the change in the microstructure of the deposit. Nanosize particles of three phases of calcium phosphates were obtained at t_{on} and t_{off} 60 s and 90 s, respectively. The amorphous phase with spherical particles with Ca/P ratio of 1.38 and a diameter of 100 nm was obtained. Santos et al. [86] used cathodic electrodeposition to deposit Ti-6Al-4V surface with HAP nano-/microsingle crystals. Large HAP-single crystals oriented along the c-axis appeared to grow up in an amorphous HAP base material. Park et al. [87] used pulse electrodeposition method to produce calcium phosphate/chitosan coatings with variation of the deposition cycle and contents of chitosan solution. The electrodeposition was performed by pulsing the voltage during 5–20 cycles. The composite coatings have shown a high corrosion resistance than bare Ti-6Al-4V substrate. Gopi et al. [88] electrodeposited CNTs reinforced HAP coatings on CP-Ti substrate at a potential of -1.4 V. Adhesion strength, hardness, and elastic modulus values were measured to be 15.8–22 MPa, 6.1–7.5 GPa, and 115–130 MPa, respectively. Yan et al. [89] fabricated a wear-resistive antibacterial coating on pure titanium through deposition of hydroxyapatite and nano-Ag effectuated by hybridization of chitosan (CS) with Ag+ and Ca^{2+}. Pulsed electrodeposition mode was used. Uniformly dispersed spherical nanoparticles of composite coating were obtained at optimized deposition potential and Ag-apatite concentration with excellent bio-inductive activity.

Lee [90] studied the effect of electroplated HAP-TiO_2 composite coatings on the wear and corrosion resistance in Hanks' solution on Ti-6Al-4V alloy. The anodizing treatment was conducted at 10 V at different durations of time (i.e., 40, 50, 60, 120, and 180 min). After the anodizing process, HAP-TiO_2 composite coatings were fabricated at constant current densities of 10 mA/cm^2 in the electrolyte at 80 °C for 60 min. The electrolyte was magnetically agitated at a speed of 200 rpm to produce HAP-TiO_2 coating. Microhardness of the coatings was ranging from 380.62

to 890.92 Hv, and the maximum hardness of 890.92 Hv was obtained with 180-minute anodized sample. R_a and RMS values varied from 500–650 to 750–900 nm, respectively, and higher values corresponding to the higher duration of anodizing. An increase in corrosion resistance and a decrease in coefficient of friction of the coatings with increasing anodizing duration of alloy substrate were observed. Huang et al. [91] presented the antibacterial efficacy and cytocompatibility of electroplated Cu and Zn co-substituted HAP (Zn-Cu-HAP) coating on CP-Ti. Cu^{2+} was substituted into hydroxyapatite structure to improve the antibacterial property. Cu ions released from Zn-Cu-HAP and produced strong antibacterial effect due to the high antimicrobial ratio (K > 95%). Zn^{2+} acts as secondary material into Cu-HAP to reduce Cu concentration. A dense coating of thickness approximately 10 μm was obtained. The adhesion strength of Zn-Cu-HAP and HAP coating was approximately 12 MPa and 10 MPa, respectively. Zn-Cu-HAP coating exhibit superior corrosion protection for Ti substrates. Saremi and Golshan [92] used pulsed cathodic electrodeposition to produce a film of osteoconductive and biocompatible HAP on Ti-6Al-4V biomedical implants. The process was done at different potentials of 3, 4.5, and 6 volts and different duty cycles of 0.4, 0.5, and 0.05. Nucleation process enhanced at higher potentials resulting in an increase in a number of fine grain particles. Smaller particles were produced at higher duty cycles. Grain growth has effected when the number of cycles increased, owing to the fact that more base material was generated at optimum values of pH and concentration of the electrolyte. The low temperature of electrolyte reduced the grain growth, and diffusion process was stopped. Fine particles were obtained on the polished surface due to the conversion of the nucleation process from heterogeneous to homogeneous mode. When current density increased from 1 to 4 mA/cm^2, thicker coatings were obtained.

3.2 *Electric Discharge Machining*

Electrical discharge machining (EDM) process is the only non-conventional machining process, which can machine effectively hard-to-cut materials [4]. EDM is generally a thermal and electric machining process that produces a large number of electrical sparks in a fraction of seconds. These intense sparks produce a large amount of heat to remove material from the workpiece surface [93]. Figure 10 shows the experimental and material removal mechanics during EDM machining process. Moreover, EDM can produce a biocompatible layer on the surface, which also enhanced the surface hardness and corrosion resistance of Ti-64 alloy [94]. Peng et al. used EDM to machine the Ti-64 alloy and tuned the surface characteristics as required for the osseointegration process [95]. The nanoporous layer has been synthesized by EDM, which promising to promote cell adhesion and growth. Ben et al. investigated the effect of EDM to improve the surface, chemical, and metallurgical properties of the Ti-64 alloy [96, 97]. Recently, Prakash et al. critically reviewed the utility of EDM process for machining Ti and its alloy for biomedical applications [4]. Owing to electrothermal process, the material gets thermally influenced and various surface

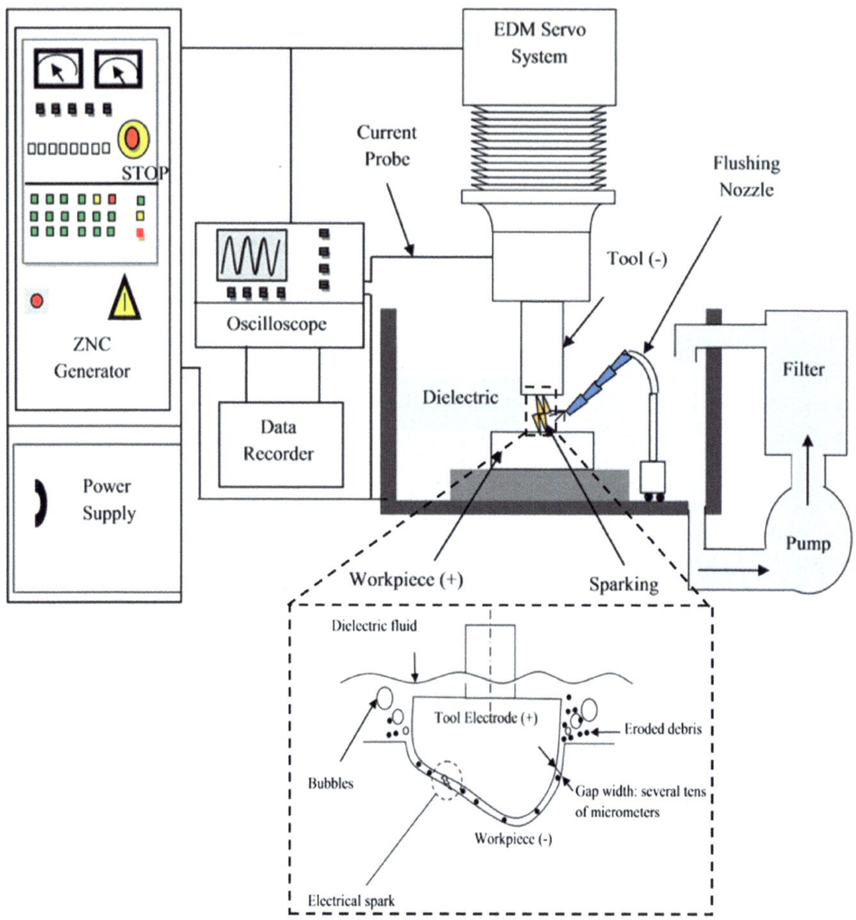

Fig. 10 Schematic illustration of the mechanism of material removal in EDM process

imperfections like high surface roughness and microcracks were developed on the machined surface. These developed surface irregularities reduced the quality and performance of machined medical implants such as despoiled corrosion performance, and fatigue performance of implant and as a result bone–implant interface failure. In order to avoid the formation of surface defects, numerous progressions/hybridization in EDM process has been carried out such as wire-EDM, ultrasonic vibration-assisted EDM (UVAEDM), rotary-assisted EDM (RAEDM), HyFlex EDM, electrodischarge coating/surface modification by composite or green tool electrode, near-dry EDM, and powder-mixed EDM (PM-EDM) [98].

Most of the researchers used powder-mixed dielectric to deposit workpiece surface [99, 100]. Prakash et al. investigated the effect of Si-mixed EDM to alter the surface characteristics to improve the biocompatibility, mechano-corrosion, and wear

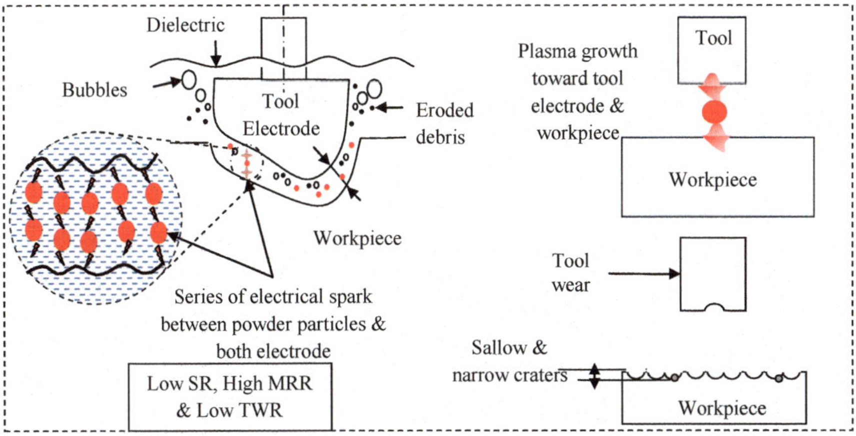

Fig. 11 Mechanism of material removal from the workpiece in PM-EDM process [98]

resistance properties of a specially designed Ti-35Nb-7Ta-5Zr (β-phase) for orthopedic applications [101]. Xie et al. reported that the surface hardness of 45-C steel has been increased from 415 to 1420 Hv using graphite-mixed EDM [102]. Arun et al. synthesized a hard layer of Ni-W coating on tool-steel by Ni&W-mixed EDM to improve the tribological performance [103]. Ekmekci et al. reported that hydroxyapatite (HA)-enriched bioceramic layer can be successfully deposited on Ti-64 surface using HA-mixed EDM process [104]. Ou and Wang used EDC to deposit HA-enriched layer on Ti alloy to enhance the biocompatibility of base material [105]. Recently, Prakash et al. uncovered the ability of powder-mixed EDM process to deposit nano-HA layer to enhance the mechanical, corrosion, bioactivity of Mg-based biodegradable implants [106]. Figure 11 shows the experimental and material removal mechanics during the PM-EDM machining process.

3.3 Thermal Spray Coating Process

Thermal spraying is a common way of coating the materials, for industrial benefits, bypassing the deposition material (powder or wire) through a heating zone where it is melted. Finally, the coatings are developed because of the propelling of the melted or semi-melted particles toward the substrates [107]. Since the mid-1980s, thermal spraying is considered the most promising technique to apply bioactive films on metallic implants for orthopedic application [108]. Thermal sprayed coating processes for biomedical applications include plasma spray deposition, HVOF spraying, detonation-gun (D-GUN) spaying), warm spraying, cold spraying/cold gas dynamic spraying (CGDS), and flame spraying. In thermal spray process, the materials primarily as a powder provided from the feedstock into the plasma jet and get deposited

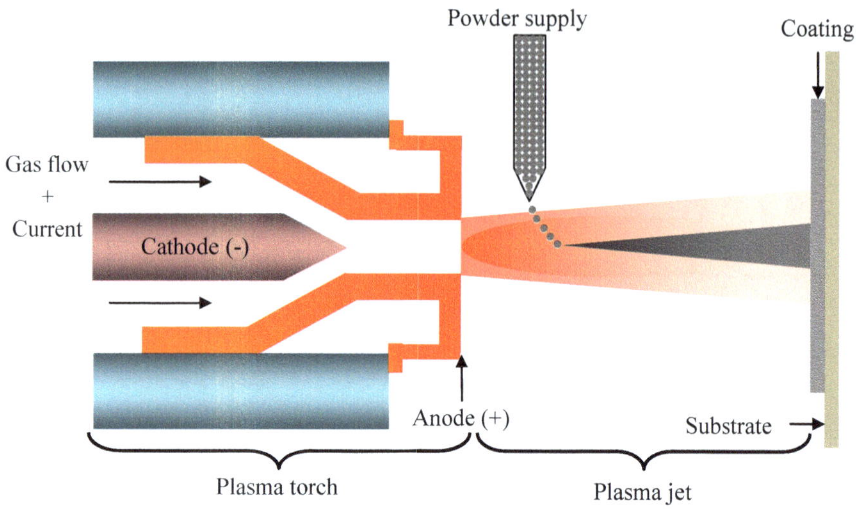

Fig. 12 Schematic representation of Thermal Spray Deposition process

on the substrate. In the jet stream, the temperature rises up to the 10,000 °K, and the material in the form of powder get melted and propelled toward the surface of the substrate. On the surface of substrate, molten droplets strike the surface and flatten after that rapidly solidify and form a coat. The schematic of the thermal spray process is shown in Fig. 12. This system is consisting of the torch, plasma jet, powder, injector, and the substrate. Plasma spray process is a high-temperature process, in which the jet stream temperature varies in the range from 8000 to 15,000 °C, which allows to deposit of a large number of biomaterials [109, 110].

Li et al. (2004) deposited pure hydroxyapatite coating (HA) and 50%HA + 50%TiO_2 coating (HT) on Ti substrate by plasma spraying deposition process. The effect of heat treatment followed by the coating on the mechanical properties has been studied. The heat treatment enhanced the coating bond strength and removed the residual stress developed during the coating process [111]. Sarao et al. (2012) deposited the composite coating of hydroxyapatite (HA) along with titanium oxide (TiO_2) in the ratio of 50:50 in weight percentage by a thermal spray deposition process. The cell culture investigation of the coated layer presents the conceivable applications in the biomedical field. The electrochemical examination demonstrated that HA-TiO_2 coatings enhanced the corrosion resistance of Ti alloy after immersion in human body host conditions [112]. Sidhu et al. (2018) enhance the mechanical properties and corrosion resistance of SS 304 L alloy by depositing HA + 15% Al_2O_3 coating using plasma spraying deposition [113]. Galdino et al. (2018) deposited HA-TiO_2 (50–50% volume) coating on the surface of Ti-6Al-4V alloy by a plasma spray technique. Microstructure and morphology of coated surface at different HA-TiO_2 composition were investigated, and it has been found that best range for the HA-TiO_2 composition is somewhere in the range of 47 and 53%, respectively [114].

Hu et al. (2018) deposited HA/TiO_2 and HA/ZrO_2 coating on stainless steel substrates by high-velocity suspension flame spray (HVSFS) deposition process for biomedical applications. The HA/TiO_2 coating possessed higher mechanical properties than HA/ZrO_2 [115]. Hameed et al. (2019) successfully deposited hydroxyapatite using on Ti-6Al-4V alloy using axial suspension plasma spray technique. The mechanical, in vitro corrosion and bioactivity analysis results of the coating revealed that exhibit excellent cell adhesion, superior adhesion strength, and better corrosion resistance [116]. Hazoor Singh et al. (2018) studied the corrosion and bioactivity of HA/TiO_2 composite coating deposited by plasma spray deposition process. The $50\%HA/50\%TiO_2$ composite coating exhibited excellent corrosion and bioactivity [117].

3.4 Surface Finishing of Implants

Surface finishing processes are used to transform/alter the surface characteristics which primarily occur because of preceding manufacturing process like casting or forming process [118]. Low surface roughness of a part is important to limit erosion powers and to upgrade wear resistance and mechanical properties, for example, fatigue life and durability. The goal of the present work is to survey distinguishing divisions of nanofinishing processes applicable to freeform/sculptured surfaces. Conventional techniques for accomplishing high surface complete incorporate lapping, grinding, turning, honing, polishing, furthermore, and burnishing. Figure 13 presents conventional and advanced finishing processes.

Conventional finishing processes like lapping, hopping, polishing, and ball burnishing are the oldest methods to get the fine surface finish of about 0.08 to 0.25 μm [119, 120]. In the lapping process, free rough particle in the form of the concentrated slurry is utilized to rub against the workpiece. The finishing process then deals between lapping plate, abrasive, and workpiece. The work surface was abraded in the form of microchips and smoother surface was produced [121, 122]. In this process, there is no change to any dimensional accuracy because of very low material removal rate. Figure 14 shows the schematic representation of the lapping process. In the grinding process, abrasive particle are bounded on a plate or wheel. The rotating wheel or plate is press against the linear traveling workpiece for fast removal of stock. The surface finish of about 0.9–5 μm is achievable by grinding process. Low grinding ratio is seen with the prominent wearing of the wheel while grinding of hard-brittle materials. Besides, while grinding debris tried to block the wheel. This will now influenced the uniform distribution of abrasive grain in grinding wheel [123, 124].

Today the polishing of bio-implant is usually joined with high precision CNC procedure till the operative proficiency to affirm the uniform evacuation of material over entire complex surfaces. The smoothness of polished spherical cups and balls has obtained up to nanoscale, with the progress of refined polishing machines [123]. Cheung et al. examined the factor that influences the nanosurface generation in ultra-

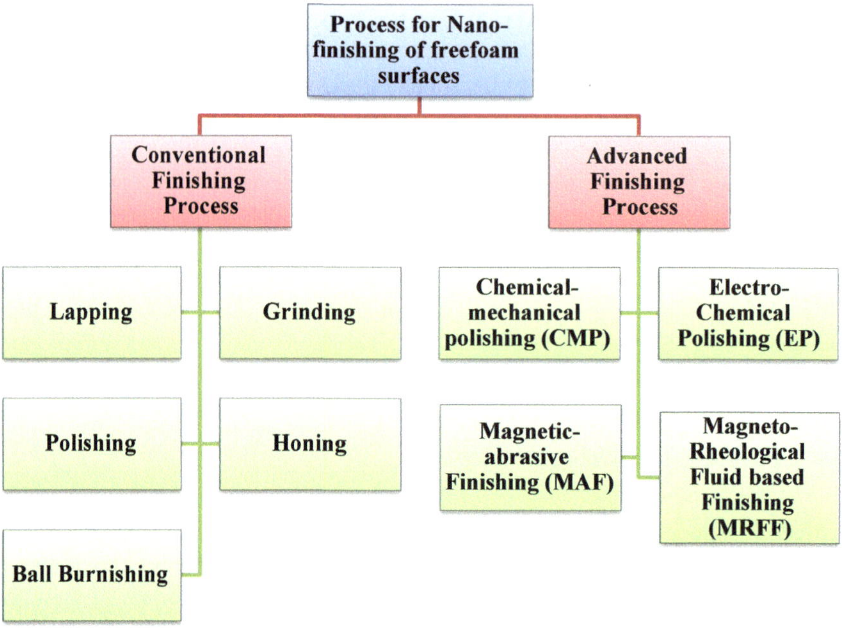

Fig. 13 Various types of process to finish the freeform surfaces

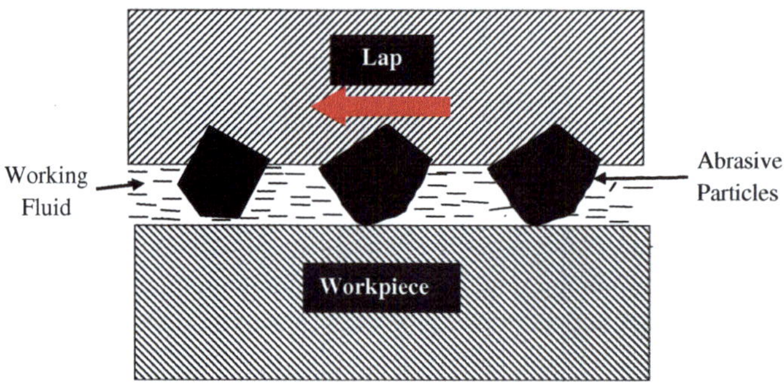

Fig. 14 Scheme of the lapping process

precision polishing. The investigation offered reasonable techniques for upgrading the surface finish of freeform components. Better surface finish was procured including/plus less cost and operative time. An arrangement of ideal polishing conditions for machining Ti-based knee implants was then presented and uncertainly confirmed. Super-finished orthopedic implants alongside surface smoothness of beneath 10 nm were accomplished [125]. Honing is also a finishing process used for internal finishing of workpiece. In this process, a honing stone is used which is composed of abrasive grains of very fine powder. It is basically utilized for better the geometric form of a surface, but side by side it may improve the surface texture too [121]. This process is also used in automobile sector to finish the internal cylinder walls of the engine. Burnishing process is also used for internal finishing of holes or tubes. It provides better results if compared with the honing process. In this process, a hardened steel balls or rolls are pressed on to the surface of workpiece and imparting a feed motion to the same surface [126]. These conventional finishing processes have some limitations such as low surface finish and dimensional accuracy, which restrict their applications. That's why these processes are not suitable on newly developed harder and tougher materials. There are following disadvantages of the conventional finishing processes:

(a) Very expensive when machining harder materials up to higher precision.
(b) More energy utilization.
(c) Normal force applied on workpiece surface is more, which further can harm the finished surface.
(d) Unfit to finish complicated 3-D shapes and time-consuming.

To deal with above-stated issues in conventional grinding and polishing process, modern and advanced finishing processes such as electrolytic in-process dressing (ELID) grinding, chemical–mechanical polishing (CMP), and magnetic abrasive finishing (MAF) processes were developed [126–128]. ELID is a hybrid process in which electrolytic process for dressing the grinding wheel and grinding process for metal removal [129]. The hard and fine abrasive particles bounded in grinding wheel which performed material removal. Figure 15 shows the schematic representation of ELID grinding process. Ohmori et al. [130] studies stated that a smooth surface with nanoscale roughness on hard-brittle materials impressively produces by ELID grinding. Accordingly, ongoing researchers have demonstrated the interesting suitability of ELID grinding on manufacturing metallic and ceramic-based bio-implants. Kotani et al. [131] employed present same technology to finish Co–Cr–Mo alloys. Mirror-like surface finish that compares positively with polished ones was accomplished. An oxide layer formed on the surface of the products which is considered in resisting the corrosion was found to be improving by rising/amplifying the electrolytic current used in machining. But then also fear/danger of peeling of inducing abrasive grain on decline/decay the surface finish will be there. Kotani et al. [132] proposed another ELID grinding system to machine inside the surface of a hemispherical cup. The hybrid process unite ELID grinding for surface finishing of sliding or rotating parts.

The magnetic abrasive finishing (MAF) process has been widely adopted for the finishing of engineering components [133, 134]. MAF is a modern precision machining procedure in which the finishing is carried out with the help of magnetic force and abrasive particle [135]. Recently, magnetorheological fluid-based finishing (MRFF) process developed by Jain and co-authors was reported a very potential technique for biomedical applications [136]. The MRFF process has several advantages upon conventional finishing processes. MRFF is an unconventional finishing process, in which finishing action was carried out by utilizing magnetic field and magnetic abrasives particles in the working gap. Figure 16 presents the schematic representation of the magnetorheological fluid-based finishing process.

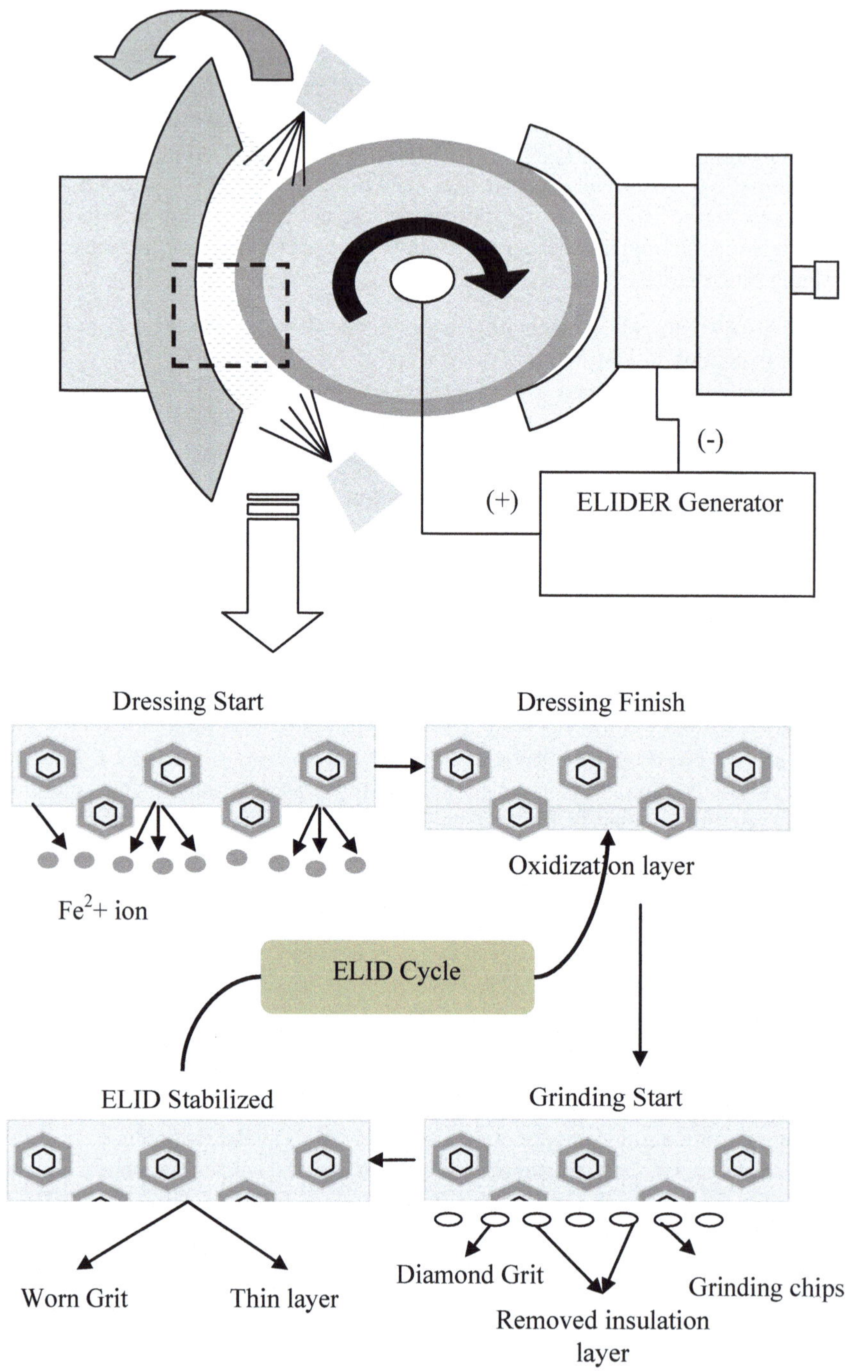

Fig. 15 Schematic representation of ELID setup and finishing process

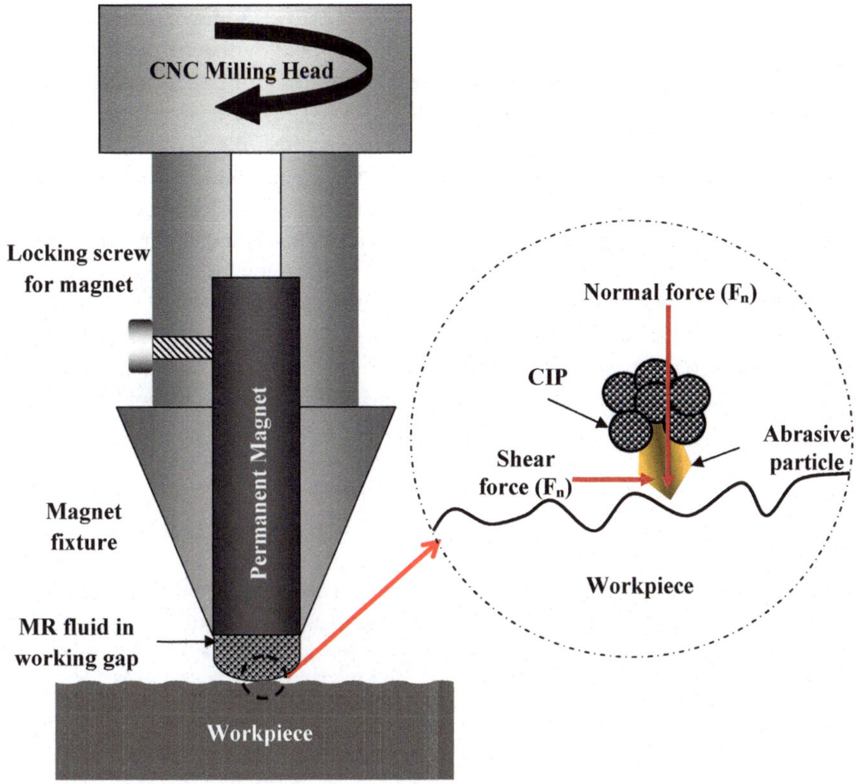

Fig. 16 Schematic representation of magnetorheological fluid-based finishing (MRFF) process

A homogeneous mechanical blend of rough and iron particles is set up for the magnetic abrasive finishing process. This blend is called magnetic abrasive powder and the particles are, therefore, called magnetic abrasive particles (MAPs). The MAPs are used in magnetic abrasive finishing of flat and cylindrical (external and internal) surfaces. Many times the abrasive particles are bonded by a sintering process, to the magnetic particles. The particles are then called bounded magnetic abrasive particles (BMAPs). Lubricating oil is sometimes added to the blend to acquire holding strength, and then it is called as loosely bounded magnetic abrasive particles (LMAPs) or it might be a basic mechanical blend of the two, called as unbounded magnetic abrasive particles (UMAPs). A magnet is brought close to the surface of the workpiece maintaining some gap between the two. The shape of the magnet depends upon the shape of the work surface which is to be finished. The magnetic field generator can be either electromagnetic coils or permanent magnets. The gap between the magnet and workpiece is known as finishing gap or machining gap or working gap. When a magnetic field is produced by the magnet, the iron particles in the mixture get magnetized and get aligned along the lines of magnetic force. The abrasive particles in the mixture get trapped in the iron particles matrix or, get sandwiched between the iron particles and thereby, the abrasive particles are held by the magnetic force too. Overall, this alignment of the magnetic abrasive particles forms a brush-like structure known as flexible magnetic abrasive brush (FMAB). The brush is termed "Flexible" due to its capacity to take the shape of any of surface profile irregularities, if present on the work surface. The normal magnetic force acting on the abrasive particles that are in contact with the workpiece produces indentation into the workpiece. The relative movement between the induced rough particles of the FMAB and workpiece produces the essential shearing action at the abrasive–workpiece interface to expel material from the workpiece as smaller than expected chips. MRFF tool was developed for processing complex freeform surfaces, in order to attain nanoscale level surface roughness. According to the literature, MRFF for finishing of knee implants and nanolevel finish has been achieved [137]. Nagdeve et al. further modified MRFF and developed a new rotational–magnetorheological abrasive flow finishing process (R-MRFF) for the knee implant applications. In this inverse replica of the workpiece, a finishing fixture is recommended for finishing of freeform surfaces up to nanolevel roughness [138]. Kumar et al. also used R-MRFF for finishing of knee implants [139]. Barman et al. reported that not only abrasive particles were sufficient for the finishing process, but also MRFF fluid composition also affects the surface integrity. To finish the titanium alloy at the nanometer level, two types of acidic base medium are prepared and specific composition of the various acidic base medium is very important to generate required surface morphology on the implant [140].

4 Conclusion and Future Work Scope

Development of biomedical implants is a biomechanical engineering process, which comprised and integrated both mechanical and bioengineering for material selection, fabrication, surface finishing, in vivo and in vitro evaluation of implants. Thus, the chapter presented the major biomechanical processes. As per the biomaterial requirement and selection, metallic biomaterials were widely used. A number of attempts were made to develop a new generation of alloys and composites for biomedical applications for load-bearing applications. The major challenge experience for metallic biomaterials is stress shielding due to their high elastic modulus. In order to deal with this problem, the implants with porous structure were developed and believed to be the most potential approach. The porous structure not only reduced the elastic modulus but also stimulated the bone growth around the implant in the host body. In this regard, a number of strategies were developed such as powder metallurgy, additive manufacturing, and electrospinning which shows huge potential. With the critical interest from the general public and industry, proposing very productive and practical manufacture approaches with diminished follow-up procedures ought to be the primary need. After the development of medical implants, surface engineering like modification, treatment, processing, and finishing were required for the successful implantation of the artificial organ. This includes machining, coating, and finishing, etc., as per the specific application of implant. Surface modification by material removal/deposition/treatment also played a very important and crucial role in the stability of the implant in the host body. Surface modification by coating and deposition processes leads to develop wear- and corrosion-resistant surfaces, and surface treatment enhanced the fatigue performance of implants. Similarly, surface finishing processes are also essential for implant processing for a specific application like knee and acetabular head. In such type of applications, defect-free and high-quality surfaces are required for long-term performance. For high-quality and precision finishing, a number of finishing processes were developed and used for knee implant applications. Recently, the application of magnetorheological abrasive flow finishing process was found most suitable and potential for the finishing of knee implant and can be used for finishing of acetabular head components. There is huge scope in the development, processing, and finishing of implants, and this chapter provides an insight for the researchers to select material, process technique, and surface engineering technique for biomedical applications.

References

1. Natarajan S (2016) Biomimetic, bioresponsive, and bioactive materials edited by Matteo Santin and Gary J. Phillips. Mater Manuf Process 31(7):976–977
2. Geetha M, Singh AK, Asokamani R, Gogia AK (2009) Ti-based biomaterials, the ultimate choice for orthopaedic implants—a review. Prog Mater Sci 54(3):397–425
3. Gao C, Peng S, Feng P, Shuai C (2017) Bone biomaterials and interactions with stem cells. Bone Res 5:1–33. https://doi.org/10.1038/boneres.2017.59
4. Prakash C, Kansal HK, Pabla BS, Puri S, Aggarwal A (2016) Electric discharge machining—a potential choice for surface modification of metallic implants for orthopedic applications: a review. Proc Inst Mech Eng Part B: J Eng Manuf 230(2):331–353
5. Liu X, Chu PK, Ding C (2004) Surface modification of titanium, titanium alloys, and related materials for biomedical applications. Mater Sci Eng R: Rep 47(3–4):49–121
6. Bartolo P, Kruth JP, Silva J, Levy G, Malshe A, Rajurkar K, Mitsuishi M, Ciurana J, Leu M (2012) Biomedical production of implants by additive electro-chemical and physical processes. CIRP Ann Manuf Technol 61(2):635–655
7. Kang CW, Fang FZ (2018) State of the art of bioimplants manufacturing: part I. Adv Manuf 6(1):20–40
8. Kang CW, Fang FZ (2018) State of the art of bioimplants manufacturing: part II. Adv Manuf 6(1):137–154
9. Prakash C, Kansal HK, Pabla BS, Puri S (2017) On the influence of nanoporous layer fabricated by PMEDM on β-Ti implant: biological and computational evaluation of bone-implant interface. Mater Today Proc 4(2):2298–2307
10. Zhao B, Gain AK, Ding W, Zhang L, Li X, Fu Y (2018) A review on metallic porous materials: pore formation, mechanical properties, and their applications. Int J Adv Manuf Technol 95(5–8):2641–2659
11. Raza MR, Sulong AB, Muhamad N, Akhtar MN, Rajabi J (2015) Effects of binder system and processing parameters on formability of porous Ti/HA composite through powder injection molding. Mater Des 87:386–392
12. Ryan G, Pandit A, Apatsidis DP (2006) Fabrication methods of porous metals for use in orthopaedic applications. Biomaterials 27(13):2651–2670

13. Zhang L, He ZY, Zhang YQ, Jiang YH, Zhou R (2016) Rapidly sintering of interconnected porous Ti-HA biocomposite with high strength and enhanced bioactivity. Mater Sci Eng, C 67:104–114
14. Torres-Sanchez C, McLaughlin J, Fotticchia A (2018) Porosity and pore size effect on the properties of sintered Ti35Nb4Sn alloy scaffolds and their suitability for tissue engineering applications. J Alloy Compd 731:189–199
15. Cook SD, Walsh KA, Haddad JR (1985) Interface mechanics and bone growth into porous Co-Cr-Mo alloy implants. Clin Orthop Relat Res 193:271–280
16. Camron HU, Pilliar RM, Macnab I (1976) The rate of bone ingrowth into porous metal. J Biomed Mater Res 10(2):295–302
17. Hofmann AA, Bloebaum RD, Bachus KN (1997) Progression of human bone ingrowth into porous-coated implants: rate of bone ingrowth in humans. Acta Orthop Scand 68(2):161–166
18. Bobyn JD, Pilliar RM, Cameron HU, Weatherly GC (1980) The optimum pore size for the fixation of porous-surfaced metal implants by the ingrowth of bone. Clin Orthop Relat Res 150:263–270
19. Hulbert SF, Young FA, Mathews RS, Klawitter JJ, Talbert CD, Stelling FH (1970) Potential of ceramic materials as permanently implantable skeletal prostheses. J Biomed Mater Res 4(3):433–456
20. Barth E, Ronningen H, Solheim LF, Saethren B (1986) Bone ingrowth into weight-bearing porous fiber titanium implants. Mechanical and biochemical correlations. J Orthop Res 4(3):356–361
21. Tsuruga E, Takita H, Itoh H, Wakisaka Y, Kuboki Y (1997) Pore size of porous hydroxyapatite as the cell-substratum controls BMP-induced osteogenesis. J Biochem 121(2):317–324
22. Clemow AJT, Weinstein AM, Klawitter JJ, Koeneman J, Anderson J (1981) Interface mechanics of porous titanium implants. J Biomed Mater Res 15(1):73–82
23. Bobyn JD, Stackpool GJ, Hacking SA, Tanzer M, Krygier JJ (1999) Characteristics of bone ingrowth and interface mechanics of a new porous tantalum biomaterial. J Bone Joint Surg 81-B(5):907 (British Volume)
24. Vasconcellos LMR, Leite DO, Oliveira FN, Carvalho YR, Cairo CAA (2010) Evaluation of bone ingrowth into porous titanium implant: histomorphometric analysis in rabbits. Braz Oral Res 24:399–405
25. Taniguchi N, Fujibayashi S, Takemoto M, Sasaki K, Otsuki B, Nakamura T, Matsushita T, Kokubo T, Matsuda S (2016) Effect of pore size on bone ingrowth into porous titanium implants fabricated by additive manufacturing: an in vivo experiment. Mater Sci Eng, C 59:690–701
26. Wen CE, Yamada Y, Shimojima K, Chino Y, Asahina T, Mabuchi M (2002) Processing and mechanical properties of autogenous titanium implant materials. J Mater Sci Mater Med 13(4):397–401
27. Singh R, Singh BP, Gupta A, Prakash C (2017). Fabrication and characterization of Ti-Nb-HA alloy by mechanical alloying and spark plasma sintering for hard tissue replacements. In: IOP Conference Series: Materials Science and Engineering, vol 225, no 1. IOP Publishing, p 012051
28. Sharma N, Kumar K (2018) Mechanical characteristics and bioactivity of porous Ni50− x Ti50Cu x (x= 0, 5 and 10) prepared by P/M. Mater Sci Technol 34(8):934–944
29. Sharma B, Vajpai SK, Ameyama K (2016) Microstructure and properties of beta Ti–Nb alloy prepared by powder metallurgy route using titanium hydride powder. J Alloy Compd 656:978–986
30. Roseti L, Parisi V, Petretta M, Cavallo C, Desando G, Bartolotti I, Grigolo B (2017) Scaffolds for bone tissue engineering: state of the art and new perspectives. Mater Sci Eng, C 78:1246–1262
31. Zhu SL, Yang XJ, Hu F, Deng SH, Cui ZD (2004) Processing of porous TiNi shape memory alloy from elemental powders by Ar-sintering. Mater Lett 58(19):2369–2373
32. Zhang L, He ZY, Zhang YQ, Jiang YH, Zhou R (2016) Rapidly sintering of interconnected porous Ti-HA biocomposite with high strength and enhanced bioactivity. Mater Sci Eng, C 67:104–114

33. Prakash C, Singh S, Gupta M, Mia M, Królczyk G, Khanna N (2018) Synthesis, characterization, corrosion resistance and in-vitro bioactivity behavior of biodegradable Mg–Zn–Mn–(Si–HA) composite for orthopaedic applications. Materials 11(9):1602
34. Prakash C, Singh S, Pabla BS, Sidhu SS, Uddin MS (2018) Bio-inspired low elastic biodegradable Mg-Zn-Mn-Si-HA alloy fabricated by spark plasma sintering. Mater Manuf Process 1–12
35. Prakash C, Singh S, Abdul-Rani AM, Uddin MS, Pabla BS, Puri S (2019) Spark plasma sintering of Mg-Zn-Mn-Si-HA alloy for bone fixation devices: fabrication of biodegradable low elastic porous Mg-Zn-Mn-Si-HA alloy. In: Handbook of research on green engineering techniques for modern manufacturing. IGI Global, pp 282–295
36. Yazdimamaghani M, Razavi M, Vashaee D, Moharamzadeh K, Boccaccini AR, Tayebi L (2017) Porous magnesium-based scaffolds for tissue engineering. Mater Sci Eng, C 71:1253–1266
37. Kirkland NT, Kolbeinsson I, Woodfield T, Dias GJ, Staiger MP (2011) Synthesis and properties of topologically ordered porous magnesium. Mater Sci Eng, B 176(20):1666–1672
38. Witte F, Ulrich H, Rudert M, Willbold E (2007) Biodegradable magnesium scaffolds: part 1: appropriate inflammatory response. J Biomed Mater Res A 81:748–756
39. Uddin MS, Hall C, Murphy P (2015) Surface treatments for controlling corrosion rate of biodegradable Mg and Mg-based alloy implants. Sci Technol Adv Mater 16:053501
40. Uddin MS, Rosman H, Hall C, Murphy P (2017) Enhancing the corrosion resistance of biodegradable Mg-based alloy by machining-induced surface integrity: influence of machining parameters on surface roughness and hardness. Int J Adv Manuf Technol 90:2095–2108. https://doi.org/10.1007/s00170-016-9536-x
41. Hedayati R, Ahmadi SM, Lietaert K, Tümer N, Li Y, Amin Yavari S, Zadpoor AA (2018) Fatigue and quasi-static mechanical behavior of bio-degradable porous biomaterials based on magnesium alloys. J Biomed Mater Res Part A
42. Bakhsheshi-Rad HR, Hamzah E, Staiger MP, Dias GJ, Hadisi Z, Saheban M, Kashefian M (2018) Drug release, cytocompatibility, bioactivity, and antibacterial activity of doxycycline loaded Mg-Ca-TiO_2 composite scaffold. Mater Des 139:212–221
43. Ghomi H, Emadi R (2018) Fabrication of bioactive porous bredigite ($Ca_7MgSi_4O_{16}$) scaffold via space holder method. Int J Mater Res 109(3):257–264
44. Singh S, Bhatnagar N (2018) A survey of fabrication and application of metallic foams (1925–2017). J Porous Mater 25(2):537–554
45. Singh S, Bhatnagar N (2018) A novel approach to fabricate 3D open cellular structure of $Mg_{10}Zn$ alloy with controlled morphology. Mater Lett 212:62–64
46. Dutta S, Devi KB, Roy M (2017) Processing and degradation behavior of porous magnesium scaffold for biomedical applications. Adv Powder Technol 28(12):3204–3212
47. Bose S, Ke D, Sahasrabudhe H, Bandyopadhyay A (2018) Additive manufacturing of biomaterials. Prog Mater Sci 93:45–111
48. DebRoy T, Wei HL, Zuback JS, Mukherjee T, Elmer JW, Milewski JO, Beese AM, Wilson-Heid A, De A, Zhang W (2018) Additive manufacturing of metallic components–process, structure and properties. Prog Mater Sci 92:112–224
49. Wang Y, Shen Y, Wang Z et al (2010) Development of highly porous titanium scaffolds by selective laser melting. Mater Lett 64(6):674–676
50. Pattanayak DK, Fukuda A, Matsushita T et al (2011) Bioactive Ti metal analogous to human cancellous bone: fabrication by selective laser melting and chemical treatments. Acta Biomater 7(3):1398–1406
51. Weißmann V, Bader R, Hansmann H et al (2016) Influence of the structural orientation on the mechanical properties of selective laser melted Ti_6Al_4V open-porous scaffolds. Mater Des 95:188–197
52. Mueller B (2012) Additive manufacturing technologies: rapid prototyping to direct digital manufacturing. Assem Autom 32(2):151–154
53. Shipley H, McDonnell D, Culleton M, Lupoi R, O'Donnell G, Trimble D (2018) Optimisation of process parameters to address fundamental challenges during selective laser melting of Ti-6Al-4V: a review. Int J Mach Tools Manuf

54. Khorasani A, Gibson I, Awan US, Ghaderi A (2019) The effect of SLM process parameters on density, hardness, tensile strength and surface quality of Ti-6Al-4V. Add Manuf 25:176–186
55. Zhao B, Wang H, Qiao N, Wang C, Hu M (2017) Corrosion resistance characteristics of a Ti-6Al-4V alloy scaffold that is fabricated by electron beam melting and selective laser melting for implantation in vivo. Mater Sci Eng, C 70:832–841
56. Ran Q, Yang W, Hu Y, Shen X, Yu Y, Xiang Y, Cai K (2018) Osteogenesis of 3D printed porous Ti6Al4V implants with different pore sizes. J Mech Behav Biomed Mater 84:1–11
57. Yan C, Hao L, Hussein A, Wei Q, Shi Y (2017) Microstructural and surface modifications and hydroxyapatite coating of Ti-6Al-4V triply periodic minimal surface lattices fabricated by selective laser melting. Mater Sci Eng, C 75:1515–1524
58. Subbiah T, Bhat GS, Tock RW, Parameswaran S, Ramkumar SS (2005) Electrospinning of nanofibers. J Appl Polym Sci 96(2):557–569
59. Reneker DH, Yarin AL (2008) Electrospinning jets and polymer nanofibers. Polymer 49:2387–2425
60. He J, Wan YQ, Yu JY (2005) Scaling law in electrospinning: relationship between electric current and solution flow rate. Polymer 46:2799–2801
61. He W, Horn SW, Hussain MD (2007) Improved bioavailability of orally administered mifepristone from PLGA nanoparticles. Int J Pharm 334:173–178
62. Chew SY, Wen Y, Dzenis Y, Leong KW (2006) The role of electrospinning in the emerging field of nanomedicine. Curr Pharm Des 12:4751–4770
63. Chew SY, Hufnagel TC, Lim CT, Leong KW (2006) Mechanical properties of single electrospun drug-encapsulated nanofibres. Nanotechnology 17:3880–3891
64. Liang D, Hsiao BS, Chu B (2007) Functional electrospun nanofibrous scaffolds for biomedical applications. Adv Drug Deliv Rev 59:1392–1412
65. Bhardwaj N, Kundu SC (2010) Electrospinning: a fascinating fiber fabrication technique. Biotechnol Adv 28(3):325–347
66. Doustgani A, Vasheghani-Farahani E, Soleimani M (2013) Aligned and random nanofibrous nanocomposite scaffolds for bone tissue engineering. Nanomed J 1(1):20–27
67. Raeisdasteh Hokmabad V, Davaran S, Ramazani A, Salehi R (2017) Design and fabrication of porous biodegradable scaffolds: a strategy for tissue engineering. J Biomater Sci 28(16):1797–1825 (Polymer edition)
68. Yao J, Bastiaansen CW, Peijs T (2014) High strength and high modulus electrospun nanofibers. Fibers 2(2):158–186
69. Petrík S (2011) Industrial production technology for nanofibers. Nanofibers–production, properties and functional applications, p 1
70. Esmaeilzadeh I et al (2015) A feasibility study on semi industrial nozzleless electrospinning of cellulose nanofiber. Int J Ind Chem 6(3):193–211
71. Dubský M et al (2012) Nanofibers prepared by needleless electrospinning technology as scaffolds for wound healing. J Mater Sci Mater Med 23(4):931–941
72. Brown TD et al (2012) Design and fabrication of tubular scaffolds via direct writing in a melt electrospinning mode. Biointerphases 7(1):13
73. Muerza-Cascante ML et al (2014) Melt electrospinning and its technologization in tissue engineering. Tissue Eng Part B: Rev 21(2):187–202
74. Mota C et al (2013) Melt electrospinning writing of three-dimensional star poly (ϵ-caprolactone) scaffolds. Polym Int 62(6):893–900
75. Gazzarri M et al (2013) Fibrous star poly (ε-caprolactone) melt-electrospun scaffolds for wound healing applications. J Bioact Compatible Polym 28(5):492–507
76. Li X et al (2012) Preparation and characterization of PLLA/nHA nonwoven mats via laser melt electrospinning. Mater Lett 73:103–106
77. Kim SJ et al (2010) Fabrication and characterization of 3-dimensional PLGA nanofiber/microfiber composite scaffolds. Polymer 51(6):1320–1327
78. Asri RIM, Harun WSW, Hassan MA, Ghani SAC, Buyong Z (2016) A review of hydroxyapatite-based coating techniques: Sol–gel and electrochemical depositions on biocompatible metals. J Mech Behav Biomed Mater 57:95–108

79. Gurrappa I, Binder L (2008) Electrodeposition of nanostructured coatings and their characterization—a review. Sci Technol Adv Mater 9(4):043001
80. Bicelli LP, Bozzini B, Mele C, D'Urzo L (2008) A review of nanostructural aspects of metal electrodeposition. Int J Electrochem Sci 3(4):356–408
81. Qiu D, Yang L, Yin Y, Wang A (2011) Preparation and characterization of hydroxyapatite/titania composite coating on NiTi alloy by electrochemical deposition. Surf Coat Technol 205(10):3280–3284
82. Qiu D, Wang A, Yin Y (2010) Characterization and corrosion behavior of hydroxyapatite/zirconia composite coating on NiTi fabricated by electrochemical deposition. Appl Surf Sci 257(5):1774–1778
83. Peng P, Kumar S, Voelcker NH, Szili E, Smart RSC, Griesser HJ (2006) Thin calcium phosphate coatings on titanium by electrochemical deposition in modified simulated body fluid. J Biomed Mater Res Part A: An Official J Soc Biomater. The Japanese Society for Biomaterials, and The Australian Society for Biomaterials and the Korean Society for Biomaterials 76(2):347–355
84. Kar A, Raja KS, Misra M (2006) Electrodeposition of hydroxyapatite onto nanotubular TiO_2 for implant applications. Surf Coat Technol 201(6):3723–3731
85. Saremi M, Golshan BM (2007) Microstructural study of nano hydroxyapatite coating obtained by pulse electrodeposition process on Ti–6Al–4V. Trans IMF 85(2):99–102
86. Dos Santos EA, Moldovan MS, Jacomine L, Mateescu M, Werckmann J, Anselme K, Mille P, Pelletier H (2010) Oriented hydroxyapatite single crystals produced by the electrodeposition method. Mater Sci Eng, B 169(1–3):138–144
87. Park KH, Kim SJ, Hwang MJ, Song HJ, Park YJ (2017) Pulse electrodeposition of hydroxyapatite/chitosan coatings on titanium substrate for dental implant. Colloid Polym Sci 295(10):1843–1849
88. Gopi D, Shinyjoy E, Sekar M, Surendiran M, Kavitha L, Kumar TS (2013) Development of carbon nanotubes reinforced hydroxyapatite composite coatings on titanium by electrodeposition method. Corros Sci 73:321–330
89. Yan L, Xiang Y, Yu J, Wang Y, Cui W (2017) Fabrication of antibacterial and antiwear hydroxyapatite coatings via in situ chitosan-mediated pulse electrochemical deposition. ACS Appl Mater Interfaces 9(5):5023–5030
90. Lee CK (2012) Fabrication, characterization and wear corrosion testing of bioactive hydroxyapatite/nano-TiO_2 composite coatings on anodic Ti–6Al–4V substrate for biomedical applications. Mater Sci Eng, B 177(11):810–818
91. Huang Y, Zhang X, Mao H, Li T, Zhao R, Yan Y, Pang X (2015) Osteoblastic cell responses and antibacterial efficacy of Cu/Zn co-substituted hydroxyapatite coatings on pure titanium using electrodeposition method. RSC Adv 5(22):17076–17086
92. Saremi M, Golshan BM (2007) Microstructural study of nano hydroxyapatite coating obtained by pulse electrodeposition process on Ti–6Al–4V. Trans IMF 85(2):99–102
93. Abbas NM, Solomon DG, Bahari MF (2007) A review on current research trends in Electrical Discharge Machining (EDM). Int J Mach Tools Manuf 47:1214–1228
94. Prakash C, Kansal HK, Pabla BS, Puri S (2015) Processing and characterization of novel biomimetic nanoporous bioceramic surface on β-Ti implant by powder mixed electric discharge machining. J Mater Eng Perform 24(9):3622–3633
95. Peng PW, Ou KL, Lin HC, Pan YN, Wang CH (2010) Effect of electrical-discharging on formation of nanoporous biocompatible layer on titanium. J Alloy Compd 492(1–2):625–630
96. Bin TC, Xin LD, Zhan W, Yang G (2011) Electro-spark alloying using graphite electrode on titanium alloy surface for biomedical applications. Appl Surf Sci 257(15):6364–6371
97. Prakash C, Kansal HK, Pabla BS, Puri S (2015) Potential of powder mixed electric discharge machining to enhance the wear and tribological performance of β-Ti implant for orthopedic applications. J Nanoeng Nanomanuf 5(4):261–269
98. Prakash C, Kansal HK, Pabla BS, Puri S (2017) Experimental investigations in powder mixed electric discharge machining of Ti–35Nb–7Ta–5Zrβ-titanium alloy. Mater Manuf Process 32(3):274–285

99. Prakash C, Kansal HK, Pabla BS, Puri S (2016) Multi-objective optimization of powder mixed electric discharge machining parameters for fabrication of biocompatible layer on β-Ti alloy using NSGA-II coupled with Taguchi based response surface methodology. J Mech Sci Technol 30(9):4195–4204
100. Prakash C, Kansal HK, Pabla BS, Puri S (2016) Powder mixed electric discharge machining: an innovative surface modification technique to enhance fatigue performance and bioactivity of β-Ti implant for orthopedics application. J Comput Inf Sci Eng 16(4):041006
101. Prakash C, Uddin MS (2017) Surface modification of β-phase Ti implant by hydroaxyapatite mixed electric discharge machining to enhance the corrosion resistance and in-vitro bioactivity. Surf Coat Technol 326:134–145
102. Xie ZJ, Mai YJ, Lian WQ, He SL, Jie XH (2016) Titanium carbide coating with enhanced tribological properties obtained by EDC using partially sintered titanium electrodes and graphite powder mixed dielectric. Surf Coat Technol 300:50–57
103. Arun M, Duraiselvam V, Senthilkumar R (2014) Synthesis of electric discharge alloyed Nickel-Tungsten coating on tool steel and its tribological studies. Mater Des 63:257–262
104. Ekmekci N, Ekmekci B (2015) Electrical discharge machining of Ti6Al4V in hydroxyapatite powder mixed dielectric liquid. Mater Manuf Process. https://doi.org/10.1080/10426914.2015.1090591
105. Ou SF, Wang CY (2016) Fabrication of a hydroxyapatite-containing coating on Ti–Ta alloy by electrical discharge coating and hydrothermal treatment. Surf Coat Technol 302:238–243
106. Prakash C, Singh S, Pabla BS, Uddin MS (2018) Synthesis, characterization, corrosion and bioactivity investigation of nano-HA coating deposited on biodegradable Mg-Zn-Mn alloy. Surf Coat Technol 2018(346):9–18. https://doi.org/10.1016/j.surfcoat.2018.04.035
107. Wang Y, Khor K, Chang P (1998) Thermal spraying of functionally graded calcium phosphate coatings for biomedical implants. J Therm Spray Technol 7(1):50–57
108. Khor K, Yip C, Cheang P (1997) Ti-6Al-4V/Hydroxyapatite composite coatings prepared by thermal spray techniques. J Therm Spray Technol 6(1):109–115
109. Mostaghimi J, Chandra S (2002) Splat formation in plasma-spray coating. Pure Appl Chem 74(3):441–445
110. Fantassi S, Vardelle M, Fauchais P, Moreau C (1992) Investigation of the splat formation versus different particulate temperatures and velocities prior to impact. In: Berndt CC (ed) Proceeding of 13th international thermal spray conference. ASM International, Materials Park, OH, USA, Florida, pp 755–760
111. Lu YP, Li MS, Li ST, Wang ZG, Zhu RF (2004) Plasma-sprayed hydroxyapatite+titania composite bond coat for hydroxyapatite coating on titanium substrate. Biomaterials 25(18):4393–4403
112. Sarao TPS, Sidhu HS, Singh H (2012) Characterization and in vitro corrosion investigations of thermal sprayed hydroxyapatite and hydroxyapatite-titania coatings on Ti alloy. Metall Mater Trans A 43(11):4365–4376
113. Rattan V, Sidhu TS, Mittal M (2018) Study and characterization of mechanical and electrochemical corrosion properties of plasma sprayed hydroxyapatite coatings on AISI 304L Stainless Steel. In: J Biomimetics Biomater Biomed Eng, vol 35. Trans Tech Publications, pp 20–34
114. Rocha RC, Galdino AGDS, Silva SND, Machado MLP (2018) Surface, microstructural, and adhesion strength investigations of a bioactive hydroxyapatite-titanium oxide ceramic coating applied to Ti-6Al-4V alloys by plasma thermal spraying. Mater Res 21(4)
115. Yao H-L, Hu X-Z, Bai X-B, Wang H-T, Chen Q-Y, Ji G-C (2018) Comparative study of HA/TiO_2 and HA/ZrO_2 composite coatings deposited by high-velocity suspension flame spray (HVSFS). Surf Coat Technol 351:177–187
116. Hameed P, Gopal V, Bjorklund S, Ganvir A, Sen D, Markocsan N, Manivasagam G (2019) Axial suspension plasma spraying: an ultimate technique to tailor Ti_6Al_4V surface with HAp for orthopaedic applications. Colloids Surf B: Biointerfaces 173:806–815
117. Sarao TPS, Singh H, Singh H (2018) Enhancing biocompatibility and corrosion resistance of Ti-6Al-4V alloy by surface modification route. J Therm Spray Technol 1–13

118. Sarkar M, Jain VK (2017) Nanofinishing of freeform surfaces using abrasive flow finishing process. Proc Inst Mech Eng Part B: J Eng Manuf 231(9):1501–1515
119. Ghosh G, Sidpara A, Bandyopadhyay PP (2018) Review of several precision finishing processes for optics manufacturing. J Micromanuf, p 2516598418777315
120. Jain VK (2016) Nanofinishing: an introduction. In: Nanofinishing science and technology. CRC Press, pp 23–46
121. Jain VK (ed) (2016) Nanofinishing science and technology: basic and advanced finishing and polishing processes. CRC Press
122. Li S, Wang Z, Wu Y (2008) Relationship between subsurface damage and surface roughness of optical materials in grinding and lapping processes. J Mater Process Technol 205(1–3):34–41
123. Kuhar M, Funduk N (2005) Effects of polishing techniques on the surface roughness of acrylic denture base resins. J Prosthet Dent 93(1):76–85
124. Zaborski S, Sudzik A, Wołyniec A (2011) Electrochemical polishing of total hip prostheses. Arch Civ Mech Eng 11(4):1053–1062
125. Cheung C, Ho L, Charlton P et al (2010) Analysis of surface generation in the ultraprecision polishing of freeform surfaces. Pro Inst Mech Eng Part B J Eng Manuf 224(1):59–73
126. Shiou FJ, Chen CH (2003) Freeform surface finish of plastic injection mold by using ball-burnishing process. J Mater Process Technol 140(1–3):248–254
127. Jain VK (2008) Abrasive-based nano-finishing techniques: an overview. Mach Sci Technol 12(3):257–294
128. Ohmori H, Li W, Makinouchi A, Bandyopadhyay BP (2000) Efficient and precision grinding of small hard and brittle cylindrical parts by the centerless grinding process combined with electro-discharge truing and electrolytic in-process dressing. J Mater Process Technol 98(3):322–327
129. Zhang D, Li C, Jia D et al (2014) Grinding model and material removal mechanism of medical nanometer zirconia ceramics. Recent Pat Nanotechnol 8(1):2–17
130. Ohmori H, Nakagawa T (1995) Analysis of mirror surface generation of hard and brittle materials by ELID (electronic inprocess dressing) grinding with superfine grain metallic bond wheels. CIRP Ann Manuf Technol 44(1):287–290
131. Kotani H, Komotori J, Mizutani M et al (2009) Surface finishing and modification for cobalt-chromium-molybdenum alloy by electrolytic in-process dressing (ELID) grinding. In: 5th international conference on leading edge manufacturing in 21st century, LEM 2009
132. Kotani H, Komotori J, Naruse T et al (2013) Development of a new grinding system for finishing of hemispherical inside surface. Int J Nanomanuf 9(1):77–86
133. Baghel P, Singh S, Nagdeve L, Jain VK, Sharma ND (2015) Preliminary investigations into finishing of artificial dental crown. Int J Precis Technol 5(3–4):229–245
134. Nagdeve L, Jain VK, Ramkumar J (2016) Experimental investigations into nano-finishing of freeform surfaces using negative replica of the knee joint. Procedia CIRP 42:793–798
135. Sarkar M, Jain VK (2017) Nanofinishing of freeform surfaces using abrasive flow finishing process. Proc Inst Mech Eng Part B: J Eng Manuf 231(9):1501–1515
136. Sidpara AM, Jain VK (2012) Nanofinishing of freeform surfaces of prosthetic knee joint implant. Proc Inst Mech Eng Part B: J Eng Manuf 226(11):1833–1846
137. Nagdeve L, Jain VK, Ramkumar J (2016) Experimental investigations into nano-finishing of freeform surfaces using negative replica of the knee joint. Proc CIRP 42:793–798
138. Nagdeve L, Jain VK, Ramkumar J (2018) Differential finishing of freeform surfaces (knee joint) using R-MRAFF process and negative replica of workpiece as a fixture. Mach Sci Technol 22(4):671–695
139. Kumar S, Jain VK, Sidpara A (2015) Nanofinishing of freeform surfaces (knee joint implant) by rotational-magnetorheological abrasive flow finishing (R-MRAFF) process. Precis Eng 42:165–178
140. Barman A, Das M (2018) Nano-finishing of bio-titanium alloy to generate different surface morphologies by changing magnetorheological polishing fluid compositions. Precis Eng 51:145–152

Harjit Singh graduated with B. Tech in mechanical engineering from PTU Punjab and completed his Master in engineering from PEC University of Technology Chandigarh (UT), and doing PhD from Lovely Professional University Punjab in the field of coating on biomaterials. He guided the graduate students in their capstone projects like laser engraving machine, design of an electrically actuated artificial limbs to reduce the fatigue, SMS-controlled material handling system.

Sunpreet Singh is Assistant Professor in the School of Mechanical Engineering, Lovely Professional University, Jalandhar, India. He has received Ph. D. in Mechanical Engineering from Guru Nanak Dev Engineering College, Ludhiana, India. His areas of research are additive manufacturing and application of 3-D printing for development of new biomaterials for clinical applications. He has contributed extensively in additive manufacturing literature with publications appearing in Journal of Manufacturing Processes, Composite Part: B, Rapid Prototyping Journal, Journal of Mechanical Science and Technology, Measurement, International Journal of Advance Manufacturing Technology, and Journal of Cleaner Production. He authored ten chapters and monographs. He is working with joint collaboration with Prof. Seeram Ramakrishna, NUS Nanoscience and Nanotechnology Initiative, and Prof. Rupinder Singh, Manufacturing Research Lab, GNDEC, Ludhiana.

Chander Prakash is Associate Professor and head of Industrial Engineering in the School of Mechanical Engineering, Lovely Professional University, Jalandhar, India. He has received Ph. D. in mechanical engineering from Panjab University, Chandigarh, India. His areas of research are biomaterials, rapid prototyping and 3-D printing, advanced manufacturing, modeling, simulation, and optimization. He has more than 11 years of teaching experience and 6 years research experience. He has contributed extensively to the world in the titanium- and Magnesium-based implant literature with publications appearing in Surface and Coating Technology, Materials and Manufacturing Processes, Journal of Materials Engineering and Performance, Journal of Mechanical Science and Technology, Nanoscience and Nanotechnology Letters, Proceedings of the Institution of Mechanical Engineers, Part B: Journal of Engineering Manufacture. He authored 50 research papers and 10 chapters. He is also editor of three books: “Current Trends in Biomanufacturing”, Springer Series in Advanced Manufacturing, Springer International Publishing AG, Gewerbestrasse 11, 6330 Cham, Switzerland., Dec. 2018; “3D Printing in Biomedical Engineering”, Book series Materials Horizons: From Nature to Nanomaterials, Springer International Publishing AG, Gewerbestrasse 11, 6330 Cham, Switzerland., August 2019, and “Biomaterials in Orthopaedics and Bone Regeneration—Design and Synthesis”, Book series Materials Horizons: From Nature to Nanomaterials, Springer International Publishing AG, Gewerbestrasse 11, 6330 Cham, Switzerland., March 2019. He is also guest editor of three journals: Guest Editor of Special Issue of “Functional Materials and Advanced Manufacturing”, Facta Universitatis, Series: Mechanical Engineering (Scopus Index), Materials Science Forum (Scopus Index), and Special Issue on “Metrology in Materials and Advanced Manufacturing”, Measurement and Control (SCI indexed).

Recent Advances in Additive Manufacturing of Bio-inspired Materials

Swaroop Gharde, Aarsha Surendren, Jay M. Korde, Shubham Saini, Nikit Deoray, Rajendra Goud, Sunil Nimje and Balasubramanian Kandasubramanian

Abstract The changing scenario in the rapidly developing economies and industries requires bulk-scale fabrication of materials. The production of materials on such a large scale by industries requires high quality, low-cost production, and high efficiency, in order to sustain the innovative market competition. Complexities like high initial tooling, part design restrictions, bounded degree of designing freedom, and machinery cost in traditional manufacturing have led to the need of new approaches and techniques of manufacturing. To overcome these complexities, additive manufacturing (i.e. 3D Printing) has been proven to be a paramount method, which has the potential to perform all the operations in one place such as cutting, forming, bending or transforming materials and components for further assembling in one part and in short time, due to which it is also useful in biomedical applications from medicine to anthropology. Recently, the polymers have become prime choice of the materials for additive manufacturing, and various thermoplastic materials like acrylonitrile butadiene styrene (ABS) and polylactic acid (PLA) together with thermosetting polymeric materials can be easily processed by 3D printing. This chapter discusses 3D printing of various biologically inspired structures like molluscan shell and honeycomb structure with above-mentioned matrix materials and their reinforce-

S. Gharde · R. Goud · B. Kandasubramanian (✉)
Rapid Prototyping Laboratory, Department of Metallurgical and Materials Engineering, Defence Institute of Advanced Technology (DU), Ministry of Defence, Girinagar, Pune 411025, India
e-mail: meetkbs@gmail.com

A. Surendren
Centre for Biopolymer Science and Technology, Central Institute of Plastics Engineering & Technology, Udyogamandal, Eloor, Kochi, Kerala 683501, India

J. M. Korde · N. Deoray
Department of Plastic and Polymer Engineering, Maharashtra Institute of Technology, Aurangabad, Maharashtra 431028, India

S. Saini
Dr. B. R. Ambedkar National Institute of Technology, Jalandhar, India

S. Nimje
Department of Mechanical Engineering, Defence Institute of Advanced Technology (DU), Ministry of Defence, Girinagar, Pune 411025, India

C. Prakash et al. (eds.), *Biomanufacturing*,
https://doi.org/10.1007/978-3-030-13951-3_2

ments with synthetic and natural fibres. The developed materials were characterized via Fourier-transform infrared spectroscopy (FTIR), wear test analysis and impact strength analysis (ASTM standard). Finally, the chapter concludes with a discussion on future scope of 4D printing for additive manufacturing.

Keywords 3D printing · Molluscan shell structures · Honeycomb structures · Acrylonitrile butadiene styrene (ABS) · Polylactic acid (PLA) · 4D printing

1 Introduction to Rapid Prototyping

Nature being a massive productive laboratory covers a list full of efficient details for an extensive collection of technical and scientific problems [1]. Nature has inspired the human kind with concepts for contouring a range of innovative designs for systems and high-performance materials serving from macro- to nanoscale [2, 3]. In the recent scenario, in order to sustain the market competition before introducing the product or material in the market, they need to be qualitatively assessed in the product development stage. The initial product assessment can be done by rapid prototyping of the materials, which mainly involves technologies such as 3D printing, freeze casting, layer-by-layer deposition, self-assembly and electrophoretic deposition (Table 1) [4, 5].

As compared to above-mentioned techniques, 3D printing possesses the ability to accomplish all the operations in one place such as cutting, forming, bending or transforming materials and components for further assembling in one part and in short time due to which it is also useful in biomedical applications like teeth, bone, organ scaffolding, bionic ears, skin, kidney, heart, tissue hand, leg, etc. [6–18]. Historically, in 1986, Charles Hull was the first to depict additive manufacturing (AM), also called as 3D printing, solid free-form (SFF) or rapid prototyping (RP), as a 'process of combining materials for engineering monoliths from 3D model data in a layer-by-layer mechanism' [19]. 3D printing is concurrently implemented with additive manufacturing (AM) which has the capability to fabricate complex architecture with high resolution, rapid concoction, easy prototyping, and its aptitude to assign material properties at the sub-millimetre scale, inspiring multimaterial, functionally graded designs, thus making it an attractive option for composite material development [20] (Figs. 1 and 2). Additionally, 3D printing facilitates the mimicking of molluscan shell high-strength bio-inspired structures like honeycomb [12], molluscan shell [4, 21], etc. Various well-established techniques for manufacturing of polymeric composite structures like stereolithography (SLA) [22], selective laser sintering (SLS) and extrusion-based techniques like fused deposition modelling (FDM) or fused filament fabrication (Table 2) have also been explored by researchers. Considering the end applications, metals, ceramics and polymers can be used for 3D printing. Considering the easy and rapid fabrication, flexibility, noticeable mechanical properties, chemical resistance, etc., the polymers have become the prime choice of materials for 3D printing. Thermoplastic polymeric materials like acrylonitrile butadiene styrene

Table 1 Advantages and disadvantages of rapid prototyping techniques [4, 5]

Synthesis technique	Advantages	Disadvantages
Freezing casting	– Variety of materials can be used – Controlled interphase behaviour – Good control over the thickness of the various composite	– Energy-consuming process – Difficult to control the segmented overlap minerals
Layer-by-layer deposition	– Good control over the nanodimension thickness – Ability to manufacture homogenous film – Good control of the layered structure	– Time-consuming – Mechanical property reduces at higher concentration of additives
Electrophoretic deposition	– Simple process – Large film forming capability	– Mechanical properties are low
Mechanical assembly	– Economically feasible and relatively faster process	– Nanodimension thickness control is not easy
Chemical self-assembly	– Molecular-level assembly is possible to bulk material construction	– Difficult to fabricate large-scale layered composite
3D printing	– Quick production, printing of various intricate structure, on-demand printing, capability to print various biological inspired architecture or structures with higher mineral concentration, exploited as an efficient predictive tool for enhanced synthetic materials	– Printing of limited number of materials, inability to combine nanoscale printing with macroscale design, control over surface quality

(ABS) and polylactic acid (PLA) together with thermosetting polymeric materials like epoxy resins can be easily processed by 3D printing [23, 24]. The modelling of the 3D printed layered of the various structures was contoured in the standard tessellation language (STL) format with the aid of designing software for the further meshing into the G-code by slicing it with the aid of Repetier-Host computational software with specific resolution for a meticulous mimicking orientation. 3D printing process involves polymer-based filament, which passes through heating nozzles where it melts and extrudes into a semi-liquid state. The diameter of the polymer-based filament defines the resolution for the building platform in a layer-by-layer fashion, wherein the layers amalgamate and solidify into final parts owing to the post-extrusion air convection process [25]. Concurrently, FDM has found utilization

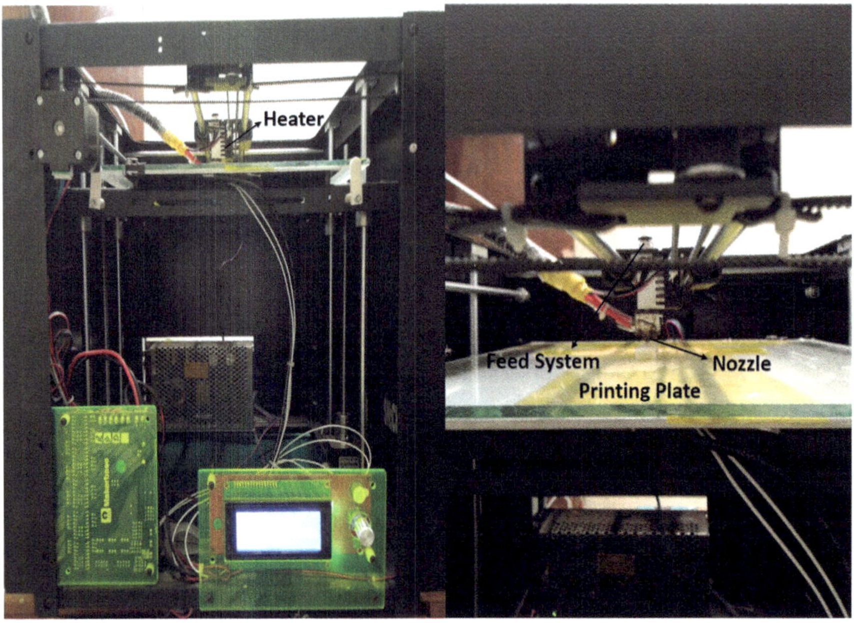

Fig. 1 Experimental set-up of 3D printing machine

Table 2 Advantages and disadvantages of different 3D printing methods [4]

Method	Principle/materials	Advantages	Disadvantages
SLS	Laser-induced sintering of powder materials (metals, ceramics and polymers)	– High-strength object – Self-supporting process – An extensive range of materials	– Only for the powder form of materials – Shrinkage of materials due to thermal stress
SLA	laser scanning and UV induced curing Material: liquid photopolymer resin	High printing resolution Available for biological materials	– Expensive – Time-consuming – Cytotoxicity
FDM	Extrusion deposition of materials Material: thermoplastics	– Low cost – Good strength – An extensive range of materials	– Anisotropy, chordal defects, nozzle clogging

for prototyping of ceramics and metals with the aid of polymeric binder materials [26, 27].

Though there is a widely available literature on 3D printing for the specific applications such as biomedical, structural, aerospace, automobile and defence applications, there is still a need for consolidation of widely used conventional and recently advanced works on 3D printing for a broader understanding of the scientific community. The chapter discusses process defects, which occur during the 3D printing of

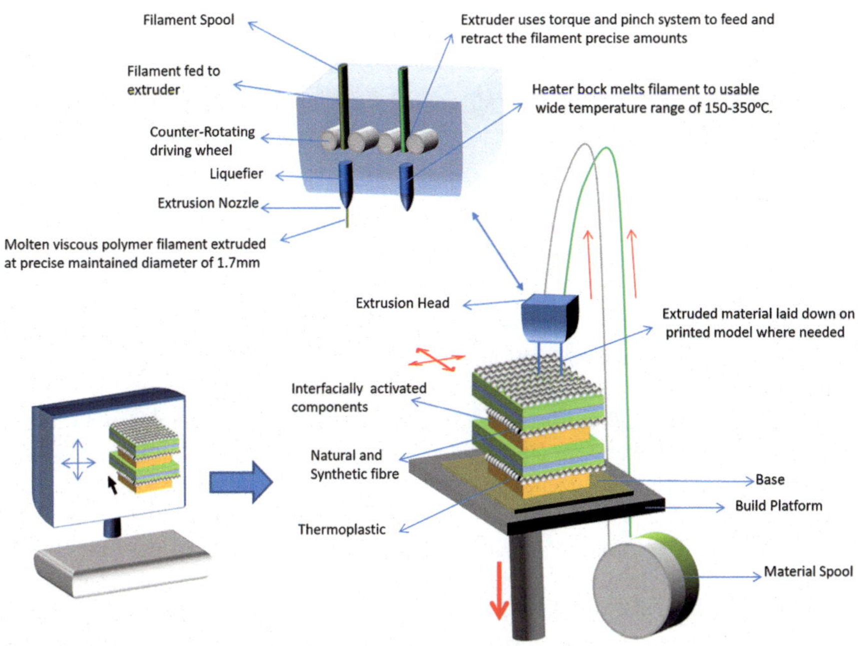

Fig. 2 Working mechanism of 3D printing process

materials such as voids in layers, weak interfacial forces and interlocking of nano-asperities. These defects cause the breakage of the object when a specific amount of stress or load is applied, thereby reducing service applicability of the 3D printed materials or products. The weak interfacial forces can also lead to delamination of the adjacent printed layer. Also, the extrusion temperature variation can also lead to the poor melt viscosity of the polymers, thus resulting in shrinkage of the printed layers. The chapter will give an overview on 3D printing of biologically inspired molluscan shell structures with indigenously fabricated in-laboratory extruded engineering thermoplastic materials such as ABS and PLA reinforced with natural and synthetic fibres such as human hair, sheep wool, coir, jute, cotton fabrics, polyester fabrics, nylon fabrics, PLA and ABS microfibres. Further, the effect of nanoclay and functionalization methods like silane coupling, sodium lauryl sulphate treatment, ethanol treatment, sodium hydroxide (NaOH) treatment and maleic anhydride treatment has been discussed in detail. The chapter also discusses the effect of the above-mentioned matrix materials and their reinforcements on 3D printed products, characterized via impact analysis (ASTM D256-06 standard, Tinius Olsen IT 503) and Fourier-transform infrared spectroscopy (FTIR). Finally, the chapter concludes with the future scope of 4D printing for additive manufacturing.

2 3D Printing of Bio-inspired Structures

In the following section, we have discussed the fabrication and biomimicking of different bio-inspired structures such as honeycomb and molluscan shell structures.

2.1 *Honeycomb Structures*

The honeycomb structure is the most common paradigm for artificial schemes for cellular materials mimicked from the prototypes elucidated in nature. Honeycomb structures have had the researchers and the designers conjointly enticed owing to its low density and high-performance mechanical properties coupled with a high stiffness which find extensive engineering applications in the sections of aerospace, transportation, chemical, mechanical, biomedical and architecture [28–30]. Many researchers have proposed flexible shaped honeycombs from an array of contouring methodologies like extrusion, folding techniques [31], thermoforming [32] and 3D printing [33, 34]. Extensive studies have been carried out for elucidating and understanding conjointly the mechanical behaviour [35–38] and the thermal behaviour [39–41] of 3D printed honeycomb composite structures (Fig. 3); however, the structural applications have been limitedly explored. The mechanical behaviour of hierarchical honeycomb monoliths was investigated by Ajdari et al. [42], wherein their elucidations exhibited that the first- and second-order hierarchical honeycombs stiffness is 2.0 and 3.5 times, respectively, higher in comparison with honeycomb with regular geometry showed 1.0–0.28 stiffness. The compressive strength of honeycomb structures exhibiting chirality was investigated by Scarpa et al. [43], where they elucidated that the hexagonal honeycombs with chirality demonstrated nearly 3.6 times higher normalized collapse stress consistently. The hexagonal geometry for the cellular monoliths can be elucidated conjointly in the 'Plantae and Animalia' kingdoms as the hexagonal shape exhibits the most stability. The geometry of honeycomb structure permits the minimization of the material for realizing minimum weight while simultaneously attaining maximum strength, owing to which it has proven to be an excellent innovation in the composite sector. The natural cellular structure of a cork was discovered to be similar to that of a honeycomb structure with hexagonal geometry by Robert Hook in 1665. The first structural application of the honeycomb structure was patented by Hofler and Renyi in 1914 [44].

2.1.1 Honeycomb Structures with Fabrics

We validated the contouring of 3D printed multilayered fabric-sandwiched honeycomb composite structures with augmented impact properties, which is an impregnated structure of honeycomb composites engineered from acrylonitrile butadiene styrene (ABS) fused with numerous fabrics like polyester, nylon and cotton with

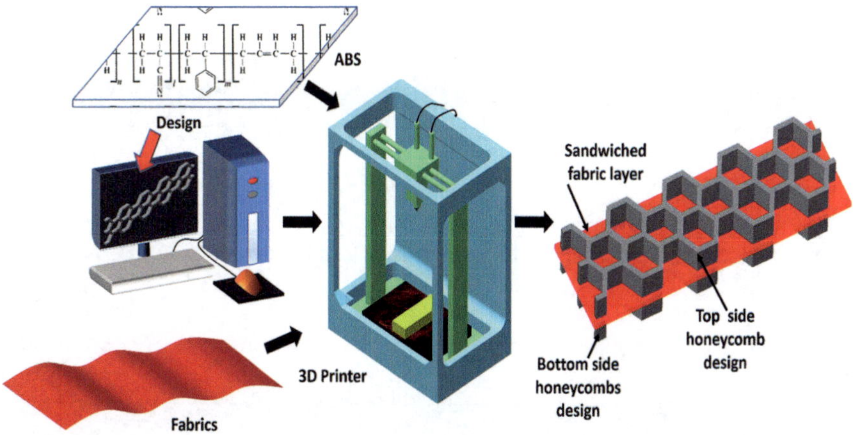

Fig. 3 Schematic representation of 3D printing of honeycomb structure

the aid of fused deposition modelling methodology. Comparative studies have been investigated between the pristine ABS honeycomb composite structure with the superficially maleated and non-maleated polyester, nylon and cotton fabrics. Functionalization of the surface of the fabrics with maleic anhydride was confirmed from FTIR characterization. The improvement of the impact properties of the 3D printed multilayered fabric-sandwiched honeycomb composite structure was revealed by impact analysis. Lightweight, facile engineering methodology and augmented impact properties exhibit 3D printed layered fabric-sandwiched honeycomb composite as a potential aspirant for mechanical applications in the mass transport and avionics sectors. Acrylonitrile butadiene styrene (ABS) (Absolac 920) granules were initially preheated at 80 °C for 3 h before being extruded into filament form at 250 °C having a diameter of 1.75 mm for 3D printing from a 20-mm twin screw extruder. Polyester, nylon and cotton fabrics were firstly washed with methanol in an ultrasonicator bath for 30 min and then washed again with deionized water for 30 min in the ultrasonicator bath for removing dirt and cleaning the fabric thoroughly. The cleansed fabrics were dried in an air convection-type oven at 120 °C until an equilibrium weight was elucidated. The maleation process was adopted for optimizing the interfacial adhesiveness between fabrics and the ABS matrix. The cleansed fabric was then immersed in 2% maleic anhydride (MA) solution at a constant temperature of 65 °C for a total period of 18 h following which they were further dried in an air convection-type oven at 85 °C until an equilibrium weight was attained.

Designing and Contouring of Honeycomb Structure

The hexagonal geometry was chosen owing to it being the most paradigm cellular structure. The height and the length of the hexagonal modelled design due to its

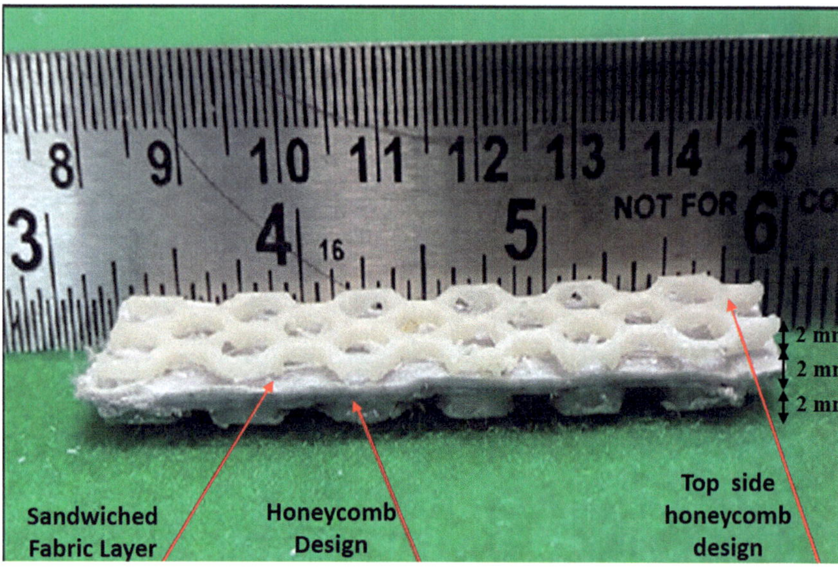

Fig. 4 3D printed layered fabric-sandwiched honeycomb composite structure

isotropic nature were equal with a dimension of 2 mm having a thickness of 1 mm. The angle between all the nodes of the cell of the honeycomb structure was kept symmetric in nature. The proposed 3D model design of the layered fabric-sandwiched honeycomb structure exhibited an overall dimension of 63.5 mm × 12.7 mm × 6 mm similar to that of the pristine ABS honeycomb design model, but the new proposed design honeycomb structure exhibited an alteration in the model having a sandwiched design where the top section and the bottom section exhibited a thickness of 2 mm, respectively, with a throughout hollow cavity of 2 mm, wherein the layered fabrics are to be sandwiched in the honeycomb structure (Fig. 3). The maleated and non-maleated polyester, cotton and nylon fabrics were multilayered for obtaining a thickness of 2 mm by the hand lay-up technique, wherein the adhesive used for bonding the fabrics was 30 wt% ABS solution. The sandwiched honeycomb composite structure with multilayers has been shown in Fig. 4.

Results and Discussions

Analysis of Impact Properties: For scrutinizing the impact properties of the 3D printed superficially grafted layered fabric honeycomb composite structure in comparison with the non-superficially grafted layered fabric honeycomb composite structure engineered via 3D printing, the composite samples were subjected to Izod impact test (ASTM D256-06 standard) for notched samples with a notch radius of 0.25 mm and depth of 2.54 mm, as depicted in Fig. 5. The sandwiching of layered textiles in

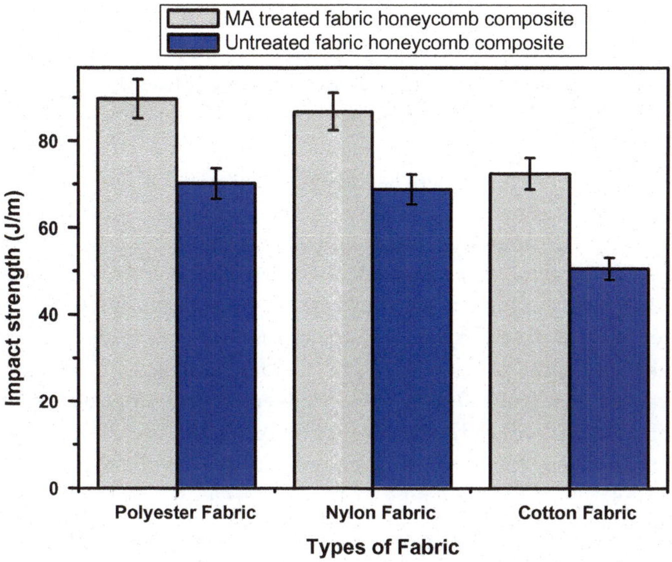

Fig. 5 Impact strength of different fabric-sandwiched honeycomb composites engineered by fused deposition modelling

the ABS honeycomb structure exhibited a value of 50.63 J/m in comparison with pure ABS honeycomb composite structure exhibiting a value of 13.24 J/m owing to the meshing of the fabrics in the composite structure. For 3D printed superficially maleated and non-maleated layered polyester fabric honeycomb composite structures, the superficially maleated layered polyester fabric honeycomb structures exhibited the highest impact value of 89.72 J/m in comparison with 70.22 J/m of the superficially non-maleated multilayered polyester fabric honeycomb structures. In case of 3D printed superficially maleated and non-maleated layered nylon fabric honeycomb composite structure, the superficially non-maleated layered Nylon fabric honeycomb structures exhibited an impact value of 68.87 J/m to that of 86.79 J/m impact energy of the superficially maleated layered nylon fabric-sandwiched honeycomb composite structure. For the 3D printed superficially maleated and non-maleated layered cotton fabric-impregnated honeycomb composite structures, the superficially maleated stacked cotton fabric layered honeycomb composite structure exhibited an impact value of 72.49 J/m, whereas the superficially non-maleated multilayered cotton fabric-sandwiched honeycomb composite structure showed an impact value of 50.63 J/m.

The successful interfacial grafting between the maleic anhydride and the fabrics exhibited augmented impact values by 27, 26 and 43% for the polyester, nylon and cotton fabrics sandwiched honeycomb composite structures, respectively, in comparison with the 3D printed superficially non-maleated multilayered fabric-impregnated honeycomb composite structures. It can be elucidated that the impact strength of the honeycomb structure increased with the sandwiching of layered fabric in the composite honeycomb structure in comparison with that of the pristine ABS honeycomb structure. The maleated fabrics sandwiched honeycomb structures exhibited a higher impact strength in comparison with the non-maleated fabric-sandwiched honeycomb structures owing to the strongly bonded interfacial adhesion between the fabrics and the ABS matrix, which was augmented on the treatment of the fabrics with the Maleic anhydride.

Analysis of FTIR Spectra: For scrutinizing the interaction between the maleic anhydride with the fabrics, the maleated fabrics were subjected to FTIR spectroscopy in comparison with the non-maleated fabrics for the region from 4000 to 500 cm^{-1}. The infrared spectra of (a), (b), (c), (d) are shown in Fig. 6. In the spectrum of MA (Fig. 6a), the absorption bands at 3736 cm^{-1} and 3612 represent characteristic O–H stretch, while the bands at 1779 cm^{-1} and 1698 cm^{-1} represent a characteristic C=O stretch of carboxylic acid [45]. The absorption bands at 1277, 1232 and 1039 cm^{-1} are owing to the C–O stretching of the carboxylic acids, while the bands at 891 and 827 cm^{-1} depict the C=C stretching of the anhydride [45, 46]. In the spectrum of cotton fabric (Fig. 6b), the absorption bands at 1320, 1154, 1113 and 989 cm^{-1} were elucidated conjointly in the maleated and non-maleated fabrics owing to the C–O bending vibrations [47]. The absorption band at 3268 cm^{-1} corresponds to the H-bonded O–H stretching, whereas the absorption bands at 3735, 2933 and 1437 cm^{-1} correlate to the free hydroxyl O–H stretching of an alcohol group, C–H stretching of alkanes and C–C in-ring stretching of aromatic groups, respectively, for the cotton fabric. New absorption bands at 1712 and 1632 cm^{-1} elucidated in the maleated cotton fabrics were associated with C=O stretching of carboxylic acid testifying the grafting of maleic anhydride on the surface of cotton fabrics [48].

In the spectrum of polyester fabric (Fig. 6c), the absorption bands at 1341, 1238 and 1020 cm^{-1} were elucidated in both the maleated and non-maleated polyester fabric corresponding to the stretching of the C–O ether linkages, while the weak absorption bands at 1521 and 1400 cm^{-1} correspond to C–C in-ring stretching of the aromatic groups. The strong absorption band at 1705 cm^{-1} corresponds to the stretching of C=O ester group, whereas the band at 714 cm^{-1} associates with the heterocyclic aromatic ring of the polyester fabric [49]. From Figs. 6a, c it is apparent that the boost of the intensities of major of the characteristic absorption bands is owing to the interfacial interaction between the anhydride and the fabric, corresponding to the grafting of the maleic anhydride on the polyester fabric surface. In the spectrum of nylon fabric (Fig. 6d), the weak absorption bands at 1339 and 1022 cm^{-1} conjointly with strong absorption bands at 1234 and 1086 cm^{-1} corresponding to the stretching of C–N group were observed in both the maleated and non-maleated nylon fabric [49]. The strong absorption band at 1707 cm^{-1} represents the C=O group, whereas the absorption band at 1516 cm^{-1} is owing to the combination of bending of N–H and stretching of C–N groups [50]. It is noticeably evident from Fig. 6a, d that the intensities of most of the absorption bands of the maleated Nylon fabrics augmented with a slight decreasing in a few absorption bands associated the grafting of the maleic anhydride onto the surface of nylon fabric owing to the interfacial interaction between the maleic anhydride and the fabric [50].

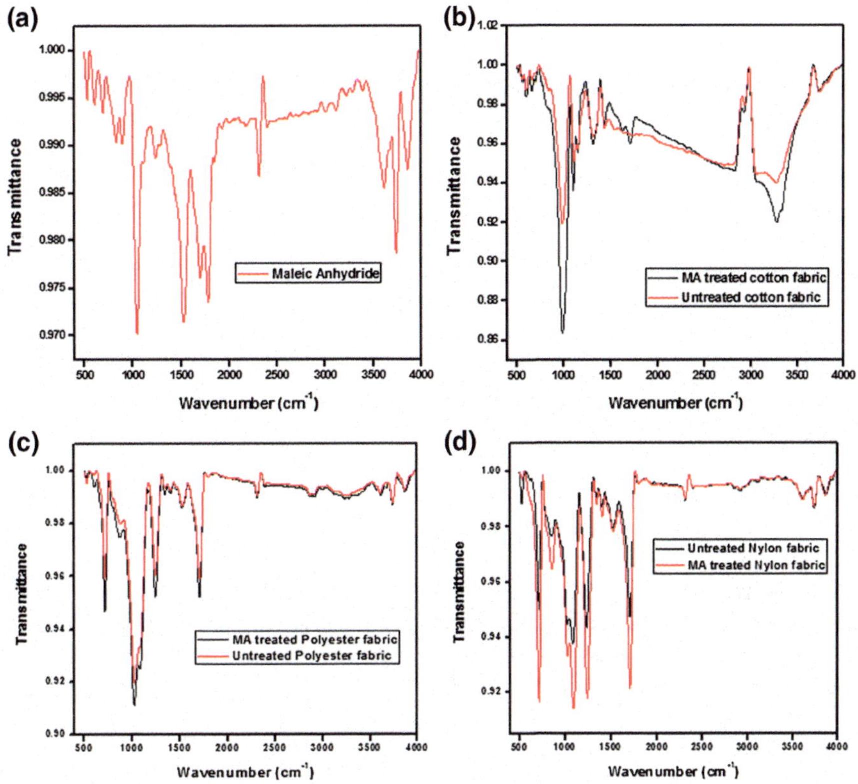

Fig. 6 FTIR spectra of: **a** maleic anhydride; **b** cotton fabrics; **c** polyester fabrics; **d** nylon fabrics

2.2 Molluscan Shell Structures

Molluscan shells, a natural biocomposite with hierarchical structures having various rough interfaces of individual tablets, contribute to the strength and toughness of mollusc shell-inspired materials such as mineral bridges and nano-asperities. It has a role to protect the soft body from the killers and from the environment. It contains >95% of calcium carbonate and 0.1–5% of organic matrices, the organic matrix as both adhesive and cohesive bond glue together the stacked polygonal mineral tablets due to that the enhanced microstructures, mechanical strength and fracture toughness it is attractive as natural composites materials. In this section, we have studied 3D printing of different molluscan structures with the incorporation of different natural fibres and films [51, 52].

2.2.1 Molluscan Shell with Films

We validated the contouring of 3D printed multilayered film-sandwiched composite structures explored through notched Izod impact test and the aggrandized impact property achieved by amalgamating solution-casted films of pristine ABS and different wt% (i.e. 0.5, 1, 1.5 wt%) loaded organoclay-functionalized ABS nanocomposite films in between 3D printed layers. The impact characteristics of the 3D printed samples were evaluated according to the ASTM D256-06 standard dimension (63.5 mm × 12.7 mm × 6 mm), the rapidly engineered layers' grafted composites were characterized by FTIR analysis, and the tribological possessions were premeditated by wear analysis.

Preparation of Solution-Casted Films

Acrylonitrile butadiene styrene (ABS) (Absolac 920) granules were initially preheated at 80 °C for 3 h before being extruded into filament form at 250 °C having a diameter of 1.75 mm for 3D printing from a 20-mm twin screw extruder. In the solution casting, the thin film is made by the evaporation of solvent from the solution, which is casted on a substrate, without applying any external stress or energy. The primary homogeneous solution is primed by dissolving 20 wt% ABS in 10 mL DCM solution and organically modified MMT with high-purity, surface-compatibilized montmorillonites having d-spacing of 1.8–2.2 nm and contains silane coupling agent as nanoclay edge treatment agent to enhance the dispersion of polymer resin is loaded in different concentrations such as 0.5, 1.0, 1.5 wt% in DCM solution and sonicated for 20 min with a frequency of 20 kHz and the probe tip (13 mm in diameter) with titanium alloy and homogeneously dispersed in the primary solution by continuous stirring of laboratory magnetic stirrer at 250 rpm for 1–2 h in room temperature. The homogenized solutions were poured into a hydrophobic glass plate, and the solvent is evaporated at ambient temperature and pressure. Films of desired dimensions (63.5 mm × 12.7 mm) with 2 mm thickness were expurgated using razor blade from the resultant casted films (Fig. 7). The temperature of the forming film, entrapped gas bubbles in casting solution and non-uniform dispersion of organoclay particles would greatly affect the properties of films.

Characterization Techniques: Impact specimen of ASTM standard dimension (63.5 mm × 12.7 mm × 6 mm) was fabricated by 3D printing multilayer thick laminates of 2 mm thickness integrated in between 2-mm-thick solution-casted films of pristine ABS and different wt% loaded organoclay/ABS nanocomposites with a notch radius of 0.25 mm and depth of 2.54 mm. Fourier-transformed infrared spectra of organoclay and solution-casted films were recorded at a wavenumber range of 500–4000 cm^{-1} having eco Zn–Se ATR crystal with a gold-coated mirror. The films were investigated in transmittance mode at a resolution of 4 cm^{-1} with 25 scans at room temperature. Specimen of 27 mm length, 10 mm width and 6 mm thickness is expurgated from the FDM samples for wear test, performed on a pin on disc friction and wear test rig arrangement by dry sliding on an abrasive surface at room

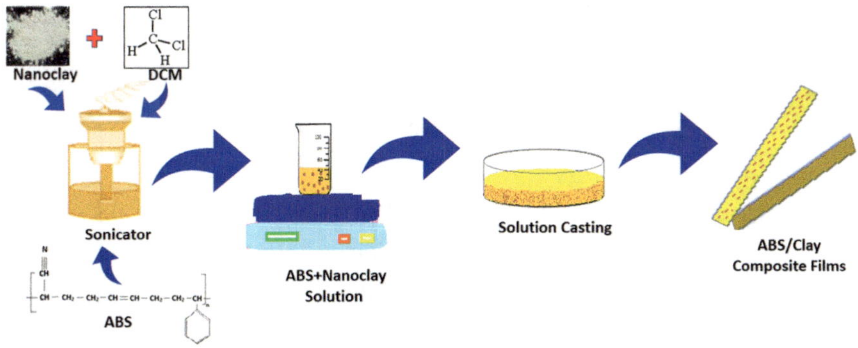

Fig. 7 Solution casting of ABS/organoclay nanocomposite (NC) films

temperature. The tests were conducted in the sliding distance of 25.13 m with the constant sliding speed of 0.418 m/s, a load of 9.8 N, track radius of 16 at 50 rpm for about 5 min.

Results and Discussion

Analysis of Impact Properties: Impact properties of the notched 3D printed samples are scrutinized, and from Fig. 8, solution-casted pristine ABS film-incorporated 3D printed multilayer laminate sample (51.09 J/m and 0.66 J) shows a tremendous increase in impact strength and breaking energy compared to that of fully 3D printed specimen (13.25 J/m and 0.17 J). The increment in impact properties of film integrated structured sample is due to its resistance to fracture propagation path along the sandwich structure, and the decrease of impact property in the fully 3D printed sample is due to its porous structure [53]. Considering 3D configured porous structured sample, the incorporation of solution-casted films of the non-porous structure with high material volume and more elastic behaviour gives a lamellar structure which transfers stress and thereby hinders the crack propagation path, which lead to enhancement in the toughness of the film integrated specimens [54, 55].

The impact strength and energy of the ABS-based 0.5 wt% organically modified montmorillonite-incorporated solution-casted ABS nanocomposite film-reinforced composite-integrated FDM specimen were observed as 88.73 J/m and 1.12 J, respectively. On further addition of 1 wt% (58.07 J/m and 0.75 J) and 2 wt% (35.10 J/m and 0.44 J) organoclay-reinforced ABS nanocomposite solution-casted film integrated FDM sample shows a decrease in impact strength and energy. The impact strengths of ABS/organoclay nanocomposite fabricated by 3D printing (film-sandwiched layered structure) with other ABS nanocomposites processed using other techniques are illustrated in Table 3. Maximum impact strength was depicted in solution-casted 0.5 wt% clay-loaded ABS nanocomposite film-incorporated 3D printed structure (85% increase) than other processing techniques [56–59]. This shows that 3D printed

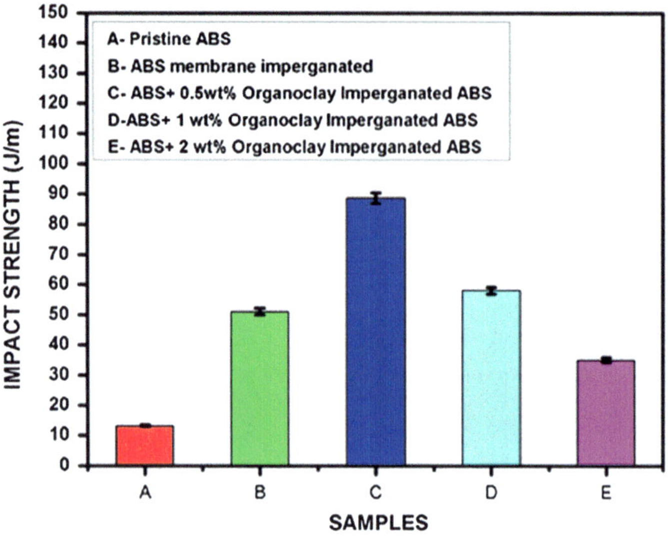

Fig. 8 Impact strength of different samples

Table 3 Impact strengths of 3D printed film-sandwiched structure of ABS composites with other processing techniques

Processing technique	Polymer	Reinforcement	Percentage increase/decrease in impact strength
Extruded composite Injection moulded samples	ABS	Organoclay (5 wt%)	84% decrease
Extruded composite Injection moulded samples	ABS	NPCC (nanosized precipitated $CaCO_3$) (2 wt%)	13% increase
Extruded composite Injection moulded samples	ABS	Calcite-filled (11.7 vol.%)	69% decrease
Extruded composite Injection moulded samples	ABS	Short glass fibre (15 vol.%)	38% decrease
Two roll mill	ABS	Organoclay (5 wt%)	90% decrease
3D printed film sandwiched	ABS	Organoclay (0.5 wt%)	85% increase

film-sandwiched layered structure with critical filler loading can achieve augmented impact property.

Surface Hardness Analysis: The resistance of a material to plastic deformation is quantified by hardness measurement. Depth-sensing indentation hardness of 3D printed specimen surface was performed by Vickers microhardness tester having a minute indenter with a constant load of less than 1 kg [60]. Dimensions of indentation and load are the important parameters used for the calculation of Vickers hardness number, H_v, by following Eq. (1).

$$H_v = \frac{2F \sin \frac{\theta}{2}}{d^2} \quad (1)$$

where θ, d and F denote the face angle of a pyramidal diamond indenter (136°), arithmetic mean of the two diagonals, $d1$ and $d2$ in mm, and test load (N), respectively. H_v of 13.63 was observed for 3D printed surface with $d1$ and $d2$ with 0.1631 mm and 0.1668 mm, respectively. It shows that Vickers hardness of 13 obtained using 1 kg of load on 3D printed ABS samples.

Analysis of FTIR Spectra: The IR characteristics peaks present in the region of 1482 and 1055–1020 cm^{-1} were assigned to the aliphatic C–H vibrations owing to the organic modifier aliphatic chains and Si–O–Si bond in clay and nanocomposites, respectively [45, 61]. The asymmetric stretching, symmetric stretching and bending of C-H vibrations of methylene groups at 2920, 2849 and 1463 cm^{-1} bands show the confirmation of organoclay [62]. 3625 and 920 cm^{-1} show corresponding peaks to the stretching vibrations of OH in clay and Al–OH–Al deformation of aluminates of clay, respectively [62, 63]. The OH bond vibrations disappear on higher loading of organoclay in ABS nanocomposite, which is attributed to interaction between hydroxyl group of organically modified clay surface and nitrile group of Styrene acrylonitrile (SAN), which results in poor stress transfer at the polymer-organoclay interface and make it rigid, thereby causing the reduction in impact load resistance [64]. The characteristics peaks of ABS polymer are observed at 2280–2240 cm^{-1} (aliphatic cyanide/nitrile stretching), 3150–3020 cm^{-1} (aromatic C–H vibrations), and 3000–2800 cm^{-1} (aliphatic vibrations). An absorption peaks occurred at 1615–1580 cm^{-1} are attributed to the double bond of styrene block in the ABS terpolymer. As the addition of organically modified MMT increases, the intensities of the filler particle bands (Si–O–Si bond) become stronger in the FTIR spectra of polymer clay nanocomposite (PCN) materials, Fig. 9.

Wear Analysis: Friction and wear are the tribological characteristics of a material, and polymers and its composites give only less wear value owing to the formation of a thin protective film while sliding against a steel counterface [65]. The solution-casted ABS film-impregnated 3D printed sample displays an augmentation in wear rate and diminution in coefficient of friction as compared to that of 3D printed virgin ABS (Fig. 10). The improvement in wear rate value is due to the abrasive nature of the multilayer laminated structure of the sample. Different wt% organoclay-loaded solution-casted ABS nanocomposite film-incorporated 3D printed samples show a decrease in wear rate due to the higher load bearing capacity of hard reinforcing material [66]. Simultaneously increase in wear rate occurs on further addition of organoclay owing to its adsorptive interaction of macromolecules with adequately developed clay filler surface and polymer transition to confined molecular mobility.

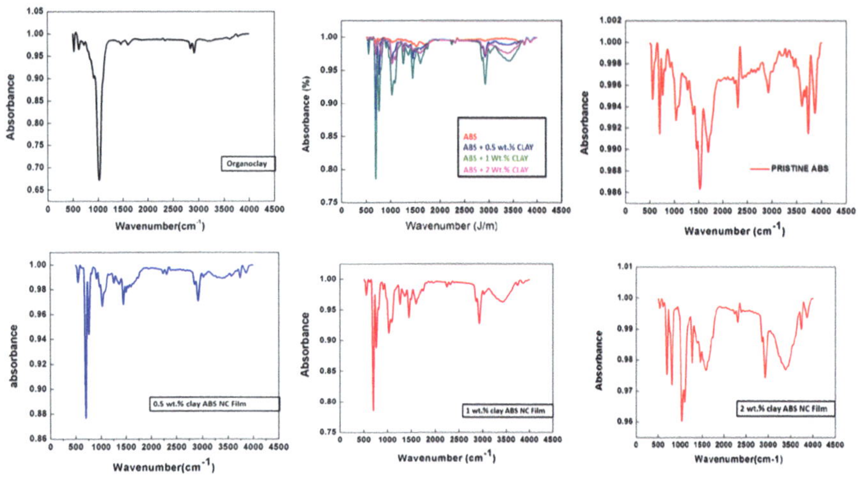

Fig. 9 FTIR spectra of 3D printed samples

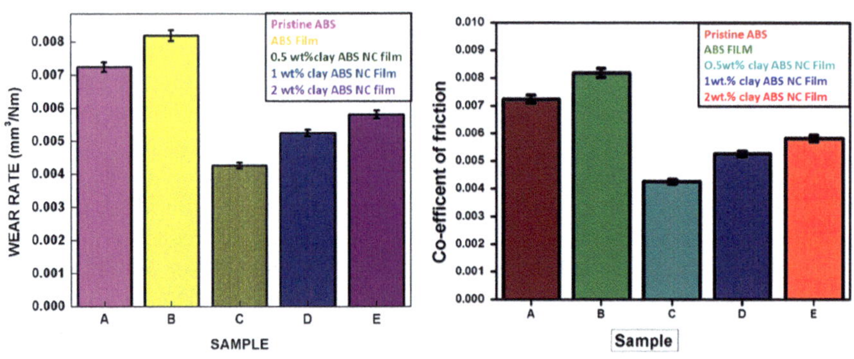

Fig. 10 Schematic representation of the wear rate and COF of 3D printed ABS

3D Printed PLA Nanocomposites

In one of our group study, we fabricated a biomimetic material with tremendous fracture toughness owing to its ingenious complex molluscan shell micro-architecture, which exhibits well-developed energy absorption characteristics stimulated as to design a novel bio mimicked molluscan shell-structured nanocomposite with augmented impact properties by 3D printing. Herein, inspired from the layered aragonite platelet/nanofibrillar chitin/protein structure of molluscan shell, artificial molluscan shell structures were primed by doctor blade coating of pristine PLA solution and different wt% organoclay-functionalized PLA solution sandwiched between 3D printed sample layers. The impact properties of the 3D printed samples were evaluated as per the standard ASTM D256-06 standard dimension (63.5 mm × 12.7 mm × 6 mm), and the rapidly engineered layers' grafted composites were characterized by FTIR

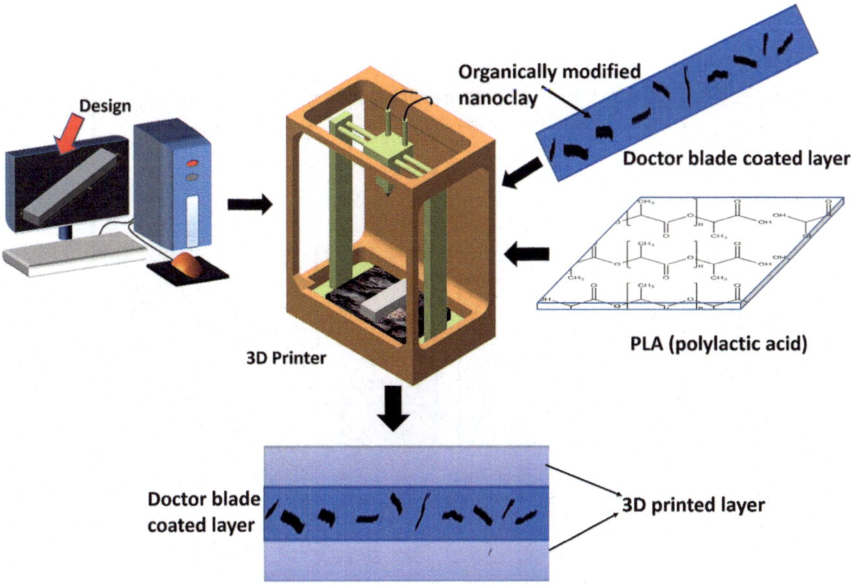

Fig. 11 Schematic representation of 3D printing of PLA/organoclay composites

analysis. Molluscan shell structure was fabricated by 3D printing technique, where it exhibited layers of 2-mm-thick laminate, which was sandwiched between two successive layers of PLA polymer. The 2-mm-thick coating of PLA and PLA/organoclay composites were contrived by the evaporation of solvent from the solution, which is coated on a substrate via doctor blade coating. In DBC, a restrained blade smears a unidirectional shear force to the viscous solution that passes through a small gap between the blade and the substrate. Capillary force is the major driving force for producing highly ordered multilayer of polymer nanocomposite [67]. The primary homogeneous solution is prepared by dissolving 20 wt% PLA in 10 mL DCM solvent. Different loadings of organically modified MMT such as 0.5, 1.0, 1.5 wt% were dissolved in DCM and sonicated for 60 min with a frequency of 20 kHz using probe tip (13 mm in diameter with titanium alloy) and homogeneously dispersed in the primary solution by continuous stirring using magnetic stirrer with 250 rpm for 1–2 h at room temperature. Final solution is coated on the 3D printed substrate of 2 mm thickness (Fig. 11).

Results and Discussion

Analysis of Impact Properties: Impact strength of the notched 3D printed samples of dimension (63.5 mm × 12.7 mm × 6 mm) is carried out, and from Fig. 12, doctor blade-coated pristine PLA-incorporated 3D printed sample (24.66 and 0.3404 J/m), biomimicking molluscan shell structure, shows tremendous increase

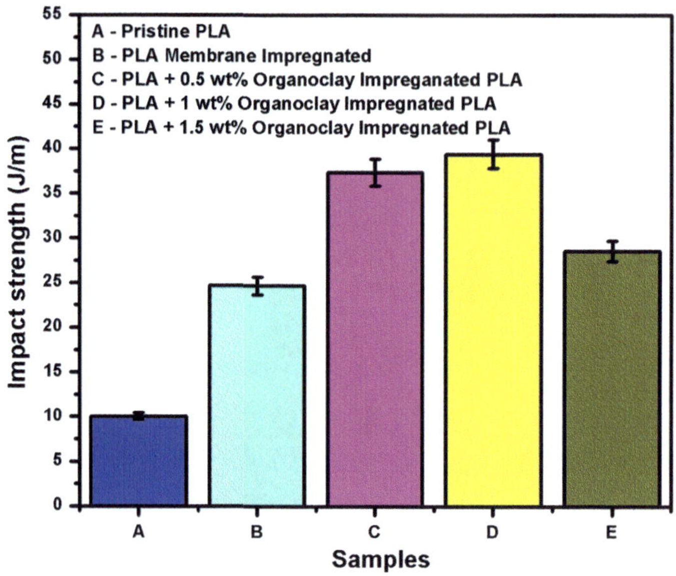

Fig. 12 Impact strength of different composites

in impact strength and breaking energy compared to that of fully 3D printed sample (10.0450 and 0.1265 J). The increase in impact properties of molluscan shell-structured sample is due to its difficult fracture propagation path along the sandwich structure, and the decrease of impact property in the fully 3D printed sample is due to its porous structure [54]. The impact property of doctor blade-coated 0.5 and 1.0 wt% organoclay-incorporated PLA solution integrated 3D printed sample displays an upsurge in impact strength and breaking energy from 37.34 J/m and 0.50 J to 39.42 J/m and 0.52 J, respectively. The improvement in impact property is due to the addition of organoclay, which has a high aspect ratio between 10:1 and 1000:1 and causes exfoliation of the composite, which is responsible for the improvement in fracture toughness [68]. Here, the organoclay particles will act as crack stoppers and form a tortuous crack propagation path which results in high impact energy at 1 wt% loading [69]. Further increment of organoclay (1.5 wt%) causes diminution in impact strength and breaking energy to 28.5779 J/m and 0.36 J, respectively. The decrease in impact property is due to the excess addition of filler leading to the non-uniform dispersion, which causes agglomeration. Due to agglomeration, stress concentration in the polymer matrix, poor bonding between the polymer chains and lower degree of clay–polymer interaction will lead to decrease in cross-linking density results in interfacial de-bonding and decrease in impact strength [69]. Among all 3D printed samples, PLA with 1 wt% organoclay solution-coated sample displayed the most effective enhancement in impact property with an impact strength of 39.42 J/m.

Analysis of FTIR Spectra: The peaks at 1748.17, 2920.19 and 3310.48 cm^{-1} were assigned to the C=O, C–H and O–H stretching of –CH(CH3)OH end group of PLA,

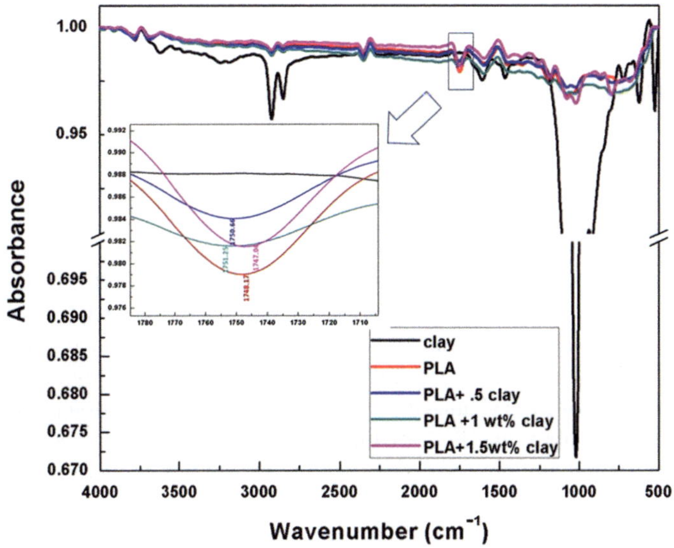

Fig. 13 FTIR spectra of organoclay, pristine PLA and different wt% clay loaded

respectively, and 1055–1020 cm^{-1} region shows the peak of Si–O–Si bond in clay and nanocomposites [45]. In the spectrum of organoclay, the peak around 3625 cm^{-1} shows O–H stretching bond of silica and the peak pertaining to the O–H disappeared in PLA nanocomposite due to the interaction between PLA and organoclay. In case of PLA/organoclay system, the CO band is shifted towards higher wavenumber, from 1748 to 1751 cm^{-1} (Fig. 13) due to the confinement of polymeric chains inside the silicate galleries (due to decreasing mobility and entanglement), which causes weakening of the polymer-polymer hydrogen bonding. The strengthening of the C=O bonds in the nanocomposites causes them to vibrate at higher frequencies which absorbs more energy and leads to higher impact properties [70]. The wavelength of CO bond again decreases to 1747 cm^{-1} in case of nanocomposite with 1.5 wt% of clay due to the collection of clay particles which causes non-uniform dispersion and hinders the polymer–polymer bond leading to decrease in impact properties [70].

2.2.2 Molluscan Shell Structures with Fibres

We have validated the contouring of 3D printed multilayered natural fibre composite structures explored through Izod impact test and the aggrandized impact property achieved by amalgamating of pristine PLA and different natural fibres (i.e. human hair and sheep wool) between 3D printed layers. The impact characteristics of the samples were evaluated according to the ASTM D256-06 standard dimension (63.5 mm × 12.7 mm × 6 mm), and the rapidly engineered layers' grafted composites were characterized by FTIR analysis.

Designing and Preparation of Composites

In this method, fabrication of composite is done by bottom-up layering by melt extruded PLA filament at 190–200 °C in a molten state through a nozzle onto a glass platform with a printing resolution of 150 μm, where the layered PLA solidifies into 2-mm-thick block. Subsequently, a 2-mm hand-laid layer of human hair or sheep wool bonded through an adhesive, i.e. 30% weight PLA solution, is impregnated and breathing time of 1 min is provided for evaporation of dichloromethane (DCM) from the adhesive solution. Ultimately, 2-mm-thick block is again printed through melt extrusion of PLA over the reinforcement layer. Initially, human hair and sheep wool were thoroughly washed with ethanol for 15 min and again washed with deionized water for the removal of dust and dirt from the fibres followed by drying. Human hair and sheep wool were silanized for the surface functionalization to obtain better adhesion between fibres and PLA by the formation of the siloxane bond. Typically, 10% weight relative to fibre weight of silane was dissolved in methanol/H_2O/acetic acid (75/8/17 v/v/v) solution at 25 °C with continuous stirring for 10 min. The fibres were moistened into silane containing mixture followed by drying at 60 °C for 25 min. Completion of the reaction was done after raising the temperature to 120 °C for 2 h. After completion, the fibres were cleaned with acetone for the removal of unattached silane and dried [71–73].

Results and Discussion

Analysis of Impact Properties: The impact properties of 3D printed PLA, superficially altered keratin fibre (human hair and sheep wool)/PLA-laminated 3D printed biocomposites and non-superficially grafted keratinous fibre 3D printed biocomposites were analysed and compared through IZOD impact strength (J/m) as depicted in Fig. 14. The ethanol-cleaned human hair-suffused PLA composite depicted a value of 140.35 J/m, whereas silanized human hair-infused PLA composite depicted a value of 208.54 J/m as compared to pristine 3D printed PLA which exhibited a value of 9.64 J/m. The augmented value of silanized human hair-infused composite is due to the better interfacial adhesion of fibre with PLA matrix owing to siloxane bond formation between human hair and PLA matrix. Similarly for silanized sheep wool-infused PLA composite, the augmented value of 63.42 J/m was recorded as compared to simply ethanol-cleansed sheep wool-infused PLA composite and pristine PLA whose values were 42.05 and 9.64 J/m owing to increased interfacial adhesion between sheep wool and PLA by silane treatment.

Analysis of FTIR Spectra: From previous studies, it is evident that the hair samples are rich in hydroxyl groups and are shown a broad peak in the range of 3200–3600 that corresponds to the –OH stretch of hydroxyl or other such groups. The PLA contributes to the peak at 1700 due to the C=O stretch that appears to be shifted due to the various treatments on the hair/PLA composites. The peak corresponding to 1500 might be attributed to the C=C stretch of aromatic rings present in the functional groups that are constituents of the hair sample. When the sample is treated

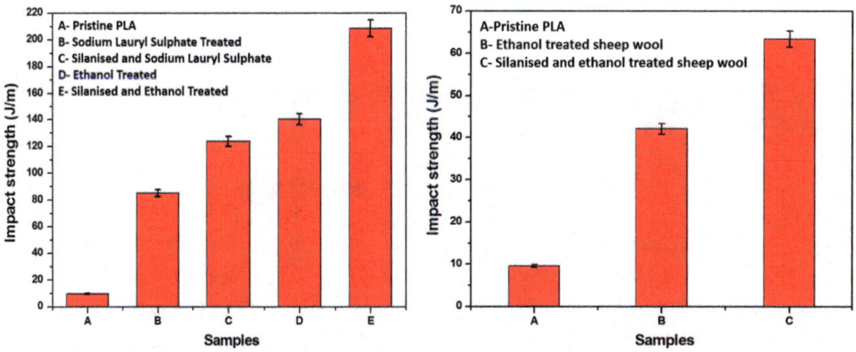

Fig. 14 Impact strength of different natural fibres (human hair and sheep wool) composites

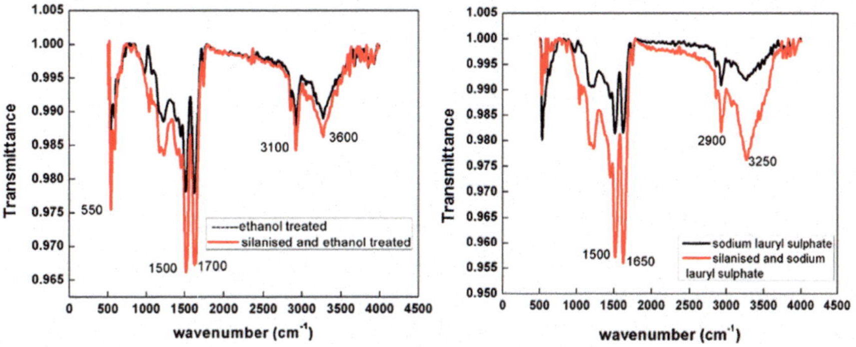

Fig. 15 FTIR spectra of human hair, ethanol-treated and silanized, and ethanol-treated

with ethanol, the hydroxyl groups form hydrogen bonds that reduce the –OH stretch intensity. The sample is further silanized and the concentration of hydroxyl increases leading to an increase in the broad peak at 3600. An interesting thing to notice is that the –C=O stretch peak has also increased in intensity. This indicates that the presence of the silanized group is affecting the polarity of the molecule and the carbon in the –C=O is becoming more electropositive. When the hair is washed with sodium lauryl sulphate, the surface is cleaned and thus the presence of any groups interfering with hydrogen bonding is removed. When this treatment is accompanied by silanization, the maximum number of hydroxyl groups appears to be silanized and an increase is visible in the intensity of the –OH peak. It can also be said that the addition of Si is affecting the –C=O groups and increasing the bond polarity as a tremendous increase is observed in the –C=O peak intensity (Fig. 15).

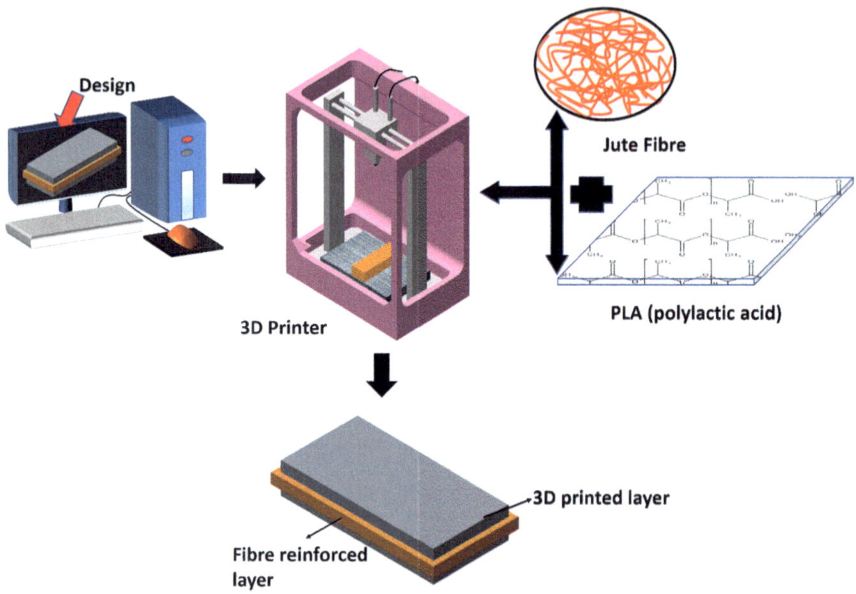

Fig. 16 Schematic representation of 3D printing of PLA/jute fibre composite

3D Printing of PLA with Jute Fibre

In one of our group research investigation, 3D printing methodology is exploited for engineering of superficially functionalized jute fibre-embedded polylactic acid sandwich pattern-laminated biocomposite. Aiming augmented mechanical properties, jute fibre surface is altered via chemical treatments such as mercerization and maleation for enhanced interfacial adhesiveness and homogeneity between jute fibre and PLA matrix. Superficial modification of jute fibre surface is analysed through Fourier-transform infrared spectroscopy (FTIR). Impact strength as per ASTM D256-06 standard dimension (63.5 mm × 12.7 mm × 6 mm) and wear resistance test have been explored for determination of augmented mechanical properties, owing to superficially altered jute fibre embedded in PLA matrix as a reinforcing agent (Fig. 16).

In this method, PLA filament was melt extruded at 190–200 °C, in semi-liquid state through nozzle in layer-by-layer fashion with printing resolution of 150 μm, onto the glass constructed platform where these layers fuse and solidify together in 2-mm-thick block upon which 2-mm-thick jute layer is deposited, bonded with a homogenous paste of 20% PLA (4 gm PLA in 20 ml DCM). Breathing time of 1 min is delivered for the evaporation of DCM from homogenous polymer paste, over which again 2-mm block is printed through melt extrusion of PLA in a semi-liquid state, considering the jute/PLA paste layer as a platform (Fig. 17).

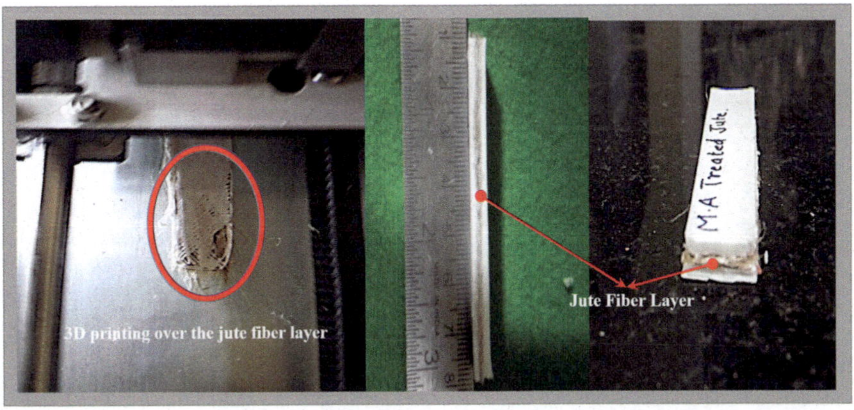

Fig. 17 Impact specimen of jute-embedded polylactic acid sandwich pattern-laminated biocomposite

Superficial Treatment: All retted jute fibre was immersed in deionized water for the duration of 15 min followed by drying process at 85 °C in air convection oven until the exclusion of moisture. They were cut into 6 mm length afore to superficial functionalization treatment. For enhancement of interfacial adhesiveness between jute fibre and polylactic acid matrix, jute fibre was superficially altered through chemical functionalization. (1) Mercerization treatment: In this process, jute fibre was immersed in 8% NaOH solution (8 gm NaOH in 100 ml deionized water) at room temperature for the duration of 12 h, further to neutralize superficial of jute fibre dried at 85 °C in an air convection oven. (2) Maleation treatment: The alkali-treated jute was immersed in 2% maleic anhydride solution (2 gm MA in 100 ml deionized water) for the duration of 18 h at a constant temperature of 65 °C, over which they are dried at 85 °C in air convection-type oven for the duration of 12 h.

Result and Discussion

Analysis of Impact Properties: The impact strengths of various PLA composite samples have been represented in Fig. 18. The notched Izod impact strength for superficially unmodified jute fibres/PLA biocomposites was augmented by 45.22% in comparison with pristine PLA due to efficient stress transfer between fibre and matrix, consistency in fibre length and homogeneity. PLA/NaOH-grafted jute fibre laminated biocomposites show 450% and 279% enhancement in impact strength values when compared against pristine PLA and unmodified jute fibre PLA, respectively. This enhancement is attributed to the mercerization, which eradicates impure elements like lignin and wax from the surface of jute fibre, thereby improving the homogeneity and interfacial adhesivenes with PLA matrix. Similarly, PLA/NaOH + MA-grafted jute-laminated biocomposites attributed to enhancement of 501, 314 and 9.23% in

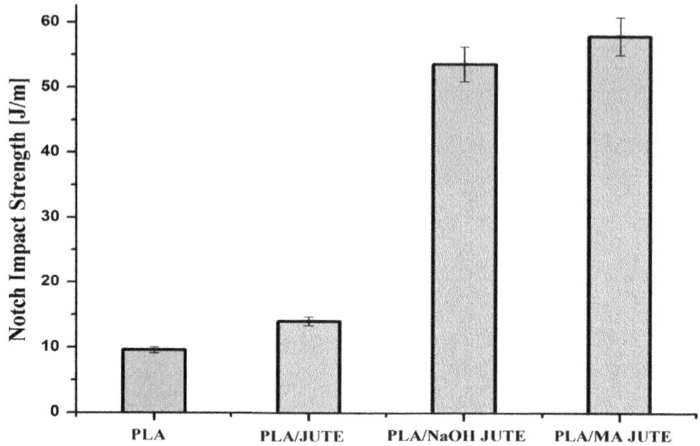

Fig. 18 Impact strength of PLA and jute fibre

values of impact strength when compared to pristine PLA, unmodified jute/PLA and NaOH-grafted jute/PLA biocomposite, respectively, owing to establishment of ester group on the superficial of jute fibre, resulting in optimum interface adhesion with the polymer matrix [74].

Analysis of FTIR Spectra: The chemical reactions of cellulose of jute fibre, with untreated, NaOH-treated and NaOH + MA-treated jute fibres samples, were confirmed using FTIR spectroscopy analysis, as represented in Fig. 19. During interpretation, strong and broad transmittance band in the region of 3200–3600 cm^{-1} is a characteristic of –OH group, which was observed similar to all spectra [75]. Concurrently, two strong peaks were elucidated for NaOH-treated and NaOH + MA-treated jute fibres in the region of 2900 cm^{-1} which conforms C–H stretching vibration of hemicellulose and cellulose which contains methylene group, while in untreated jute fibre, the same peak intensity was observed low, which has enhanced by the superficial functionalization of jute fibre by NaOH and NaOH + MA treatments, respectively, [76]. In untreated jute, a broad and sharp peak was observed in the range of 1730 cm^{-1} due to C=O stretching vibration of hemicellulose-containing ester and carboxylic acid groups. Meanwhile, in the interpretation of NaOH-treated jute fibre, a strong peak in the region of 1730 cm-1 disappears which corresponds to C=O stretching vibration of hemicellulose, which indicates the dissolution of a significant portion of uronic acid and fatty substances in hemicellulose due to mercerization treatment. Further, in the spectrum of NaOH + MA-treated jute fibre, similar broad and sharp peak indicating C=O stretching vibration reappeared in the region of 1730 cm-1 which indicates the ester functionality of MA imposed on superficial of jute fibre [77]. Another peak was interpreted for untreated jute fibre in the region of 1240 cm^{-1} characterized to C–O group of lignin with amorphous and aromatic structure, which disappeared in NaOH-treated and NaOH + MA-treated jute fibre which indicates the dissolution lignin and wax in NaOH treatment [78].

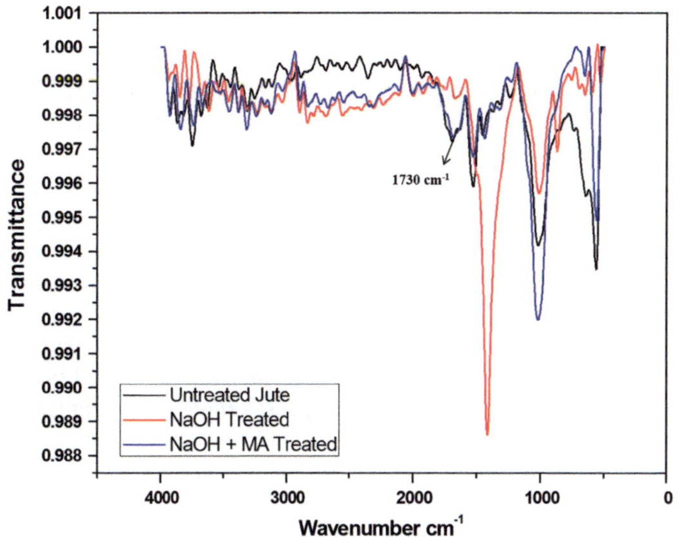

Fig. 19 FTIR spectra of untreated jute, NaOH/jute and NaOH + MA/jute

3D Printing of PLA with Coir Fibre

Coir is obtained from coconut trees which grow widely in tropical countries such as India, where owing to their durability and other advantages they find themselves being exploited for an array of functions such as rope and yarn [79]. From various surveys conducted, it has been found that only a small percentage of the potential total world production of coconut husk is consumed in these traditional coir products. Thus, numerous efforts where coir has been utilized as corroboration in polymer amalgams have been conducted for determining new functions for coir fibres [80, 81]. Similarly, in another study, we validate the contouring of the 3D printed multilayer coir fibre composite structures where chemical treatments such as mercerization and maleation for enhanced interfacial adhesiveness and homogeneity between coir fibre and PLA matrix are used. In order to analyse its mechanical properties, the comparative studies have been investigated between the pristine PLA and the coir (with mercerization- and maleation-treated) via (FTIR) and Izod impact tester as per ASTM D256-06 standard dimension (63.5 mm × 12.7 mm × 6 mm).

Results and Discussion

Analysis of Impact Properties: For analysing the impact properties of the 3D printed multilayer coir composite structure in comparison with the chemical treated composite structure engineered via 3D printing, the composite samples were subjected to Izod impact test for notched samples with a notch radius of 0.25 mm and depth of 2.54 mm. The deviation of notched Izod impact strength (J/m) between PLA and superficially altered coir/PLA-laminated biocomposites engineered via fused deposition modelling (FDM) methodology has been represented in Fig. 20. The notched Izod impact strength for superficially unmodified coir fibres/PLA biocomposites was augmented by 589% in comparison with pristine PLA due to efficient stress transfer between fibre and matrix, consistency in fibre length and homogeneity. PLA/NaOH-grafted coir fibre laminated biocomposites shows 650% and 8.88% improvement in impact strength values when compared against pristine PLA and unmodified coir fibre PLA, respectively. This improvement is attributed to the mercerization, which eradicates impure elements like lignin and wax from the surface of coir fibre, thereby enhancing the homogeneity and interfacial adhesivenes with PLA matrix. Similarly, PLA/NaOH + MA-grafted coir-laminated biocomposites attributed to the enhancement of 659, 10.19 and 1.20% in values of impact strength when compared to pristine PLA, unmodified coir/PLA and NaOH-grafted coir/PLA biocomposite, respectively.

Analysis of FTIR Spectra: The interaction between the coir fibre, untreated, NaOH-treated and NaOH + MA-treated coir fibres was analysed by FTIR spectroscopy in the region from 4000 to 500 cm^{-1}, represented in Fig. 21. Broad transmittance band was observed in the region of 3200–3500 cm^{-1} which is a characteristic of –OH, N–H and C–H stretching of amines, hydroxyl and carbon chains of various groups and C=O stretching vibration of hemicellulose-containing ester and carboxylic acid groups. The different characteristics of peaks according to their functional group, range, and bonds are given in Table 4. The peaks at about 3117.87 and 993.54 cm^{-1} for untreated, NaOH-treated and NaOH + MA-treated coir fibres/PLA conform to the methylene group present in cellulose and hemicellulose. It was witnessed that the peaks of treated coir fibre were considerably low, which can be

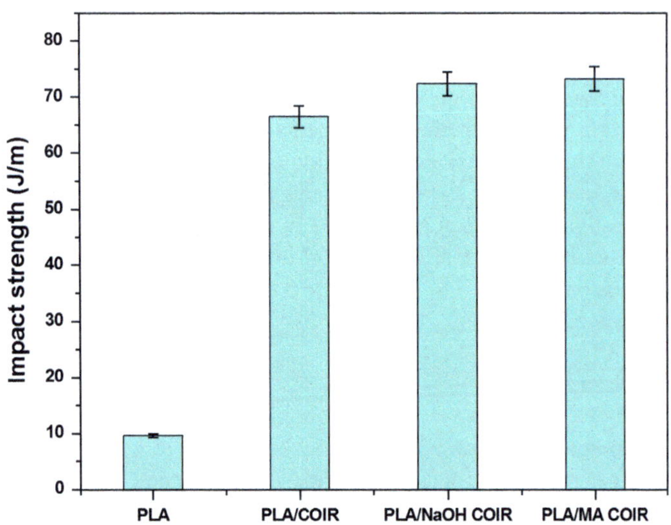

Fig. 20 Impact strength of PLA and coir fibre

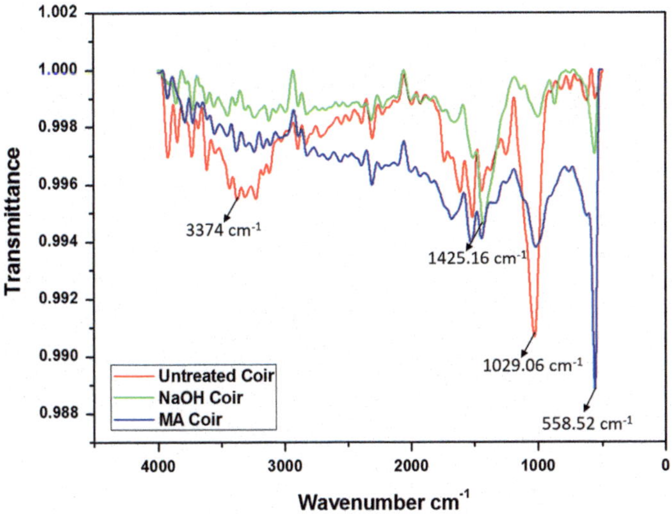

Fig. 21 FTIR spectra of untreated coir, NaOH/coir and NaOH + MA/coir

attributed to the functionalization of coir fibre by NaOH and NaOH + MA treatment [82, 83]. A peak equivalent to C=O stretching vibration of hemicellulose was witnessed in the interpretation of NaOH-treated coir fibre, and a strong peak in the region of 1730 cm^{-1} disappears which corresponds to C=O stretching vibration of hemicellulose, which indicates the dissolution of a substantial portion of uronic acid and fatty substances in hemicellulose due to mercerization treatment. Further, in the spectrum of NaOH + MA-treated coir fibre, similar broad and sharp peak indicating C=O stretching vibration reappeared in the region of 1730 cm^{-1} which indicates the ester functionality of MA imposed on superficial of coir fibre [84, 85]. Due to the increase in hydrogen bonding after treatment of the sample with NaOH, a loss in the intensity of hydroxyl peaks is observed. The nitrile denoting peaks also decrease because of hydrogen bonding in between C–N and –OH of NaOH. The hydrogen bonding between molecules results in connectivity enhancement in the fibrous structures that result in strengthening of material and impact strength augmentation. After MA of the NaOH-treated coir fibre, the impact strength is further enhanced as a result of esterification. This leads to better connectivity and thus an overall improvement in strength of the material is observed as a combined effect of hydrogen bonding and the formation of ester bonds between the fibres [86].

3 Conclusion and Future Scope

All over the world, researchers are achieving swift development in understanding the structure–property relationships of the 3D printed composite structures as 3D print-

Table 4 Peak characteristics according to their functional group, range and bonds

Wavelength (n, cm^{-1})	Compound class	Range	Assignment intensity
3611.68	Alcohols, phenol	3640–3610	O–H stretch
3374.30	1°, 2° amines, amides	3400–3250	N–H stretch
2309.19	Isocyanate	2300–2250	N=C=O stretch
1615.37	1° amines	1650–1580	N–H bend
1517.43	Nitro compounds	1550–1475	N–O asymmetric stretch
1442.10	Aromatics	1500–1400	C–C stretch
1251.63	Alkyl halides	1300–1150	C–H wag ($-CH_2X$)
1029.06	Aliphatic amines	1250–1020	C–N stretch

ing allows the engineering of composite structures having arbitrary shapes in a facile and single-step procedure. In all to summarize, we reported a superficially grafted layered fabric-sandwiched honeycomb composite from polyester, nylon and cotton fabrics via 3D printing where the interfacial adhesion between maleic anhydride and the different fabrics was confirmed via FTIR analysis. The contoured honeycomb composites were subjected for the impact analysis where the superficially grafted fabrics exhibited augmented impact strength by 27, 26 and 43% for polyester, nylon and cotton fabrics, respectively, in comparison with that of the superficially non-grafted layered textile-impregnated honeycomb composite. The pristine ABS film and surface-treated nanoclay-incorporated ABS films (ranging from 0.5 to 2 wt%) assimilated 3D printed specimens, the nanoclay-incorporated 3D structured samples showed a maximum of impact fracture energy at 0.5 wt% (85%) and on further addition caused decrease in impact value owing to agglomeration of the organoclay filler, which acts as an initial crack point and subsequently reduces the impact strength. 3D printed surface of ABS specimen displayed a hardness value of 13.63. Wear analysis of multilayered composite impregnated with virgin ABS film revealed wear properties where COF was found to be decreased as compared to 3D printed virgin sample.

Molluscan shell-structured 3D printed composites revealed improved property (74%), for organoclay loading of (ranging from 0.5 to 1.5 wt%) in PLA dispersion via doctor blade coating technique. The organoclay-incorporated 3D printed molluscan shell-structured samples showed an enhancement up to 1 wt% of loading after which the value decreased because of non-uniform dispersion due to agglomeration of the organoclay filler which acts as an initial crack point subsequently reducing the impact strength. FTIR analysis revealed strong polymer/organoclay interaction and dispersion for progressive loading of organoclay till 1 wt%, resulting in enhanced impact strength (39.4 J/m). Biocomposites prepared by 3D printing with reinforcement of natural fibres such as human hair and sheep wool amalgamated with PLA revealed an improvement in impact strength of 140.35 and 42.43 J/m, respectively, whereas the composites reinforced with silanized natural fibres exhibited impact strength values of 208.85 and 61.55 J/m as compared to pristine PLA, i.e. 15.75 J/m. It was found that aggrandized interfacial adhesion between fibres and matrix via silane coupling agent improves the impact properties. The superficially layered interfacial interaction between the PLA and keratin fibre was confirmed by FTIR analysis.

Jute fibre-infused polylactic acid-laminated biocomposite treated with mercerization treatment resulted in the dissolution of impure moieties from fibre surface, attributing it with rough and textured fibre surface for effective mechanical interlocking with polylactide matrix. However, superficial modification through maleation treatment altered the superficial condition by imposing ester moieties on the fibre surface, which improved the phase adhesiveness and mechanical interlocking between fibre and polylactide matrix. Simultaneously, impact the performance of surface-grafted jute fibre biocomposite (notched) was augmented in comparison with pristine PLA and unmodified jute fibre biocomposite. Moreover, NaOH + MA-grafted jute fibre-infused composite resulted in enhanced mechanical properties when compared NaOH-grafted jute fibre composite, which was attributed to ester group grafting on fibre surface resulting in better mechanical interlocking with enhanced adhesiveness and homogeneity in between fibre and matrix. Further, the treatment of with MA + NaOH coir fibre enhanced the impact strength of PLA/coir fibre 3D printed composites and their fibre matrix interaction was confirmed by FTIR. Recently, researchers are moving towards 4D printing, which is expected to be commercialized by 2019 with a global market growth of around 42.5% expected between 2019 and 2025 and further exhibits the ability to create objects with superior multimaterial components that can transform with respect to external stimuli caused by a change in the environment. The change in the shape structure as a response to the stimuli is time-dependent that makes time as the fourth dimension of the 4D printing, and it gives an edge over 3D printing where no such changes are observed.

References

1. Fan T-X, Chow S-K, Zhang D (2009) Biomorphic mineralization: from biology to materials. Prog Mater Sci 54:542–659. https://doi.org/10.1016/j.pmatsci.2009.02.001
2. Chen P-Y, McKittrick J, Meyers MA (2012) Biological materials: functional adaptations and bioinspired designs. Prog Mater Sci 57:1492–1704. https://doi.org/10.1016/j.pmatsci.2012.03.001
3. Aizenberg J, Fratzl P (2009) Biological and biomimetic materials. Adv Mater 21:387–388. https://doi.org/10.1002/adma.200803699
4. Yadav R, Naebe M, Wang X, Kandasubramanian B (2017) Review on 3D prototyping of damage tolerant interdigitating brick arrays of Nacre. Ind Eng Chem Res 56:10516–10525. https://doi.org/10.1021/acs.iecr.7b01679
5. Mishra N, Kandasubramanian B (2017) Biomimetic design of artificial materials inspired by iridescent nacre structure and its growth mechanism. Polym Plast Technol Eng, 1–15. https://doi.org/10.1080/03602559.2017.1326139
6. Chang CC, Boland ED, Williams SK, Hoying JB (2011) Direct-write bioprinting three-dimensional biohybrid systems for future regenerative therapies. J Biomed Mater Res Part B Appl Biomater 98B:160–170. https://doi.org/10.1002/jbm.b.31831

7. Cui H, Nowicki M, Fisher JP, Zhang LG (2017) 3D bioprinting for organ regeneration. Adv Health Mater 6:1601118. https://doi.org/10.1002/adhm.201601118
8. Gross BC, Erkal JL, Lockwood SY, Chen C, Spence DM (2014) Evaluation of 3D printing and its potential impact on biotechnology and the chemical sciences. Anal Chem 86:3240–3253. https://doi.org/10.1021/ac403397r
9. Guvendiren M, Molde J, Soares RMD, Kohn J (2016) Designing biomaterials for 3D printing. ACS Biomater Sci Eng 2:1679–1693. https://doi.org/10.1021/acsbiomaterials.6b00121
10. Ho CMB, Ng SH, Li KHH, Yoon Y-J (2015) 3D printed microfluidics for biological applications. Lab Chip 15:3627–3637. https://doi.org/10.1039/C5LC00685F
11. Kolesky DB, Truby RL, Gladman AS, Busbee TA, Homan KA, Lewis JA (2014) 3D bioprinting of vascularized, heterogeneous cell-laden tissue constructs. Adv Mater 26:3124–3130. https://doi.org/10.1002/adma.201305506
12. Korde JM, Shaikh M, Kandasubramanian B (2018) Bionic prototyping of honeycomb patterned polymer composite and its engineering application. Polym Plast Technol Eng, 1–17. https://doi.org/10.1080/03602559.2018.1434667
13. Loh QL, Choong C (2013) Three-dimensional scaffolds for tissue engineering applications: role of porosity and pore size. Tissue Eng Part B Rev 19:485–502. https://doi.org/10.1089/ten.teb.2012.0437
14. Malik A, Kandasubramanian B (2018) Flexible polymeric substrates for electronic applications. Polym Rev, 1–38 (2018). https://doi.org/10.1080/15583724.2018.1473424
15. Mandrycky C, Wang Z, Kim K, Kim D-H (2016) 3D bioprinting for engineering complex tissues. Biotechnol Adv 34:422–434. https://doi.org/10.1016/j.biotechadv.2015.12.011
16. Murphy SV, Atala A (2014) 3D bioprinting of tissues and organs. Nat Biotechnol 32:773–785. https://doi.org/10.1038/nbt.2958
17. Peltola SM, Melchels FPW, Grijpma DW, Kellomäki M (2008) A review of rapid prototyping techniques for tissue engineering purposes. Ann Med 40:268–280. https://doi.org/10.1080/07853890701881788
18. Rengier F, Mehndiratta A, von Tengg-Kobligk H, Zechmann CM, Unterhinninghofen R, Kauczor H-U, Giesel FL (2010) 3D printing based on imaging data: review of medical applications. Int J Comput Assist Radiol Surg 5:335–341. https://doi.org/10.1007/s11548-010-0476-x
19. Hull CW (1986) Apparatus for production of three-dimensional objects by stereolithography. U.S. Patent 4,575,330
20. Quan Z, Wu A, Keefe M, Qin X, Yu J, Suhr J, Byun J-H, Kim B-S, Chou T-W (2015) Additive manufacturing of multi-directional preforms for composites: opportunities and challenges. Mater Today 18:503–512. https://doi.org/10.1016/j.mattod.2015.05.001
21. Deoray N, Kandasubramanian B (2018) Review on three-dimensionally emulated fiber-embedded lactic acid polymer composites: opportunities in engineering sector. Polym Plast Technol Eng 57:860–874. https://doi.org/10.1080/03602559.2017.1354226
22. Melchels FPW, Feijen J, Grijpma DW (2010) A review on stereolithography and its applications in biomedical engineering. Biomaterials 31:6121–6130. https://doi.org/10.1016/j.biomaterials.2010.04.050
23. Wang X, Jiang M, Zhou Z, Gou J, Hui D (2017) 3D printing of polymer matrix composites: a review and prospective. Compos Part B Eng 110:442–458. https://doi.org/10.1016/j.compositesb.2016.11.034
24. Kruth J-P, Leu MC, Nakagawa T (1998) Progress in additive manufacturing and rapid prototyping. CIRP Ann 47:525–540. https://doi.org/10.1016/S0007-8506(07)63240-5
25. Ahn S, Montero M, Odell D, Roundy S, Wright PK (2002) Anisotropic material properties of fused deposition modeling ABS. Rapid Prototyp J 8:248–257. https://doi.org/10.1108/13552540210441166
26. Wu GA, Langrana N, Sadanji R, Danforth S (2002) Solid freeform fabrication of metal components using fused deposition of metals. Mater Des 23:97–105. https://doi.org/10.1016/s0261-3069(01)00079-6
27. Agarwala MK, Jamalabad VR, Langrana NA, Safari A, Whalen PJ, Danforth SC (1996) Structural quality of parts processed by fused deposition. Rapid Prototyp J 2:4–19. https://doi.org/10.1108/13552549610732034

28. Bitzer T (1997) Honeycomb technology. Springer, The Netherlands
29. Vinson JR (1999) The Behavior of Sandwich Structures of Isotropic and Composite Materials. Technomic Publishing Company Inc., Lancaster
30. Khatavkar N, Kandasubramanian B (2016) Composite materials for supersonic aircraft radomes with ameliorated radio frequency transmission—a review. RSC Adv 6:6709–6718. https://doi.org/10.1039/C5RA18712E
31. Fischer S, Drechsler K, Kilchert S, Johnson A (2009) Mechanical tests for foldcore base material properties. Compos Part A Appl Sci Manuf 40:1941–1952. https://doi.org/10.1016/j.compositesa.2009.03.005
32. Pflug J (2004) Pflug J, Verpoest I (2004) Thermoplastic folded honeycomb structure and method for the production thereof. U.S. Patent 6,726,974
33. Nygaard JV, Lyckegaard A (2007) Sandwich beam with a periodical and graded core manufactured using rapid prototyping. J Sandw Struct Mater 9:365–376. https://doi.org/10.1177/1099636207071609
34. Hou Y, Tai YH, Lira C, Scarpa F, Yates JR, Gu B (2013) The bending and failure of sandwich structures with auxetic gradient cellular cores. Compos Part A Appl Sci Manuf 49:119–131. https://doi.org/10.1016/j.compositesa.2013.02.007
35. Fíla T, Zlámal P, Jiroušek O, Falta J, Koudelka P, Kytýř D, Doktor T, Valach J (2017) Impact testing of polymer-filled auxetics using split hopkinson pressure bar. Adv Eng Mater 19:1700076. https://doi.org/10.1002/adem.201700076
36. Hancox NL, Wells H (1979) Aluminum/carbon fiber hybrid composites. Polym Eng Sci 19:917–922. https://doi.org/10.1002/pen.760191307
37. Hedayati R, Sadighi M, Mohammadi Aghdam M, Zadpoor A (2016) Mechanical properties of additively manufactured thick honeycombs. Mater Basel 9:613. https://doi.org/10.3390/ma9080613
38. Türk D-A, Brenni F, Zogg M, Meboldt M (2017) Mechanical characterization of 3D printed polymers for fiber reinforced polymers processing. Mater Des 118:256–265. https://doi.org/10.1016/j.matdes.2017.01.050
39. Yang X, Lu T, Kim T (2013) Effective thermal conductivity modelling for closed-cell porous media with analytical shape factors. Transp Porous Media 100:211–224. https://doi.org/10.1007/s11242-013-0212-4
40. Kourtides DA, Gilwee WJ, Parker JA (1979) Thermal response of composite panels. Polym Eng Sci 19:226–231. https://doi.org/10.1002/pen.760190311
41. Shafizadeh JE, Seferis JC (1999) High temperature thermal analysis characterization of honeycomb core. Polym Eng Sci 39:733–740. https://doi.org/10.1002/pen.11461
42. Ajdari A, Jahromi BH, Papadopoulos J, Nayeb-Hashemi H, Vaziri A (2012) Hierarchical honeycombs with tailorable properties. Int J Solids Struct 49:1413–1419. https://doi.org/10.1016/j.ijsolstr.2012.02.029
43. Scarpa F, Blain S, Lew T, Perrott D, Ruzzene M, Yates JR (2007) Elastic buckling of hexagonal chiral cell honeycombs. Compos Part A Appl Sci Manuf 38:280–289. https://doi.org/10.1016/j.compositesa.2006.04.007
44. Zhang Q, Yang X, Li P, Huang G, Feng S, Shen C, Han B, Zhang X, Jin F, Xu F, Lu TJ (2015) Bioinspired engineering of honeycomb structure—using nature to inspire human innovation. Prog Mater Sci 74:332–400. https://doi.org/10.1016/j.pmatsci.2015.05.001
45. Coates J (2006) Interpretation of infrared spectra, a practical approach. In: Encyclopedia of analytical chemistry. Wiley, Chichester
46. Mirone P, Chiorboli P (1962) Infrared and Raman spectra and vibrational assignment of maleic anhydride. Spectrochim Acta 18:1425–1432. https://doi.org/10.1016/0371-1951(62)80003-4
47. Gore PM, Dhanshetty MKB (2016) Bionic creation of nano-engineered Janus fabric for selective oil/organic solvent absorption. RSC Adv 6:111250–111260. https://doi.org/10.1039/c6ra24106a
48. Yang CQ, Chen D, Guan J, He Q (2010) Cross-linking cotton cellulose by the combination of maleic acid and sodium hypophosphite. Ind Eng Chem Res 49:8325–8332. https://doi.org/10.1021/ie1007294

49. Pumure I, Ford S, Shannon J, Kohen C, Mulcahy A, Frank K, Sisco S, Chaukura N (2015) Analysis of ATR-FTIR absorption-reflection data from 13 polymeric fabric materials using chemometrics. Am J Anal Chem 6:305–312. https://doi.org/10.4236/ajac.2015.64029
50. Elnagar K, Abou Elmaaty T, Raouf S (2014) Dyeing of polyester and polyamide synthetic fabrics with natural dyes using ecofriendly technique. J Text 2014:1–8. https://doi.org/10.1155/2014/363079
51. Suzuki M, Nagasawa H (2013) Mollusk shell structures and their formation mechanism. Can J Zool 91:349–366. https://doi.org/10.1139/cjz-2012-0333
52. Gu GX, Takaffoli M, Buehler MJ (2017) Hierarchically enhanced impact resistance of bioinspired composites. Adv Mater 29:1700060. https://doi.org/10.1002/adma.201700060
53. Wyzgoski MG, Yeh GSY (1974) Morphology of crazes in glassy polycarbonate. J Macromol Sci Part B 10:647–661. https://doi.org/10.1080/00222347408219412
54. Belter JT, Dollar AM (2015) Strengthening of 3D printed fused deposition manufactured parts using the fill compositing technique. PLoS ONE 10:e0122915. https://doi.org/10.1371/journal.pone.0122915
55. Bian J, Wang ZJ, Lin HL, Zhou X, Xiao WQ, Zhao XW (2017) Thermal and mechanical properties of polypropylene nanocomposites reinforced with nano-SiO_2 functionalized graphene oxide. Compos Part A Appl Sci Manuf 97:120–127. https://doi.org/10.1016/j.compositesa.2017.01.002
56. Stretz HA, Paul DR, Cassidy PE (2005) Poly(styrene-co-acrylonitrile)/montmorillonite organoclay mixtures: a model system for ABS nanocomposites. Polymer (Guildf) 46:3818–3830. https://doi.org/10.1016/j.polymer.2005.03.043
57. Jiang L, Lam YC, Tam KC, Chua TH, Sim GW, Ang LS (2005) Strengthening acrylonitrile-butadiene-styrene (ABS) with nano-sized and micron-sized calcium carbonate. Polymer (Guildf) 46:243–252. https://doi.org/10.1016/j.polymer.2004.11.001
58. Fu S-Y, Lauke B (1998) Fracture resistance of unfilled and calcite-particle-filled ABS composites reinforced by short glass fibers (SGF) under impact load. Compos Part A Appl Sci Manuf 29:631–641. https://doi.org/10.1016/S1359-835X(97)00111-5
59. Chen B, Evans JRG (2008) Impact and tensile energies of fracture in polymer–clay nanocomposites. Polymer (Guildf) 49:5113–5118. https://doi.org/10.1016/j.polymer.2008.09.024
60. Nayak SK, Yadav SN, Mohanty S (2011) Fundamentals of plastic testing. Springer, New Delhi, pp 53–62
61. Benítez JJ, San-Miguel MA, Domínguez-Meister S, Heredia-Guerrero JA, Salmeron M (2011) Structure and chemical state of octadecylamine self-assembled monolayers on mica. J Phys Chem C 115:19716–19723. https://doi.org/10.1021/jp203871g
62. Karahaliou E-K, Tarantili PA (2009) Preparation of poly(acrylonitrile-butadiene-styrene)/montmorillonite nanocomposites and degradation studies during extrusion reprocessing. J Appl Polym Sci 113:2271–2281. https://doi.org/10.1002/app.30158
63. Cai Y, Huang F, Xia X, Wei Q, Tong X, Wei A, Gao W (2010) Comparison between structures and properties of ABS nanocomposites derived from two different kinds of OMT. J Mater Eng Perform 19:171–176. https://doi.org/10.1007/s11665-009-9392-z
64. Ambre A, Jagtap R, Dewangan B (2009) ABS nanocomposites containing modified clay. J Reinf Plast Compos 28:343–352. https://doi.org/10.1177/0731684407084254
65. Boparai K, Singh R, Singh H (2015) Comparison of tribological behaviour for Nylon6-Al-Al_2O_3 and ABS parts fabricated by fused deposition modelling. Virtual Phys Prototyp 10:59–66. https://doi.org/10.1080/17452759.2015.1037402
66. Aigbodion VS, Hassan SB, Agunsoye JO (2012) Effect of bagasse ash reinforcement on dry sliding wear behaviour of polymer matrix composites. Mater Des 33:322–327. https://doi.org/10.1016/j.matdes.2011.07.002
67. Yang H, Jiang P (2010) Large-scale colloidal self-assembly by doctor blade coating. Langmuir 26:13173–13182. https://doi.org/10.1021/la101721v
68. Gatos KG, Karger-Kocsis J (2007) Effect of the aspect ratio of silicate platelets on the mechanical and barrier properties of hydrogenated acrylonitrile butadiene rubber (HNBR)/layered silicate nanocomposites. Eur Polym J 43:1097–1104. https://doi.org/10.1016/j.eurpolymj.2007.01.032

69. Baran Inceoglu A, Yilmazer U (2003) Synthesis and mechanical properties of unsaturated polyester based nanocomposites. Polym Eng Sci 43:661–669. https://doi.org/10.1002/pen.10054
70. Garofalo E, Scarfato P, Di Maio L, Incarnato L (2017) Tuning of co-extrusion processing conditions and film layout to optimize the performances of PA/PE multilayer nanocomposite films for food packaging. Polym Compos. https://doi.org/10.1002/pc.24323
71. Ahmad F, Choi HS, Park MK (2015) A review: natural fiber composites selection in view of mechanical, light weight, and economic properties. Macromol Mater Eng 300:10–24. https://doi.org/10.1002/mame.201400089
72. Wojciechowska E, Rom M, Włochowicz A, Wysocki M, Wesełucha-Birczyńska A (2004) The use of Fourier transform-infrared (FTIR) and Raman spectroscopy (FTR) for the investigation of structural changes in wool fibre keratin after enzymatic treatment. J Mol Struct 704:315–321. https://doi.org/10.1016/j.molstruc.2004.03.044
73. Boojaria A, Masrournia M, Ghorbani H, Ebrahimitalab A, Miandarhoie M (2015) Silane modified magnetic nanoparticles as a novel adsorbent for determination of morphine at trace levels in human hair samples by high-performance liquid chromatography with diode array detection. Forensic Sci Med Pathol 11:497–503. https://doi.org/10.1007/s12024-015-9702-8
74. Mwaikambo LY, Ansell MP (2002) Chemical modification of hemp, sisal, jute, and kapok fibers by alkalization. J Appl Polym Sci 84:2222–2234. https://doi.org/10.1002/app.10460
75. Gañán P, Zuluaga R, Restrepo A, Labidi J, Mondragon I (2008) Plantain fibre bundles isolated from Colombian agro-industrial residues. Bioresour Technol 99:486–491. https://doi.org/10.1016/j.biortech.2007.01.012
76. Haque MM, Hasan M, Islam MS, Ali ME (2009) Physico-mechanical properties of chemically treated palm and coir fiber reinforced polypropylene composites. Bioresour Technol 100:4903–4906. https://doi.org/10.1016/j.biortech.2009.04.072
77. Hossain MK, Dewan MW, Hosur M, Jeelani S (2011) Mechanical performances of surface modified jute fiber reinforced biopol nanophased green composites. Compos Part B Eng 42:1701–1707. https://doi.org/10.1016/j.compositesb.2011.03.010
78. Liu XY, Dai GC (2007) Surface modification and micromechanical properties of jute fiber mat reinforced polypropylene composites. Express Polym Lett 1:299–307. https://doi.org/10.3144/expresspolymlett.2007.43
79. Kulkarni AG, Satyanarayana KG, Rohatgi PK, Vijayan K (1983) Mechanical properties of banana fibres (Musa sepientum). J Mater Sci 18:2290–2296. https://doi.org/10.1007/BF00541832
80. Owolabi O, Czvikovszky T, Kovács I (1985) Coconut-fiber-reinforced thermosetting plastics. J Appl Polym Sci 30:1827–1836. https://doi.org/10.1002/app.1985.070300504
81. Khalil HPSA, Rozman HD, Ahmad MN, Ismail H (2000) Acetylated plant-fiber-reinforced polyester composites: a study of mechanical, hygrothermal, and aging characteristics. Polym Plast Technol Eng 39:757–781. https://doi.org/10.1081/PPT-100100057
82. Mofokeng JP, Luyt AS, Tábi T, Kovács J (2012) Comparison of injection moulded, natural fibre-reinforced composites with PP and PLA as matrices. J Thermoplast Compos Mater 25:927–948. https://doi.org/10.1177/0892705711423291
83. Hirschmann RP, Kniseley RN, Fassel VA (1965) The infrared spectra of alkyl isocyanates. Spectrochim Acta 21:2125–2133. https://doi.org/10.1016/0371-1951(65)80228-4
84. Coates J (2000) Interpretation of infrared spectra, a practical approach. Wiley, Chichester
85. Ebele CC, Metu Chidiebere S, Ojukwu Martin C (2016) Fourier transform infrared (FTIR) spectroscopy study on coir fibre reinforced polyester (CFRP) composites. Int J Civ Mech Energy Sci 2:20–28
86. Le Questel J-Y, Berthelot M, Laurence C (2000) Hydrogen-bond acceptor properties of nitriles: a combined crystallographic and ab initio theoretical investigation. J Phys Org Chem 13:347–358. https://doi.org/10.1002/1099-1395(200006)

Mr. Swaroop Gharde is currently working as Junior Research Fellow (JRF) in DIAT, Pune, and holds master's degree in computer-aided analysis and design (CAAD) from Manipal University, Jaipur. His research interests include design and analysis of composite structures, 3D printing, vacuum casting and polymer composite for different applications.

Mrs. Aarsha Surendren is currently working as Junior Research Fellow (JRF) in Institute of Plastics Technology, CIPET, Cochin, and holds master's degree in biopolymer science from the same institute. Her research interests are in biopolymers, 3D printing and applications of biopolymers in biomedical field.

Mr. Jay M. Korde is currently pursuing his master's degree in corrosion technology from DIAT, Pune. He has completed his bachelor's degree in plastics and polymer engineering from MIT, Aurangabad. His research interests include polymer composites, biomaterials and advanced materials for energy and environmental applications.

Mr. Shubham Saini is currently working as Graduate Engineer Trainee (GET) in Reliance Industries Limited. He has completed his bachelor's degree from NIT Jalandhar. His research interests include polymer composites, energy, environmental and industry applications.

Mr. Nikit Deoray is currently working as Operation Head in Crennovations. He has completed his bachelor's degree in plastics and polymer engineering from MIT Aurangabad. His research interests include polymer composites, biomaterials and 3D printing.

Mr. Rajendra Goud has completed his master's degree in materials science and technology from DIAT, Pune. He has completed his bachelor's degree in mechanical engineering from Biju Patnaik University of Technology, Odisha. His research interests include additive manufacturing, bio-inspired materials and renewable energies.

Mr. Sunil Nimje is Assistant Professor in Department of Mechanical Engineering, DIAT, Pune. His research interests include analysis and design of composite structures, modelling of FRP plates and shells, adhesively bonded and mechanically fastened joints of laminated FRP composite structures, functionally graded structures, finite element analysis, fracture mechanics and machine design.

Prof. Balasubramanian Kandasubramanian is highly acclaimed for his contribution towards the polymer processing and fabrication for various applications including antibacterial, smart textiles, hydrophobic coatings, ablative materials, fire-retardant fabric wastewater treatment, biomimicking and polymer nanocomposite for defence applications and has 14 Indian patents with more than 200 papers in peer-reviewed journals. He is Head of Department of Metallurgical and Materials Engineering, associated with DIAT (DU) from 2010 till date. He was actively involved in the development of various research laboratories including polymer processing laboratory and characterization laboratories (FESEM, HRTEM, SAXS). His contribution towards the wastewater management was acknowledged and perceived by National Water Academy, Pune, and Board of Research in Nuclear Science (BRNS), Mumbai. He has been recognized by Royal Society of Chemistry (RSC) for one of the most cited authors of the year for his erudite work on superhydrophobicity. He has gleaned the attention of Naval Postgraduate School, California, USA, and Naval Materials Research Laboratory (NMRL), DRDO Ambernath, Ministry of Defence, India, for the development of hydrophobic, antifungal and anticorrosive coating for ship hulls. He is recently awarded with the prestigious award 'Technology Innovation in Petrochemicals and Downstream Plastic Processing Industry' from Ministry of Chemical and Fertilizers, Government of India, for his contribution in polymer technology.

Poly-lactic-Acid: Potential Material for Bio-printing Applications

Sunpreet Singh, Chander Prakash, Manjeet Singh, Guravtar Singh Mann, Munish Kumar Gupta, Rupinder Singh and Seeram Ramakrishna

Abstract Exclusive research efforts, made across the world, in the area of material science have resulted into development of a wide range of materials which could be successfully used for numerous biomedical applications. Poly-lactic-acid (PLA) is one of these developments which could be brought in direct contact of the tissues/organs, as a medical device and support structure. For the benefit of the research scholars, this chapter is structured to review the prospective biomedical implications of PLA material, explored in the last 20 years. Further, the efficacy of PLA with different types of three-dimensional printing (3DP) technologies, especially for fused deposition modeling, is also highlighted in response of the mechanical, biological, and topological characteristics of resulting parts. Further, the printing of waste natural fiber embedded PLA structures has experimented, as a case study, via fused deposition modeling.

Keywords Biomedical · Composites · Fused deposition modeling · Poly-lactic-acid · Properties · Three-dimensional printing · Tissue engineering

---

S. Singh (✉) · C. Prakash · M. Singh · G. S. Mann
Mechanical Engineering, Lovely Professional University, Phagwara, India
e-mail: snprt.singh@gmail.com

M. K. Gupta
Mechanical Engineering, National Institute of Technology, Hamirpur, India

Mechanical Engineering, Ludhiana College of Engineering and Technology, Ludhiana, India

R. Singh
Production Engineering, Guru Nanak Dev Engineering College, Ludhiana, India

S. Ramakrishna
Mechanical Engineering, National University of Singapore, Singapore, Singapore

C. Prakash et al. (eds.), *Biomanufacturing*,
https://doi.org/10.1007/978-3-030-13951-3_3

1 Introduction

In the recent past, the requirement of biodegradable and bioactive materials has been enormously increased, especially for the biomedical applications, owing to their superior performances within the human body environment. The various types of synthetic polymers material resulted from such a vast demand are poly-lactic-co-glycolic acid (PLGA) and poly-lactic-acid (PLA) [1]. The PLA offers exclusive biodegradability, biocompatibility, ease of processing, and environmental friendliness that makes it a potential candidate of the medical industry. It has been utilized both as a surgical implant material and for delivering the drug. Moreover, the porous scaffolds could be easily produced by using PLA, in order to enhance the growth of neo-tissue [2, 3]. The use of PLA for biomedical and tissue engineering (TE) applications is not only because of its biodegradability but is majorly because of the combination of excellent properties at reasonably low cost [4]. The degradation pace PLA and PLGA is a function of numerous factors such as the configuration of the structure, copolymer ratio, crystallinity, morphology, stresses, and a fraction of residual monomer, structural porosity, and site of implantation [5].

As of now, a wide range of medical products has been produced using PLA, including degradable structures, drug releasing microparticles/nanoparticles, drug delivery capsules, and porous scaffolds for TE applications [6]. Researchers have also combined PLA with numerous other materials such as bio-ceramics, fibers, polymers, and metallic, nonmetallic for improving the plethora of desirable characteristic [7–9]. The surface topography of the resulting structures plays an utmost important role to determine the possible applications, especially for biomaterials, biocompatibility, cell adhesion, osseointegration, and mechanical properties. Since, the PLA implants and scaffolds are temporarily implanted and installed, respectively, and will disappear steadily upon degradation; hence, a postsurgical operation could be easily eliminated. Faludi et al. prepared and characterized the PLA-based composites, using three corncob fractions and a single fraction of wood fiber. Two consecutive micromechanical deformation mechanisms were detected in the fabricated composites which contained a high fraction of corncob. The fracture mechanisms were notated as soft and hard particles; however, the reason behind was the segregation of the composite elements [10].

Cui et al. [11] experimented with a hybrid manufacturing process resulted by combining porogen leaching and melt-molding. This hybrid process is highly useful to obtaining interconnected porous architecture with the resulting sample. At the same time, the complexity of the external anatomical shape can also be maintained. Yun et al. [12] characterized the expansion pattern of the genes during osteogenic differentiation of different stem cells on PLGA/nano-HA fibrous scaffolds. Sanders et al. studied the PLA microfibers (ranged 4–15 mm) for TE by using rat subcutaneous implantation [13]. Kellomaki et al. observed the constructional features and fabrication techniques of various types of bio-absorbable scaffolds for guided bone regeneration and generation. It has been found by the research group that among the

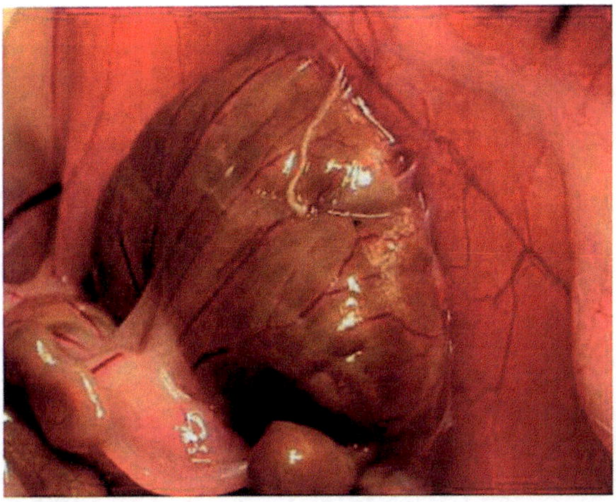

Fig. 1 Adhesion of the PLA film with wound interface [16]

various constructions, the self-reinforced PLA rods were used as scaffolds, successfully, for bone formation in muscle by free tibial periosteal grafts [14].

Hoveizi et al. enhanced the cell adherence characteristics and improved the in vitro compatibility of gelatin modified PLA scaffold. The PLA/gelatin fibers were produced in two different proportions of 3/7 and 7/3 via electro-spinning technique. The properties of the blending nano-fibers were thereby investigated by Fourier-transform infrared spectroscopy (FTIR) and scanning electron microscopy (SEM). The composition, 7/3, was found to be more suitable for fibroblasts attachment and viability as compared to the pure PLA and gelatin nano-fiber [15]. Researchers have evaluated the efficacy of an absorbable PLA film in preventing postoperative intra-abdominal adhesions in a rat model by placing it between the cecal and abdominal wall defects (refer Fig. 1). They found that the strategic placement of PLA film during abdominal surgery was associated with a noticeably reduced rate and severity of postoperative intra-abdominal adhesion in the model [16]. Li et al. investigated the performance standards of PLA ureteral stents for treating ureteral war injuries. The outcomes of their study represented PLA stents as very promising [17].

Further, in [18], scientists used PLA based on polymer blending with poly-tri-methylene carbonate via solvent casting method in order to prevent postoperative adhesions. Mi et al. have performed an extensive characterization on twin screw blended TPU/PLA scaffolds for TE applications. It has been found that the morphological representation of virgin PLA and TPU was different, significantly, and the structure of the pores was also variable. Comparatively, it was outlined by the research team that the pore size of TPU was much larger than the same of PLA. Figure 2 shows the cell viability and morphological representations of day 3 and day 10. The attachment of fibroblasts cells and its uniform distribution can be clearly

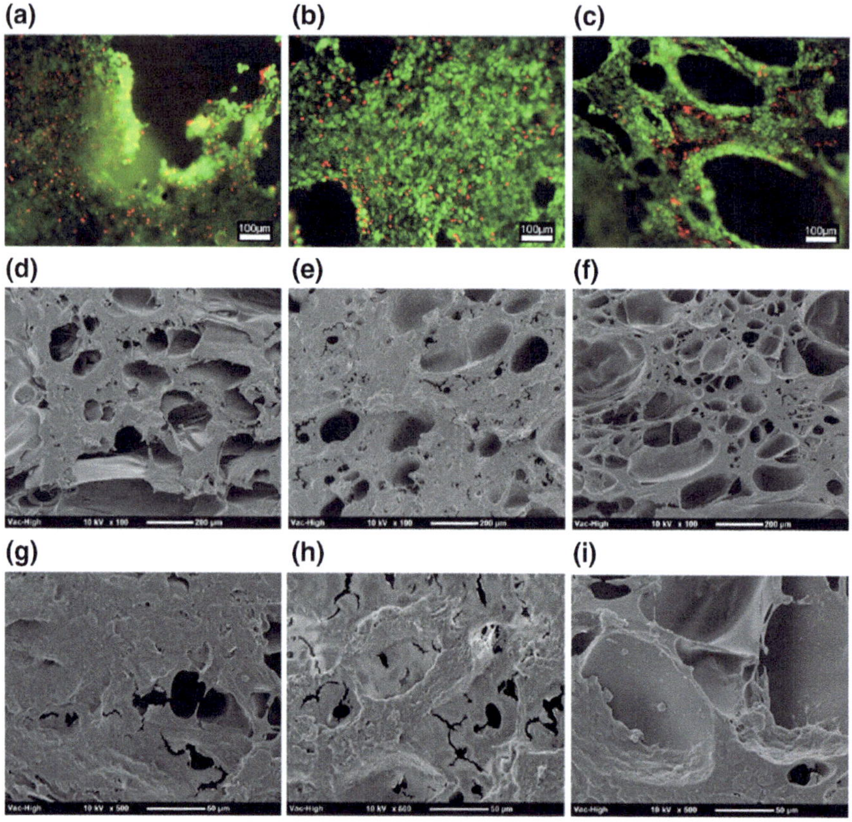

Fig. 2 Cell culture results of microcellular injection molded PLA (**a**, **d**, **g**), PLA at 50% proportion (**b**, **e**, **h**), and TPU (**c**, **f**, **i**) scaffolds [19]

seen on the scaffolds [19]. Table 1 lists the various novel benefits of PLA material for different medical applications.

Moreover, the orthopedics applications of PLA and PLGA are enormously explored in various publications, about two decades ago, demanding much tougher and mechanically stronger polymeric materials. Van Sliedregt et al. [27] performed an in vitro cell compatibility examination on PLA by using three different types of cells, including rat epithelial cells, human fibroblasts, and osteosarcoma cells. It has been found by the authors that PLA resulted in good biocompatibility. In a similar study conducted by Majola et al. [28], the PLA rods showed reasonable biocompatibility in the cancellous bone of samples. And, Böstman et al. [29] reported the biocompatibility of PLA copolymers is suitable in the human body. These materials could be used for making numerous devices, specifically in the field of sports medicine, such as interference screws, suture anchors, meniscal repair devices, and simple fracture fixation [30].

Table 1 A list of breakthroughs of PLA in biomedical

S. no.	Specific application	Material/Grade	Approach/method used	Outcomes	References
1	Cartilage TE	Chondrocyte and PLGA microspheres (30–80 μm in diameter)	Microspheres were injected at desired location with various gauges of needles	The implantation of resulting materials highlighted the formation of cartilage tissue	[20]
2	Drug releasing devices	PLA nano-fibers combined with TiO_2 nanoparticles	Electro-spinning process and solvent casting	Anticancer drug daunorubicin can be readily released to accumulate on leukemia K562 cells	[21]
3	Drug delivery technology	Poly(D,L-lactic acid) and Kirschner-wires	Solvent casting technique	Poly(D,L-lactic acid) coating of implants can be highly beneficial against the infections	[22]
4	Drug delivery system	PLA nanoparticles and fluorescent die	Solvent casting	The polymeric material showed its ability to be a good carrier for hair folli-cle/sebaceous gland targeting	[23]
5	Guided bone/tissue regeneration	PLA (ATRISORB® Free-Flow™)	Surgery	The PLA device may induce bone resorbing foreign body reactions	[24]
6	Scaffolds for wound healing	PLA nano-fibers and ibuprofen	–	The scaffolds containing 20 wt% ibuprofen promoted human skin cell viability and proliferation and reduced wound contraction	[25]
7	Orthopedic implant	PLA and magnesium particles	Solvent casting	The study elucidated that the incorporation of Mg to PLA matrix affected the degradation and osteogenesis	[26]

The emergence of three-dimensional printing (3DP) technologies is marked as the biggest breakthrough for the biomedical and tissue engineering (BTE) applications. These new classes of technologies are welcoming the wide and extensive range of materials and their combinations which were not feasible, earlier. Moreover, now, with the help of different printing technologies more stable and tailor-made geometries could be easily made with the assistance of computer-aided design and systems. The 3DP technologies enable the researchers to fabricate scaffolds, implants, or organs with a controlled internal/external micro-architecture, good accuracy, intricate internal pores, and complex geometries. The following section is dedicated to discussing the utility and implication of PLA and PLGA in 3DP focusing BTE solutions.

2 Bio-composites of PLA

In the contemporary world of medicine, one of the major tasks is to restore the broken human tissues and to restore the missing organs because of the medical history of the patient. In response to this, there is a rapid inclination in the demand of various healthcare systems as well as medical devices. Medical therapies and surgical treatments of the present world are being supported by both existing biomaterials and newly introduction of novel and innovative materials combinations. For example, traditional PLA copolymers are the potential candidates for the treatments of diseases, traumas, and disabilities. Being a thermoplastic material, PLA possesses high-strength, high-modulus, and biocompatibility [31]. Researchers prepared the composite scaffolds with porous PLLA and PLLA/hydroxyapatite (HAP) using phase separation technique. It was the porous foams prepared by a thermally induced phase separation methods exhibited good mechanical properties and found suitable for osteoblast seeding and growth [32]. Materials such as PLA, poly-caprolactone (PCL), and wollastonite (W) can be used as the construction of biomedical scaffold using PLA/PCL ratio of 85/15 using injection molding process with CO_2 and was the physical blowing agent and bioactive filler in order to induce cell attachment differentiation and proliferation [33]. Figure 3 shows the regeneration of the cells resulted in maximum cell growth. Basalt fiber-reinforced PLA was also used for hard tissue repair using solution blending and freeze drying methods as blast fibers can be uniformly dispersed in PLLA matrix which helps to improve mechanical as well as hydrophobicity of the PLLA matrix as shown in Table 2, which were osteoblast compatible due to which it has promising applications in hard tissue repair [34].

The rapidly growing market demand for biodegradable materials is primarily because these materials produce no foreign body reaction as these get a gradual absorption and therefore in later are replaced by natural tissue. Therefore in long run, fatigue characteristics of such materials might be compromised. For instance, PLGA fibers are being used for fabricating sutures and fixation devices due to their well-documented biomedical characteristics. Cooper et al. developed a novel PLGA scaffold based on a 3D fibrous hierarchical design by using custom braiding tech-

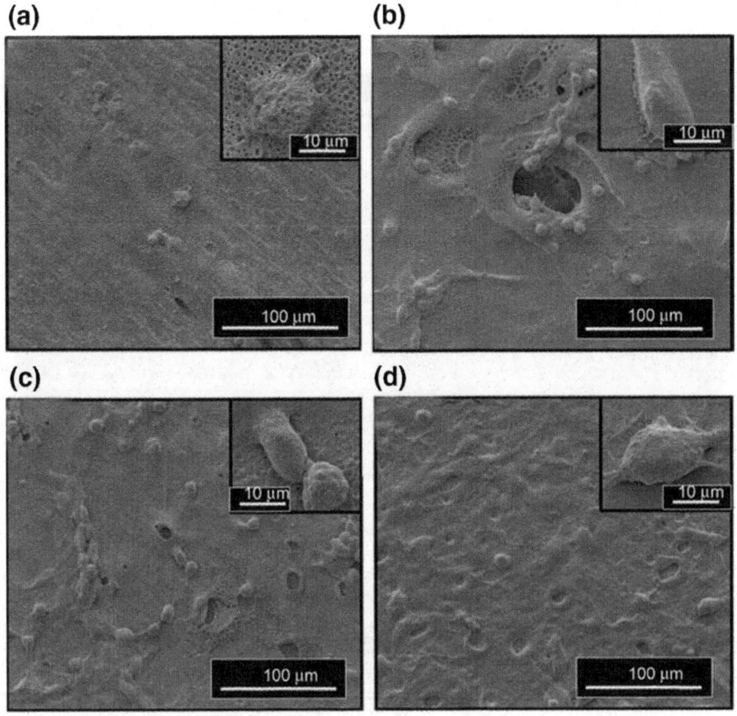

Fig. 3 SEM images of cell growth confirmation [33]

niques, which permitted the precise synthesis of substrates with the required pore diameter, porosity, mechanical properties, and architecture. Such a scaffold would provide the newly regenerating tissue with a temporary site for cell attachment, proliferation, and mechanical stability [35]. In addition, PLA copolymers were used as carriers for bone morphogenetic proteins (BMPs), biologically active molecules with the tendency of bone regeneration, refer Fig. 4, and to promote bone repair [36].

Zhang et al. [37] fabricated PLA/Octa-decylamine (ODA) based composites for use in TE through solution method. The PLA solution and functionalized nanodiamond (ND) dispersions were mixed, in different proportions, and after evaporation of the solvent thin films of the composites were obtained. The results showed that the

Table 2 Hydrophobicity of the films expressed as contact angle measurement [34]

Samples	Contact angle
PLLA	80.41 ± 0.88
5 wt% BF	75.18 ± 0.73
10 wt% BF	65.74 ± 1.02
15 wt% BF	54.03 ± 3.08

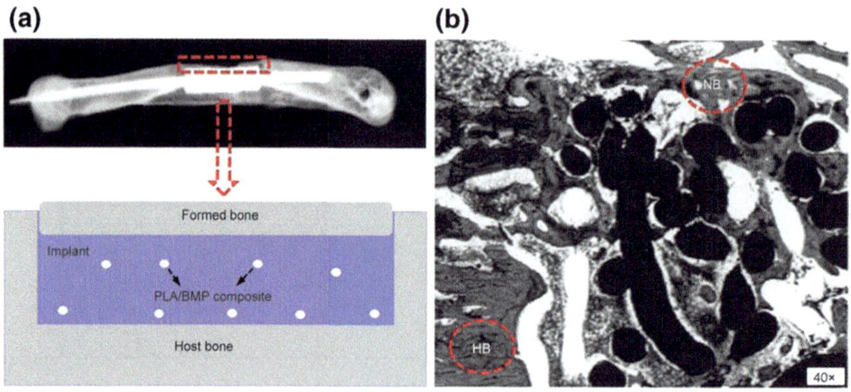

Fig. 4 Bone defect treatment using a composite of PLA and BMP (**a**) and exuded composite from the implant's pores [36, 37]

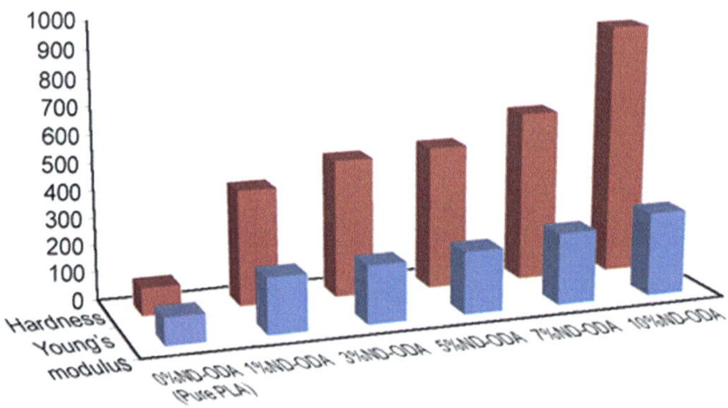

Fig. 5 Mechanical results of PLA/ND-ODA polymer composites in comparison of pure PLA [37]

mechanical properties of PLA were improved (refer Fig. 5). PLA and its copolymers were used in a range of applications related to wound management. Qin et al. [18] used PLA polymer blends with poly-tri-methylene carbonate (PTMC) in order to prevent postoperative adhesions. In their study, the polymeric solution was poured into a Teflon plate and solvent was evaporated, thereafter.

Historically, the PLA polymer was used to produce numerous biodegradable devices, which include screws and fixation pins, plates, and suture anchors. These devices gained widespread clinical use, especially where marginal mechanical stiffness or strength was desired. Pertinent orthopedic areas might include the knee [38–40], shoulder, foot and ankle [29, 41, 42], hand, wrist [43], elbow [44], pelvis, and zygomatic fractures. Although in some cases, high-performance PLA is required to improve the mechanical properties of the resulting devices, for instance, impact, tensile, and flexural strength in fracture fixation. One of the most important goals of using biodegradable materials is to avoid a second surgical procedure to remove

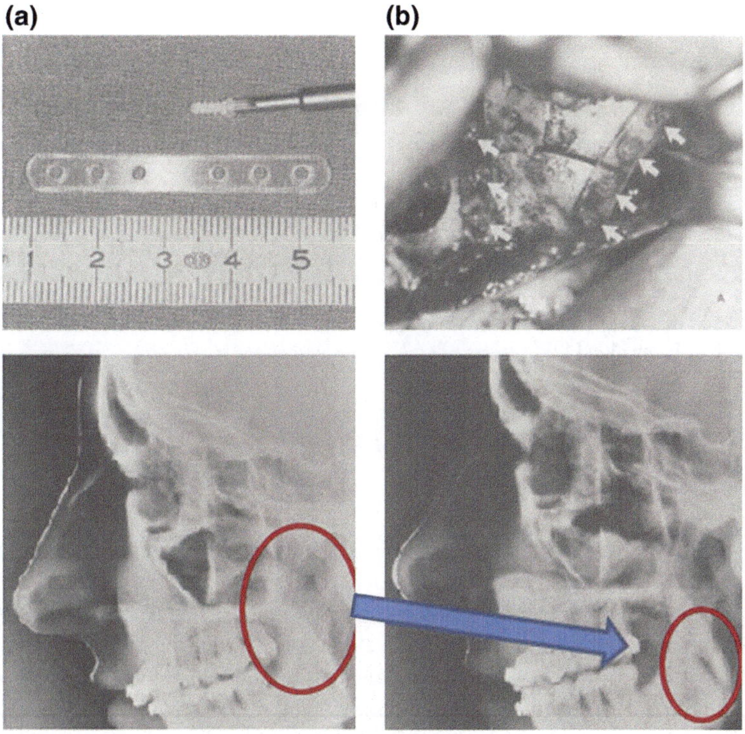

Fig. 6 **a** PLA screws and plate **b** implanted plates and screws in situ **c** and lateral cephalogram [45]

unnecessary hardware (refer Fig. 6) [45]. Novel chemical and mechanical properties can be generated by using effective materials, such as chitin and chitosan, for future-based biomedical applications owing to their remarkable biocompatibility, biodegradability, and non-toxicity. Furthermore, these materials can be processed into gels, sponges, membranes, beads, and scaffolds forms, effortlessly, in order to form nano-fibers, nanoparticles, and nano-composites [46]. Shalumon et al. made nano-fibers with electrospun processing of water-soluble carboxy-methyl chitin/poly-vinyl-alcohol (CMC/PVA) blended in different ratios (0–100%). The fibers were made water-insoluble by cross-linking with glutaraldehyde vapors followed by thermal treatment. It was found that the prepared nano-fibers exhibited reasonable bioactivity and biocompatibility. The cytotoxicity and cell attachment studies were also conducted on the fibrous scaffold by using human mesenchymal stem cells (hMSCs) by the MTT assay (refer Fig. 7) [47].

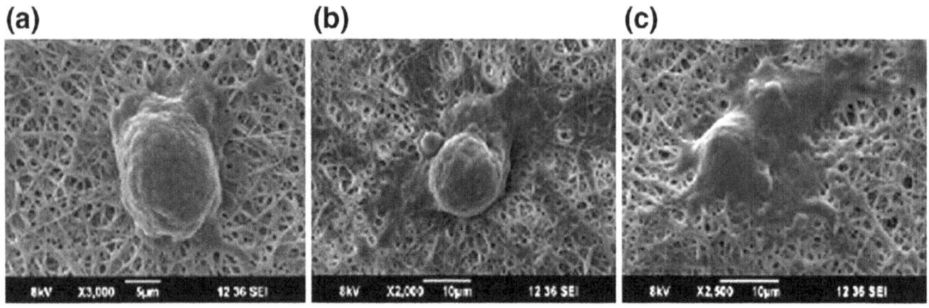

Fig. 7 SEM images of hMSCs attached on the surfaces of CMC/PVA scaffolds after **a** 12 h **b** 24 h and **c** 48 h of incubation [47]

3 PLA and Three-Dimensional Printing

In the recent growing industrial world, the globalization in the manufacturing sector has been increased. With the latest manufacturing process revolutions, the rapid prototyping is considered as the new future leading-edge technology that builds the complex physical object very rapidly and accurately [48, 49]. The various 3DP processes such as stereolithography (SLA), selective laser sintering (SLS), or laminated object manufacturing (LOM), and fused deposition modeling (FDM) are existing in mass, around the world [50]. The complete details of these subjected processes are well presented in the study of Yan and Gu and Pham and Gault, respectively [51, 52]. Albeit each of the 3DP processes has their own pros and cons, still the FDM technique is represented as a promising technology and studied extensively in assessing its viability in commercial applications such as electronics (resistors and sensors), optical (antennas), medical (artificial hip joints, tissue scaffolds, and bone structures), automotive, communication and aerospace industries (turbines engines and thermal insulation coatings) [53]. Further, printing of PLA required very less time and the only 16 g of freshly extruded PLA required only \$0.46 of PLA for common applications [54]. Moreover, by using 3DP, the organic electronic devices could be fabricated on biodegradable PLA substrates, to overcome issues of fabrication costs reduction and disposability of old, non-functional organic devices [55]. Anatomically, accurate PLA replicas can be easily created by a consumer-grade 3D printer in a fast and cost-effective manner, promoting their routine in-house use [56]. Ge et al. examined the mechanical properties and microenvironment of 3D printed PLGA scaffolds and, also, the proliferation and differentiation of human fetal osteoblasts were evaluated after 3 weeks of in vitro culture on the scaffolds. It has been found that the scaffolds had mechanical properties quite similar to trabecular bone but, otherwise, was weaker than cortical bone. Furthermore, the resulted porous interconnected tunnel like cylindrical PLGA scaffolds was of 6 mm in diameter and height, as shown in Fig. 8 [57].

Serra et al. printed 3D composite scaffolds based on PLA, glass (G5), and 5% PEG. It has been outlined by the researchers that the 3DP technology used enabled

Fig. 8 **a** PLGA scaffold, **b** SEM plot of micropore, and **c** SEM at high resolution [57]

the fabrication of highly porous scaffolds with mechanical properties considerably higher than other conventional methods such as solvent casting and phase separation [58]. Moreover, the quality of architecture produced is highly precise and tailor controlled, as shown in Fig. 9. The possibility of building PLA/nano-HA composites by FDM was systematically evaluated by producing an in-house developed feedstock for a low-cost printer was produced by the solvent-free procedure. The morphological analysis and compression testing on printed samples showed that nano-HA was uniformly dispersed within the PLA matrix and improved the PLA mechanical properties without changing the rheological performance [59]. Further, in another study, 3D printed PLA/HA scaffolds exhibited good biocompatibility and bioactive, in vitro, as compared to β-tri-calcium-phosphate (TCP). In vivo evaluation by a critical size rat, calvarial defect model showed that printed PLA/HA scaffolds had the features of little inflammation response, relatively larger resorption rate and superior osteoinductive activity to enhance bone formation [60]. Furthermore, poly-dopamine can be coated on the PLA scaffolds prints to improve the adhesion, proliferation, and cell growth up to a significant level [61].

The PLA and PLGA samples help to control the spatial morphology and distribution of cells on synthetic biomaterials which could have significant applications in TE [62]. Wang et al. performed cold atmospheric plasma (CAP) treatment of 3D printed

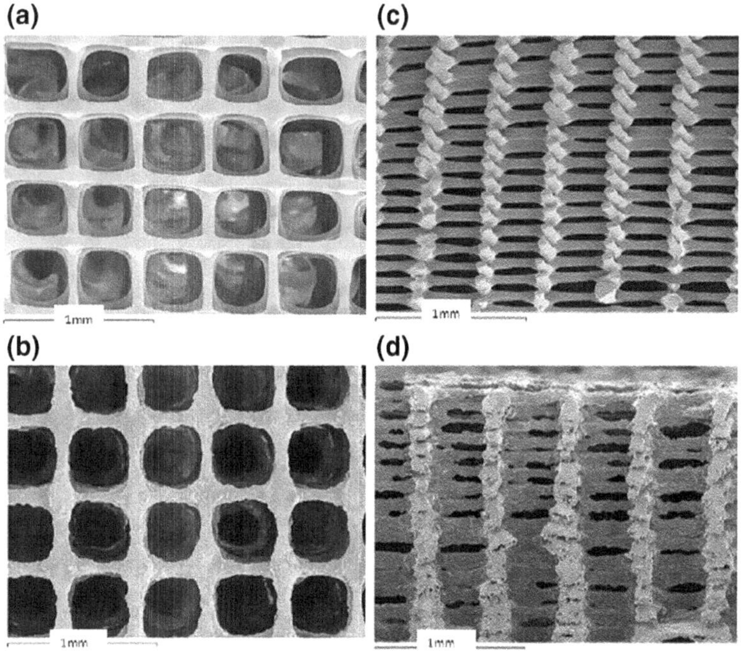

Fig. 9 SEM micrographs of 3-D printed scaffolds: **a**, **c** PLA/PEG; **b**, **d** PLA/PEG/G5; **a**, **b** top view; **c**, **d** cross-sectional view [58]

PLA scaffolds for tissue regeneration applications. It has been found in Fig. 10 that the osteoblast cell proliferation on the different CAP treated PLA scaffolds for 1, 3, and 5 days of proliferation, indicated that CAP treatment of the 3D printed PLA scaffolds improved bone cell growth [63]. Senatov et al. presented work on the development of 3D printed porous scaffolds by using HA reinforced PLA polymer. The required composite feedstock was developed via a screw extrusion process, which further undergone through mechanical, structural, and shape memory effect (SME) investigations. The mean size of the pore and average porosity level of the printed scaffold was observed as 700 μm and 30 vol.%, respectively. It has been outlined by the research group that with the dispersion of HA particles, additional rigid fixed phases were formed which reduced the mobility of the polymer mixture. Additionally, the ceramic HA particles instilled the growth of cracks during compression heating cycles, indicated the self-healing capability [64]. Castilho et al. treated the HA/TCP scaffolds with PLGA solution and enhanced the mechanical properties by a factor of 8 [65].

Almeida et al. unrevealed the effect of 3D printed structures of PLA on inflammation [66]. In their work, the fabrication process included materials such as chitosan, PLA/calcium phosphate glass, and pure PLA. Further, cytokine secretion and cell morphology of human peripheral blood monocytes were also used in order to distinguish the various types of matrices. It was observed that the differently pre-

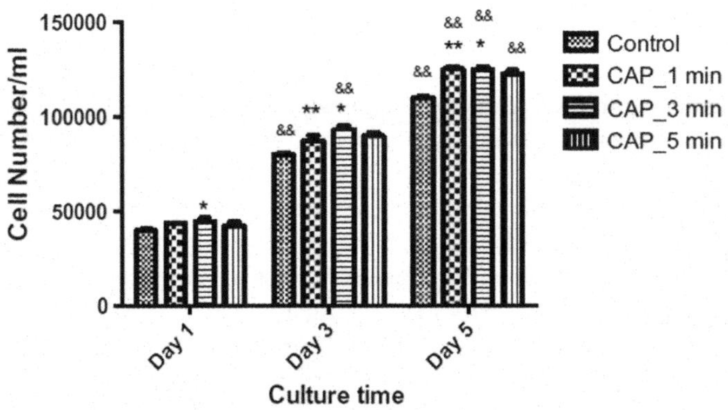

Fig. 10 Osteoblast cells proliferation on CAP treated samples [63]

pared scaffolds encouraged the monocyte/macrophage adhesion and production of a stimulated cytokine. In a comparison of PLA-based and chitosan scaffolds, a striking difference was observed with chitosan eliciting that increased the secretion of tumor necrosis factor. On the other hand, the PLA-based scaffolds induced the much higher formation of interleukin (IL)-6, IL-12/23, and IL-10. Overall, the chitosan scaffolds having larger pores and wider angles leading to higher secretion of these pro-inflammatory cytokines.

Structural anisotropy is always acted as a critical barrier in the widespread utilization of 3DP in the medical sector, however, reportedly the ionization radiation method can be a great help to control or minimizing the default anisotropy of the PLA structures produced via 3DP [67]. An inhibited biofilm colonization can be resulted by 3D printing when antimicrobial nitrofurantoin was preliminarily incorporated in PLA [68]. This could benefit various medical applications, especially wherein the degradation rate of PLA is occurring at a higher pace.

Narayan et al. created a nano-fibrous bio-ink for bio-printing soft tissue constructs [69]. Results from the aB assay showed a significant difference in % aB reduction between the 3D bio-plotted Alg-hASC and Alg-NfhASC strands over 16 days in chondrogenic differentiation media culture, which indicates that the presence of PLA nano-fibers in the alginate matrix aided hASC metabolic activity and proliferation. Tiersch and Monroe reported that 3D printed PLA is ideally suited for cryobiology device development, which can be designed in op-source electronically distributed for increasing the affordability of existing printers and to accelerate the cryobiology findings and improve reproducibility of results [70]. Yeon et al. printed PLA/HA/Silk composites for the development of biological internal fixation of the bone fractures in order to avoid secondary fracture occurrence owing to the use of conventional plates, bolts, and screws [71]. The major outcomes of their work highlighted several advantages: (i) noninvasive, (ii) personalized design to shorten the surgical time, improve surgical accuracy, and reduce postoperative complications, (iii) mechanically stable for numerous fragmented fractures, and (iv) high biocompatibility.

4 Summary

The PLA and its combinations with other biomaterials (such as glass, ceramics, metals, fibers, and polymers) have been nominated by numerous researchers as potential materials to meet with different BTE applications and provide a broad range of mechanical, morphological, and biological characteristics, when processed through 3DP systems. The 3DP technologies can help in the development of customized, precise, controlled architecture and morphology, near to bone properties, and desired mechanical strength to meet with body's structural needs. Being very flexible and readily adaptable materials it allows pre-/post-processing treatments, providing further scope for the betterment.

On the top, PLA is one of the thermoplastic materials that are cost-effective, easy to process, and eco-friendly, owing to its biodegradability. Till now, PLA has been used for various applications, which includes artificial scaffolds and sutures, bone regeneration and healing, drug delivery vessels, and orthopedic devices. Moreover, the shape memory effect and self-healing characteristics of this material are beneficial for various medical applications, especially for developing the scaffolds and soft structures.

Some PLA grades are not approved for use in human body tissue or fluids; therefore, the incorporation of protective biomaterials and films is must before in vivo practices. In the literature, different physical and chemical modification techniques have been proposed to improve the biological properties of PLA. Noticeably, the use of PLA in the biomedical sector was boosted after the advent of 3DP technologies, by which the required devices could be made quickly and accurately as compared to conventional fabrication methods.

In the last decade, many efforts were made on the development of new material combinations, on design and parametric optimization, on the exploration of 3DP technologies in various direct and indirect medical services, and on the characterization (including mechanical, morphological, biological, in vitro, and in vivo) of the produced geometries. Most of the practices were focused on the fewer material combinations only, isolating the potential of both PLA feedstock and 3DP setups. Thus, dedicated efforts are required for the synthesis and development of novel biomaterials and methodologies, respectively, in order to overcome the existing drawbacks of PLA and 3DP technologies.

References

1. Middleton JC, Tipton AJ (2000) Synthetic biodegradable polymers as orthopedic devices. Biomaterial 21:2335–2346
2. Gupta B, Revagade N, Hilborn J (2007) Poly(lactic acid) fiber: an overview. Prog Polym Sci 34:455–482
3. Yamane H, Sasai K (2003) Effect of the addition of poly(D-lactic acid) on the thermal property of poly(L-lactic acid). Polymer 44:2569–2575

4. Drumright RE, Gruber PR, Henton DE (2000) Polylactic acid technology. Adv Mater 12:1841–1846
5. Gunatillake PA, Adhikari R (2003) Biodegradable synthetic polymers for tissue engineering. Eur Cell Mater. 5(1):1–6
6. Lasprilla AJ, Martinez GA, Lunelli BH, Jardini AL, Maciel Filho R (2012) Poly-lactic acid synthesis for application in biomedical devices—a review. Biotechnol Adv 30(1):321–328
7. Wang X, Jiang M, Zhou Z, Gou J, Hui D (2001) 3D printing of polymer matrix composites: a review and prospective. Compos B Eng 110:442–458
8. Ku H, Wang H, Pattarachaiyakoop N, Trada M (2011) A review on the tensile properties of natural fiber reinforced polymer composites. Compos B Eng 42(4):856–873
9. Ramakrishna S, Mayer J, Wintermantel E, Leong KW (2001) Biomedical applications of polymer-composite materials: a review. Compos Sci Technol 61(9):1189–1224
10. Faludi G, Dora G, Renner K, Móczó J, Pukánszky B (2013) Biocomposite from polylactic acid and lignocellulosic fibers: structure–property correlations. Carbohyd Polym 92(2):1767–1775
11. Cui Y, Liu Y, Cui Y, Jing X, Zhang P, Chen X (2009) The nanocomposite scaffold of poly (lactide-co-glycolide) and hydroxyapatite surface-grafted with L-lactic acid oligomer for bone repair. Acta Biomater 5(7):2680–2692
12. Yun YP, Kim SE, Lee JB, Heo DN, Bae MS, Shin DR, Lim SB, Choi KK, Park SJ, Kwon IK (2009) Original paper: comparison of osteogenic differentiation from adipose-derived stem cells, mesenchymal stem cells, and pulp cells on PLGA/hydroxyapatite nanofiber. Tissue Eng Regenerat Med 6(1):336–345
13. Sanders JE, Bale SD, Neumann T (2002) Tissue response to microfibers of different polymers: polyester, polyethylene, polylactic acid and polyurethane. J Biomed Mater Res 62:222–227
14. Kellomaki M, Niiranen H, Puumanen K, Ashammakhi N, Waris T, TormaLa P (2000) Bioabsorbable scaffolds for guided bone regeneration and degeneration. Biomaterials 21:2495–2505
15. Hoveizi E, Nabiuni M, Parivar K, Rajabi-Zeleti S, Tavakol S (2014) Functionalisation and surface modification of electrospun polylactic acid scaffold for tissue engineering. Cell Biol Int 38(1):41–49
16. Avital S, Bollinger TJ, Wilkinson JD, Marchetti F, Hellinger MD, Sands LR (2005) Preventing intra-abdominal adhesions with polylactic acid film: an animal study. Dis Colon Rectum 48(1):153–157
17. Li G, Wang Z-X, Fu W-J, Hong B-F, Wang X-X, Cao L, Xu F-Q, Song Q, Cui F-Z, Zhang X (2011) Introduction to biodegradable polylactic acid ureteral stent application for treatment of ureteral war injury. BJU Int 108:901–906
18. Qin Y, Yuan M, Li L, Guo S, Yuan M, Li W, Xue J (2006) Use of polylactic acid/polytrimethylene carbonate blends membrane to prevent postoperative adhesions. J Biomed Mater Res Part B Appl Biomater Off J Soc Biomater Jpn Soc Biomater Aust Soc Biomater Korean Soc Biomater 79(2):312–319
19. Mi HY, Salick MR, Jing X, Jacques BR, Crone WC, Peng XF, Turng LS (2013) Characterization of thermoplastic polyurethane/polylactic acid (TPU/PLA) tissue engineering scaffolds fabricated by microcellular injection molding. Mater Sci Eng C 33(8):4767–4776
20. Kang SW, Jeon O, Kim BS (2005) Poly (lactic-co-glycolic acid) microspheres as an injectable scaffold for cartilage tissue engineering. Tissue Eng 11(3–4):438–447
21. Chen C, Lv G, Pan C, Song M, Wu C, Guo D, Wang X, Chen B, Gu Z (2007) Poly (lactic acid) (PLA) based nanocomposites—a novel way of drug-releasing. Biomed Mater 2(4):L1
22. Gollwitzer H, Ibrahim K, Meyer H, Mittelmeier W, Busch R, Stemberger A (2003) Antibacterial poly (D, L-lactic acid) coating of medical implants using a biodegradable drug delivery technology. J Antimicrob Chemother 51(3):585–591
23. Rancan F, Papakostas D, Hadam S, Hackbarth S, Delair T, Primard C, Verrier B, Sterry W, Blume-Peytavi U, Vogt A (2009) Investigation of polylactic acid (PLA) nanoparticles as drug delivery systems for local dermatotherapy. Pharm Res 26(8):2027–2036
24. Polimeni G, Koo KT, Pringle GA, Agelan A, Safadi FF, Wikesjö UM (2008) Histopathological observations of a polylactic acid-based device intended for guided bone/tissue regeneration. Clin Implant Dent Relat Res 10(2):99–105

25. Mohiti-Asli M, Saha S, Murphy SV, Gracz H, Pourdeyhimi B, Atala A, Loboa EG (2017) Ibuprofen loaded PLA nanofibrous scaffolds increase proliferation of human skin cells in vitro and promote healing of full thickness incision wounds in vivo. J Biomed Mater Res B Appl Biomater 105(2):327–339
26. Zhao C, Wu H, Ni J, Zhang S, Zhang X (2017) Development of PLA/Mg composite for orthopedic implant: tunable degradation and enhanced mineralization. Compos Sci Technol 28(147):8–15
27. Van Sliedregt A, Radder AM, De Groot K, Van Blitterswijk CA (1992) In vitro biocompatibility testing of polylactides part I proliferation of different cell types. J Mater Sci Mater Med 3(5):365–370
28. Majola A, Vainionpää S, Vihtonen K, Mero M, Vasenius J, Törmälä P, Rokkanen P (1991) Absorption, biocompatibility, and fixation properties of polylactic acid in bone tissue: an experimental study in rats. Clin Orthop Relat Res 268:260–269
29. Böstman O, Hirvensalo E, Vainionpää S, Mäkelä A, Vihtonen K, Törmälä P, Rokkanen P (1989) Ankle fractures treated using biodegradable internal fixation. Clin Orthop Relat Res 238:195–203
30. Maurus PB, Kaeding CC (2004) Bioabsorbable implant material review. Oper Techn Sports Med 12(3):158–160
31. Hartmann MH (1998) High molecular weight polylactic acid polymers. In: Biopolymers from renewable resources. Springer, Berlin, pp 367–411
32. Ma PX, Zhang R, Xiao G, Franceschi R (2001) Engineering new bone tissue in vitro on highly porous poly (α-hydroxyl acids)/hydroxyapatite composite scaffolds. J Biomed Mater Res Off J Soc Biomater Jpn Soc Biomater 54(2):284–293
33. Goswami J, Bhatnagar N, Mohanty S, Ghosh AK (2013) Processing and characterization of poly (lactic acid) based bioactive composites for biomedical scaffold application. Exp Polym Lett 7(9):767–777
34. Chen X, Li Y, Gu N (2010) A novel basalt fiber-reinforced polylactic acid composite for hard tissue repair. Biomed Mater 5(4):044104
35. Saito N, Takaoka K (2003) New synthetic biodegradable polymers as BMP carriers for bone tissue engineering. Biomaterialsm 24(13):2287–2293
36. Murakami N, Saito N, Horiuchi H, Okada T, Nozaki K, Takaoka K (2002) Repair of segmental defects in rabbit humeri with titanium fiber mesh cylinders containing recombinant human bone morphogenetic protein-2 (rhBMP-2) and a synthetic polymer. J Biomed Mater Res Off J Soc Biomater Jpn Soc Biomater Aust Soc Biomater Korean Soc Biomater 62(2):169–174
37. Zhang Q, Mochalin VN, Neitzel I, Knoke IY, Han J, Klug CA, Zhou JG, Lelkes PI, Gogotsi Y (2011) Fluorescent PLLA-nanodiamond composites for bone tissue engineering. Biomaterials 32(1):87–94
38. Barber FA, Elrod BF, McGuire DA, Paulos LE (1995) Preliminary results of an absorbable interference screw. Arthroscopy 11(5):537–548
39. Matsusue Y, Nakamura T, Suzuki S, Iwasaki R (1996) Biodegradable pin fixation of osteochondral fragments of the knee. Clin Orthop Relat Res 322:166–173
40. Stähelin AC, Weiler A, Rüfenacht H, Hoffmann R, Geissmann A, Feinstein R (1997) Clinical degradation and biocompatibility of different bioabsorbable interference screws: a report of six cases. Arthroscopy 13(2):238–244
41. Lavery LA, Higgins KR, Ashry HR, Athanasiou KA (1994) Mechanical characteristics of poly-L-lactic acid absorbable screws and stainless steel screws in basilar osteotomies of the first metatarsal. J Foot Ankle Surg Off Publ Am Coll Foot Ankle Surg 33(3):249–254
42. Bucholz RW, Henry S, Henley BM (1994) Fixation with bioabsorbable screws for the treatment of fractures of the ankle. JBJS 76(3):319–324
43. Casteleyn PP, Handelberg F, Haentjens P (1992) Biodegradable rods versus Kirschner wire fixation of wrist fractures. A randomised trial. J Bone Joint Surg Br 74(6):858–861
44. Hope PG, Williamson DM, Coates CJ, Cole WG (1991) Biodegradable pin fixation of elbow fractures in children. A randomised trial. J Bone Joint Surg Br 73(6):965–968

45. Haers PE, Suuronen R, Lindqvist C, Sailer H (1998) Biodegradable polylactide plates and screws in orthognathic surgery. J Craniomaxillofac Surg 26(2):87–91
46. Jayakumar R, Menon D, Manzoor K, Nair SV, Tamura H (2010) Biomedical applications of chitin and chitosan based nanomaterials—a short review. Carbohyd Polym 82(2):227–232
47. Shalumon KT, Binulal NS, Selvamurugan N, Nair SV, Menon D, Furuike T, Tamura H, Jayakumar R (2009) Electrospinning of carboxymethyl chitin/poly (vinyl alcohol) nanofibrous scaffolds for tissue engineering applications. Carbohyd Polym 77(4):863–869
48. Onuh SO, Yusuf YY (1999) Rapid prototyping technology: applications and benefits for rapid product development. J Intell Manuf 10(3–4):301–311
49. Yang S, Leong K-F, Du Z, Chua C-K (2002) The design of scaffolds for use in tissue engineering. Part II. Rapid prototyping techniques. Tissue Eng 8:1–11. https://doi.org/10.1089/107632702753503009
50. Garg A, Tai K, Lee CH, Savalani MM (2013) A hybrid M5-genetic programming approach for ensuring greater trustworthiness of prediction ability in modelling of FDM process. J Intell Manuf https://doi.org/10.1007/s10845-013-0734-1
51. Yan X, Gu P (1996) A review of rapid prototyping technologies and systems. Comput Des 28:307–318. https://doi.org/10.1016/0010-4485(95)00035-6
52. Pham DT, Gault RS (1998) A comparison of rapid prototyping technologies. Int J Mach Tools Manuf 38:1257–1287. https://doi.org/10.1016/S0890-6955(97)00137-5
53. Sugavaneswaran M, Arumaikkannu G (2014) Modelling for randomly oriented multi material additive manufacturing component and its fabrication. Mater Des 54:779–785. https://doi.org/10.1016/j.matdes.2013.08.102
54. Yang S, Leong KF, Du Z, Chua CK (2001) The design of scaffolds for use in tissue engineering. Part I. Traditional factors. Tissue Eng 7(6):679–689
55. Rankin TM, Giovinco NA, Cucher DJ, Watts G, Hurwitz B, Armstrong DG (2014) Three-dimensional printing surgical instruments: are we there yet? J Surg Res 189(2):193–197
56. Mattana G, Briand D, Marette A, Quintero AV, de Rooij NF (2015) Polylactic acid as a biodegradable material for all-solution-processed organic electronic devices. Org Electron 17:77–86
57. Ge Z, Wang L, Heng BC, Tian XF, Lu K, Tai Weng Fan V, Yeo JF, Cao T, Tan E (2009) Proliferation and differentiation of human osteoblasts within 3D printed poly-lactic-co-glycolic acid scaffolds. J Biomater Appl 23(6):533–547
58. Serra T, Planell JA, Navarro M (2013) High-resolution PLA-based composite scaffolds via 3-D printing technology. Acta Biomater 9(3):5521–5530
59. Esposito Corcione C, Gervaso F, Scalera F, Montagna F, Sannino A, Maffezzoli A (2017) The feasibility of printing polylactic acid–nanohydroxyapatite composites using a low-cost fused deposition modeling 3D printer. J Appl Polym Sci 134(13):44656
60. Zhang H, Mao X, Du Z, Jiang W, Han X, Zhao D, Han D, Li Q (2016) Three dimensional printed macroporous polylactic acid/hydroxyapatite composite scaffolds for promoting bone formation in a critical-size rat calvarial defect model. Sci Technol Adv Mater 17(1):136–148
61. Kao CT, Lin CC, Chen YW, Yeh CH, Fang HY, Shie MY (2015) Poly (dopamine) coating of 3D printed poly (lactic acid) scaffolds for bone tissue engineering. Mater Sci Eng, C 56:165–173
62. Lin CC, Ho CC (2005) Micropatterning proteins and cells on polylactic acid and poly (lactide-co-glycolide). Biomaterials 26(17):3655–3662
63. Wang M, Favi P, Cheng X, Golshan NH, Ziemer KS, Keidar M, Webster TJ (2016) Cold atmospheric plasma (CAP) surface nanomodified 3D printed polylactic acid (PLA) scaffolds for bone regeneration. Acta Biomater 46:256–265
64. Senatov FS, Niaza KV, Zadorozhnyy MY, Maksimkin AV, Kaloshkin SD, Estrin YZ (2016) Mechanical properties and shape memory effect of 3D-printed PLA-based porous scaffolds. J Mech Behav Biomed Mater 57:139–148
65. Castilho M, Moseke C, Ewald A, Gbureck U, Groll J, Pires I, Teßmar J, Vorndran E (2014) Direct 3D powder printing of biphasic calcium phosphate scaffolds for substitution of complex bone defects. Biofabrication 6(1):015006

66. Almeida CR, Serra T, Oliveira MI, Planell JA, Barbosa MA, Navarro M (2014) Impact of 3-D printed PLA-and chitosan-based scaffolds on human monocyte/macrophage responses: unraveling the effect of 3-D structures on inflammation. Acta Biomater 10(2):613–622
67. Shaffer S, Yang K, Vargas J, Di Prima MA, Voit W (2014) On reducing anisotropy in 3D printed polymers via ionizing radiation. Polymer 55(23):5969–5979
68. Sandler N, Salmela I, Fallarero A, Rosling A, Khajeheian M, Kolakovic R, Genina N, Nyman J, Vuorela P (2014) Towards fabrication of 3D printed medical devices to prevent biofilm formation. Int J Pharm 459(1–2):62–64
69. Narayanan LK, Huebner P, Fisher MB, Spang JT, Starly B, Shirwaiker RA (2016) 3D-bioprinting of polylactic acid (PLA) nanofiber–alginate hydrogel bioink containing human adipose-derived stem cells. ACS Biomater Sci Eng 2(10):1732–1742
70. Tiersch TR, Monroe WT (2016) Three-dimensional printing with polylactic acid (PLA) thermoplastic offers new opportunities for cryobiology. Cryobiology 73(3):396–398
71. Yeon YK, Park HS, Lee JM, Lee JS, Lee YJ, Sultan MT, Seo YB, Lee OJ, Kim SH, Park CH (2018) New concept of 3D printed bone clip (polylactic acid/hydroxyapatite/silk composite) for internal fixation of bone fractures. J Biomater Sci Polym Ed 29(7–9):894–906

Dr. Sunpreet Singh is assistant professor in the School of Mechanical Engineering, Lovely Professional University, Jalandhar, India. He has received Ph.D. in Mechanical Engineering from Guru Nanak Dev Engineering College, Ludhiana, India. His area of research is additive manufacturing and application of 3D printing for development of new biomaterials for clinical applications. He has contributed extensively in additive manufacturing literature with publications appearing in Journal of Manufacturing Processes, Composite Part: B, Rapid Prototyping Journal, Journal of Mechanical Science and Technology, Measurement, International Journal of Advance Manufacturing Technology, and Journal of Cleaner Production. He authored 10 book chapters and monographs. He is working with joint collaboration with Prof. Seeram Ramakrishna, NUS Nanoscience and Nanotechnology Initiative and Prof. Rupinder Singh, Manufacturing Research Lab, GNDEC, Ludhiana.

Dr. Chander Prakash is associate professor in the School of Mechanical Engineering, Lovely Professional University, Jalandhar, India. He has received Ph.D. in Mechanical Engineering from Panjab University, Chandigarh, India. His area of research is biomaterials, rapid prototyping and 3D printing, advanced manufacturing, modeling, simulation, and optimization. He has more than 11 years of teaching experience and 6 years of research experience. He has contributed extensively to the world in the titanium- and magnesium-based implant literature with publications appearing in surface and coating technology, materials and manufacturing processes, Journal of Materials Engineering and Performance, Journal of Mechanical Science and Technology, Nanoscience and Nanotechnology Letters, Proceedings of the Institution of Mechanical Engineers, Part B: Journal of Engineering Manufacture. He authored 50 research papers and 10 book chapters. He is also editor of three Books: "Current Trends in Bio-manufacturing", Springer Series in Advanced Manufacturing, Springer International Publishing AG, Gewerbestrasse 11, 6330 Cham, Switzerland., Dec. 2018; "3D Printing in Biomedical Engineering", Book series Materials Horizons: From Nature to Nanomaterials, Springer International Publishing AG, Gewerbestrasse 11, 6330 Cham, Switzerland., August 2019, and "Biomaterials in Orthopaedics and Bone Regeneration - Design and Synthesis", Book series Materials Horizons: From Nature to Nanomaterials, Springer International Publishing AG, Gewerbestrasse 11, 6330 Cham, Switzerland., March 2019. He is also guest editor of three Journals: Guest Editor of Special Issue of "Functional Materials and Advanced Manufacturing", Facta Universitatis, Series: Mechanical Engineering (Scopus Index), Materials Science Forum (Scopus Index), and Special Issue on "Metrology in Materials and Advanced Manufacturing", Measurement and Control (SCI indexed).

Prof. Rupinder Singh is professor in Department of Production Engineering, Guru Nanak Dev Engineering College, Ludhiana, India. He has received Ph.D. in Mechanical Engineering from Thapar Institute of Engineering & Technology, Patiala, India. His area of research is non-traditional machining, additive manufacturing, and development of porous biomaterials using 3D printing and rapid prototyping techniques. He has more than 18 years of teaching and research experience. He has contributed extensively to the world in additive manufacturing literature with publications appearing in Journal of Manufacturing Processes, Composite Part: B, Rapid Prototyping Journal, Journal of Mechanical Science and Technology, Measurement, International Journal of Advance Manufacturing Technology, and Journal of Cleaner Production. He authored 17 books and ~3 book chapters. He has received more than 3 crores research grants from various funding agencies such as DST-SERB, AICTE, CSIR, DAE, IE etc. He is working with Prof. Seeram Ramakrishna, NUS Nanoscience and Nanotechnology Initiative, Prof. Fernando Fraternali, Full Professor of Structural Mechanics, Department of Civil Engineering, University of Salerno.

Prof. Seeram Ramakrishna is co-director, NUS Nanoscience and Nanotechnology Initiative (NUSNNI). He has received Ph.D. from University of Cambridge. He is a global leader in electrospinning and nanostructured materials. Since 2001, Seeram's team has contributed significantly to the scientific knowledge in basic research, advances in the process and the applications of nano-fibers. Professor Seeram Ramakrishna's research resulted in ~1000 peer-reviewed articles with ~70,000 citations and ~120 H-index. He authored 5 books and ~25 book chapters. He has been recognized as a Highly Cited Researcher in Materials Science (www.highlycited.com) for the past four years. Thomson Reuters listed him among the Most Influential Scientific Minds in the World. NUS Vice-President (research strategy); Dean of Faculty of Engineering; Founding Director of NUSNNI, and Bioengineering; Founding Chair of Global Engineering Deans Council; Vice-President of International Federation of Engineering Education Societies; Board member of Asia Society for Innovation & Policy. EC/FP7 report lists him among the top four researchers of Singapore. He is recognized world's number one in nanofiber technologies. He also generated intellectual property in the form of 20 patents and licensed technologies to industry and start-ups. He mentors Singapore-based start-ups namely (a) Biomers International, (b) Insight Bio Ventures, (c) ceEntek, (d) Hyperion Core and (e) Everest Capital. Moreover, he works with companies such as HP, Schlumberger, BASF, Interplex, GE, Mann Hummel, Kaneka, Mitsubishi, Toray, Unitex, GSK, and AMT for value capture in Singapore.

Computer-Aided Design of Subject-Specific Dental Instruments for Preoperative Virtual Planning in Orthognathic Surgery

Faruk Ortes, Erol Cansiz and Yunus Ziya Arslan

Abstract Maxillofacial deformities and undesirable position of the mandible cause facial asymmetry and malocclusions. The techniques and equipment used for the maxillofacial surgeries have changed over the years. In this chapter, traditional preoperative preparations and surgical planning process in the orthognathic surgery have been summarized. In addition, we reviewed various software and workflows used for the preoperative planning and design of the miniplates. We also presented a systematic protocol for the subject-specific miniplate design as a case study. In this case, design steps, which are required to be taken for obtaining the virtual model of the patient head composed of the skull, mandible, and teeth were elucidated. Simulation of the Le Fort I osteotomy, which is considered as a safe and functional procedure to correct maxillary deformities, benefiting from a computer-aided design software to plan the actual surgery process was also carried out. It is expected that the presented virtual planning process would improve the accuracy of orthognathic surgery and patient satisfaction, and reduce the operation time and cost.

Keywords Orthognathic surgery · Virtual planning · Le Fort I osteotomy · Preoperative preparation · Computer-aided design · Surgical fixation devices · Miniplates

1 Introduction

Maxillofacial deformities can be caused by congenital or accidental traumas due to a variety of reasons [1]. Patients are adversely affected by psychological, cosmetic, and chewing dysfunctions originating from the facial asymmetry and dentoskeletal

F. Ortes · Y. Z. Arslan (✉)
Department of Mechanical Engineering, Faculty of Engineering, Istanbul University-Cerrahpasa, Avcilar, 34320 Istanbul, Turkey
e-mail: yzarslan@istanbul.edu.tr

E. Cansiz
Department of Oral and Maxillofacial Surgery, Faculty of Dentistry, Istanbul University, Capa, 34093 Istanbul, Turkey

C. Prakash et al. (eds.), *Biomanufacturing*,
https://doi.org/10.1007/978-3-030-13951-3_4

deformities [2]. The treatment of patients with maxillofacial deformities requires careful and precise surgical interventions to gain acceptable outcomes. Orthognathic surgery requires a bunch of operations for the reconstruction of the maxilla and/or mandible [3]. In the scope of the surgical procedure, one of the main purposes is to obtain a sufficient mobilization of the maxilla during the operation in order to provide a proper placement in the desired position. Generally, maxillary deformities are treated by Le Fort I osteotomy [4]. This method is defined as an osteoplastic resection of the maxilla and inferior displacement of upper jaw following a pre-defined fracture line [5]. Intermaxillary fixation follows the osteotomy procedure. Long-term stability of the surgically relocated maxilla is a necessity for the successful results of the treatment including the combination of surgical and orthodontic interventions. There are significant challenges in the surgery such as potential non-resectable lesions and close proximity to significant nervous and vascular structures [6]. Therefore, it requires a well-planned and precise implementation. Even though the treatment protocol to be applied would depend on the type of the clinical case, the main flow of surgical intervention would follow (*i*) the preoperative planning, (*ii*) osteotomy, (*iii*) repositioning of the maxilla, and (*iv*) fixation of the bones by means of miniplates. The materials used in the surgery (such as miniplates, splint, and plaster models) are as important as the preoperative planning and surgical intervention. Thus, improvement of the techniques, equipment, and materials are continuously aimed to obtain satisfactory outputs associated with both surgeons' and patients' needs. Even though the patients subjected to the Le Fort I osteotomy mostly need further surgical intervention including mandibular repositioning since open bite malocclusions may occur after the maxillary displacement, only the methods and equipment related to the Le Fort I osteotomy were focused in this chapter.

2 Preoperative Preparations and Surgical Planning

In orthognathic surgery, accurate repositioning and satisfactory stability of the bony segments are essential for obtaining successful cosmetic and functional outputs. Determining the desired position of the maxillary bone and proper osteotomy line requires a significant amount of time during the preoperative planning phase. Translational and rotational movements of the maxilla are likely necessary for repositioning task. Traditional preoperative planning in the orthognathic surgery has been used over the last decades, and it was accepted as a gold standard for the treatment of the dentofacial deformities [7]. In the traditional approach, preoperative preparations and surgery planning procedure consist of many sequential steps including (*i*) articulator mounting and face-bow transfer to obtain the geometrical data of the patient's skull and jaw and to fabricate cast plaster model, (*ii*) identifying the osteotomy line, (*iii*) partition of the plaster model, (*iv*) moving the maxillary bone to the desired final position, and (*v*) producing the intermediate surgical wafer (or splint) [8]. Wafers play a role in the determination of the final position for both maxillary and mandibular bones. If there is still an open bite after repositioning of the maxilla, the osteotomy

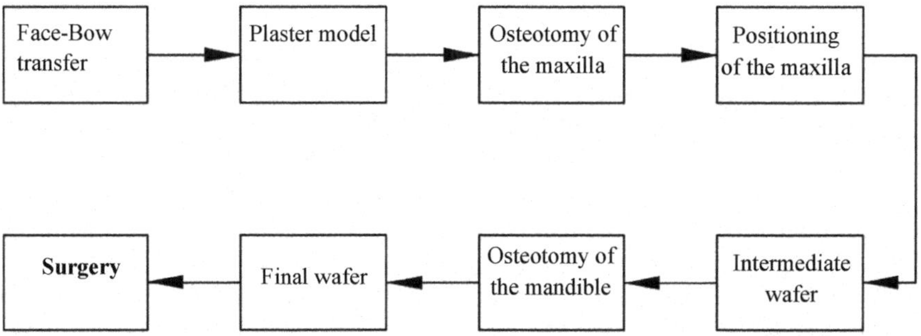

Fig. 1 Preoperative planning steps taken in the traditional treatment approach

and movement of the mandible are implemented to obtain an optimal final condition for the jaw, and a final version of the wafer is produced. In such cases, the final version of the wafer is produced after the whole segmentation and movement process is carried out on the casting plaster model. The traditional manual surgery simulation on casting models has been utilized for the fabrication of the intermediate wafers. In Fig. 1, the steps of the traditional manual surgery simulation are depicted. Accordingly, the intermediate wafer is used to reposition the maxilla to the final position, and a final wafer is employed to reposition the mandible to its final desired position. Thus, having an accurate position of the mandible depends on the accuracy of the repositioning of the maxilla. Therefore, the success of the preoperative planning for traditional orthognathic surgery still depends on the use of wafers [9]. However, using wafers may lead to unsatisfying maxillary position and occurrence of errors in the transfer of the geometrical data of the patient's jaw to the plaster models. Furthermore, this procedure is time-consuming and labor-intensive owing to multiple steps. Errors in any step of the preoperative planning cause likely significant surgical complications or malfunctions, resulting in the functional and esthetic problems. Preoperative planning takes a long time in the traditional method since the quantification of the dentoskeletal deformities needs obtaining data from various sources including photographs and cephalograms. Surgical simulation is then carried out for measuring the movement of the plaster model [10]. Aboul-Hosn Centenero and Hernández-Alfaro reported some common mistakes made in dimensioning stage which may occur (*i*) when transferring the models to the articulator, (*ii*) when drawing the vertical and horizontal reference lines in the models, and (*iii*) when repositioning, transferring, and rotating the models [11].

Three-dimensional (3D) virtual planning in orthognathic surgery is performed to mimic the actual surgical implementation. This method is also preferred for positioning the maxilla and mandible on the virtual model before the surgery by utilizing the combination of several newly recognized techniques such as obtaining the 3D model of the jaw by means of computed tomography (CT) scans and various software packages [12]. By means of virtual planning, it is possible to identify the osteotomy line on the model and examine the probable movement of the bones.

Virtual planning approach enables to detect complex problems in the surgery. CT and cone-beam computed tomography (CBCT) scans are generally processed within available software solutions that remarkably simplify diagnosis, analysis, and preoperative surgical planning. Those techniques are increasingly employed by clinicians and surgeons since they provide the 3D geometrical properties of the targeted tissues of the human body. These imaging technologies play crucial roles in virtual surgery planning that is utilized to provide a useful alternative to the traditional plaster model in the treatment planning for patients with dentofacial deformities [13].

In the orthognathic surgery planning, the mandible is mostly the targeted area. The geometry of the bone tissues is obtained in 3D space and presented in the anatomical planes, allowing observing the area and diagnosing the malformations. CT and CBCT enable to store the digitized data of the patient in the computer environment. Storing data belongs to patients allow to compare pre- and post-operative conditions and observe changes yielded by the applied treatment. Plaster models are gradually replaced with virtual 3D orthodontic models to implement surgical modifications in silico [14]. The models built in computer environment provide several advantages compared to the casting plaster models, namely computer environment (*i*) allows access to patient data easier and avoids the storage area which is required for plaster models, (*ii*) allows for many modifications, which are limited for traditional plaster casts, on the same model and taking the back up for different scenarios, (*iii*) makes the measurement of any dimension of the tissues easier, (*iv*) increases the accuracy of diagnosis, (*v*) enables the transfer of the scans of patient to anywhere and collaborating with other surgeons, and (*vi*) allows the analysis based on the objective findings [15].

Two-dimensional (2D) CT data can be converted into 3D virtual models benefiting from the various computer programs which enable surgeons to obtain 3D complex anatomical structures and to modify them consistent with surgical procedures with reduced operating time and optimal outcomes [16]. Therefore, computer-aided design (CAD) technologies have been widely used in surgical operations [17–20]. A variety of software packages has been adopted to obtain 3D models of the targeted bone structure in the human body [13, 14]. All of the preoperative planning computer programs produce the models by processing the digital imaging and communications in medicine (DICOM) data obtained from CT or CBCT devices. Digitized data existing in DICOM files are imported to the software and displayed on the computer as a first step. Surface and volumetric rendering procedures are applied by using the tissue-masking process based on the density differences of the hard and soft tissues. A pre-defined threshold based on Hounsfield unit is used to obtain different masks which can be converted to the surface models. Such programs also allow saving these masks in stereolithography (STL) format to manufacture them by means of additive manufacturing techniques such as 3D printing or laser sintering. The obtained CT or CBCT scans are converted to 3D models by taking into account the Hounsfield Unit of the corresponding tissues to distinguish the tissues from each other. However, the models mostly have a shape with a lot of artifacts, which requires the smoothing process. Artifacts, especially around the alveolar process of the mandible, cause remarkable difficulties in determination of the osteotomy line and design of implants.

In Fig. 2, the process including the steps from CT scans to the design of the miniplates are given.

In the surface reconstruction step, small holes in the geometry can be filled and sharp edges can also be smoothed. Following the surface reconstruction, the surface model is converted to a CAD model. Surface models are closed-shape with an empty volume, and they are not useful for the CAD operations such as opening holes, subtraction, and assembling with another part. CAD operations convert the surface model to a solid body model. This step also allows performing a series of modifications on the model such as drilling, bending, scaling, and boolean operations. CAD models generally take over the role of the plaster orthodontic models since they allow observing geometric and volumetric properties of the human body tissues [13]. Moreover, sectioning and movement of the maxilla and mandible can be practically simulated. After sectioning the model, the maxilla is repositioned manually in the plaster models and temporarily fixed by using wax. This traditional method is a highly error-prone and imprecise approach. Thus, using CAD models is advantageous to eliminate the probable errors.

Next steps are implementing the virtual osteotomy and replacing the maxillary bone to the desired final position. Osteotomy line is determined by executing a partition operation on the CAD model. Surgeon involvement is highly needed at this step since significant nervous and vascular structures exist around the targeted area. The

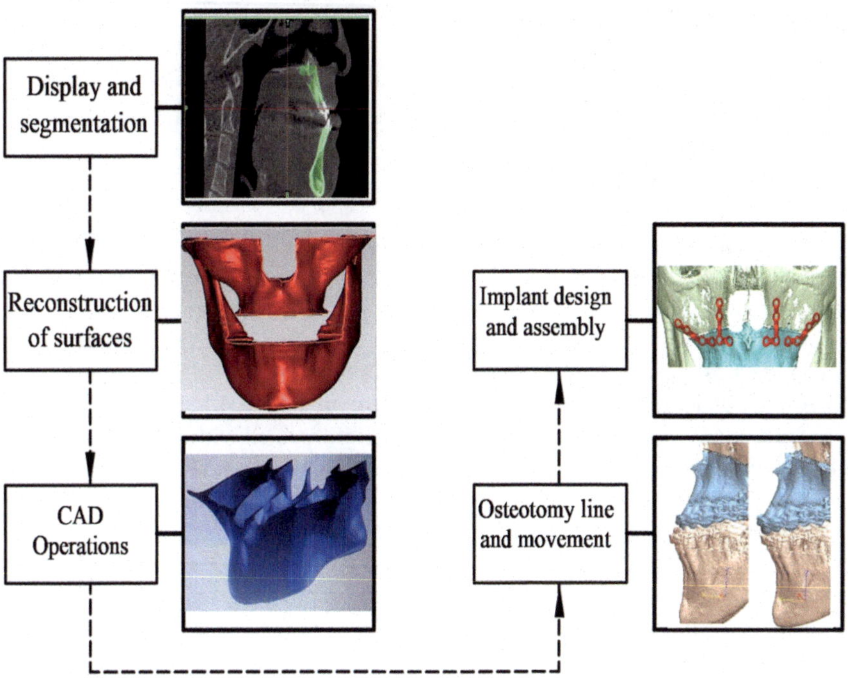

Fig. 2 Process flow of the virtual planning

CAD model is cut and separated from the osteotomy line. By using CAD software tools, which allow performing rotation and translation motions of the bone segments, the position of the maxilla is determined. Furthermore, the process enables to measure the distances between reference points located on the different segments and to identify the relative position of bony segments after the correction process. In the traditional surgical method, the same task is carried out by means of an articulator. However, in the virtual planning, the sectioning and repositioning can be performed and controlled more accurately and directly since the pre-defined points can be quantitatively determined. After repositioning of the maxilla, intermediate or final splints and wafers can be designed on the CAD model. The wafer can be used as a double-check tool to evaluate the position and orientation of the jaw after the surgery. Implant design is the last step of the virtual planning. Miniplates and screws are the fixation implants which are commonly used in orthognathic surgery to provide skeletal stability in the post-surgery period. In the traditional orthognathic surgery, the miniplates with standard shapes, dimensions, and materials are used. Surgeon changes the shape of the miniplates manually by applying external loads following the completion of the final repositioning of the maxilla. After the formation of the miniplates, the fixation is ensured by different numbers of screws. The final shape of miniplates might be still inconsistent with the topography of the bone tissues and sequential reshaping of the miniplates might be required to reach an acceptable consistency between the material and bone tissue. Benefitting from the virtual surgery planning method, subject-specific miniplates can be designed, and hence use of the dental instruments such as intermediate and final splints and wafers can be eliminated.

3 Subject-Specific Miniplate Design

Designs of the dental instruments such as surgical splints, wax bite wafers, and plasters are carried out by means of CAD technologies to determine the ideal fixation position of osteotomized maxillary segments [21–23]. Fixation is performed by means of osteosynthesis miniplates and screws following the surgical repositioning, which are mostly made of resorbable materials or titanium [24, 25]. Titanium L-type plates are commonly used in the Le Fort I osteotomy (Fig. 3).

Traditionally, the Le Fort I osteotomies are rigidly fixed with four titanium miniplates which are accepted as the standard approach for the fixation of the repositioned maxilla. Patients with a restorative need such as maxillofacial deformations start to live with these mini-implants after the surgery. Bilateral fixation of the maxilla is commonly carried out with the miniplates with standard shapes on each side of the zygomatic buttress. After tightening the screws, the miniplates maintain the position of the maxilla and keep the structural stability. The miniplates are widely made of a highly corrosion-resistant alloy Ti6AL4V and pure titanium to provide a rigid fixation, which are exposed to various dynamics forces. The use of self-tapping screws is also crucial for a stable anchorage. The screws can be used in both sturdy and

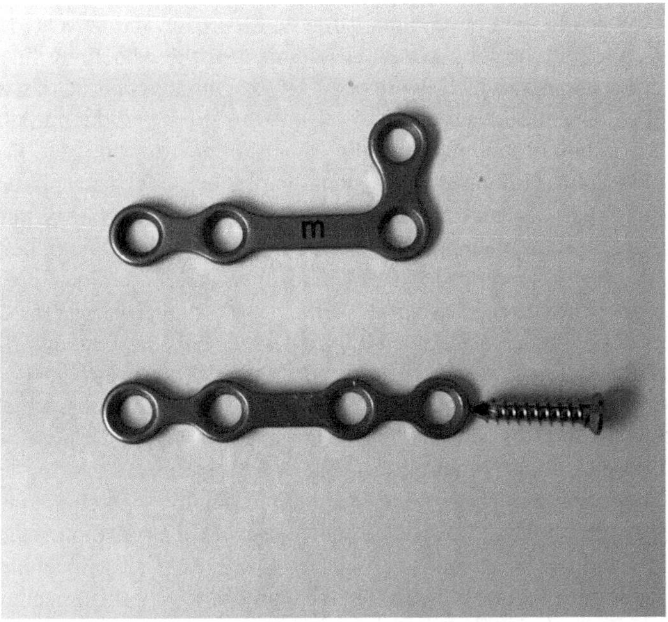

Fig. 3 Standard miniplates used in traditional orthognathic surgery

fragile facial skeleton. The inherent disadvantages associated with the design and material type of standard implants are obtrusive for surgical efficiency and patient's needs. Therefore, customized implants, which are completely adjusted to match the anatomical and morphological properties of the targeted bone tissues, became prominent with the development of new generation planning and surgical methods [26, 27]. Selection of the material for customized implants is based on various properties such as biocompatibility, corrosion resistance, formability, strength, and low price. The selected material should have a satisfactory mechanical strength and sufficient porosity [28]. Mechanical properties of the conventional Ti6A14V, which is widely used for manufacturing the standard miniplates, are improved by metallurgical process and heat treatments [29].

4 Case Study: Subject-Specific Miniplate Design for the Preoperative Virtual Planning Process

In this case study, we present a systematic protocol for the subject-specific miniplate design for the preoperative virtual planning process in the orthognathic surgery. In this methodology, the goal is to eliminate (*i*) the plaster model modifications carried out for surgery planning and (*ii*) shape adjustment procedure performed for the miniplates in the conventional surgery. Moreover, the presented method is regarded as a validation tool to test the accuracy of the osteotomy line and final position of the maxilla, which is provided by the components such as wafers and splints in the conventional surgery. We intended to show that the miniplate design using virtual planning method is a promising approach to eliminate the use of a series of intermediate equipment such as wafer, plaster model, and splint used in the orthognathic surgery. Accordingly, a series of operation is needed to be implemented in the proposed virtual planning methodology. First, the skull of a patient with the maxillary deformity is scanned with a CT scanner, and the CT data are converted to the DICOM format. Next, a virtual surface body model is generated by using DICOM data. The model is still in STL format which enables to modify the targeted structure. Le Fort I osteotomy is simulated on the 3D virtual solid model with a CAD software. Finally, a customized miniplate system is designed, which is based on the solid body model of the patient's dentoskeletal structure. Designed miniplates can be produced through rapid prototyping or additive manufacturing techniques. The subject-specific miniplates do not require pre-bending operation at the surgery phase. In the case study, CT data of an 18-year-old male patient with Class III relationships of the incisor teeth were selected to apply the virtual planning method (Fig. 4). The subject was informed about the whole details of the procedure, and the consent was taken. A detailed virtual model composed of the skull, mandible, and teeth was generated by converting the 2D CT data into the 3D model using an open-source software, 3D Slicer (or Slicer, Brigham Women's Hospital, Boston, MA) [30] (Fig. 5).

It is important to notice that a series of critical interventions such as simplifications, modifications, and smoothening were implemented to construct the model.

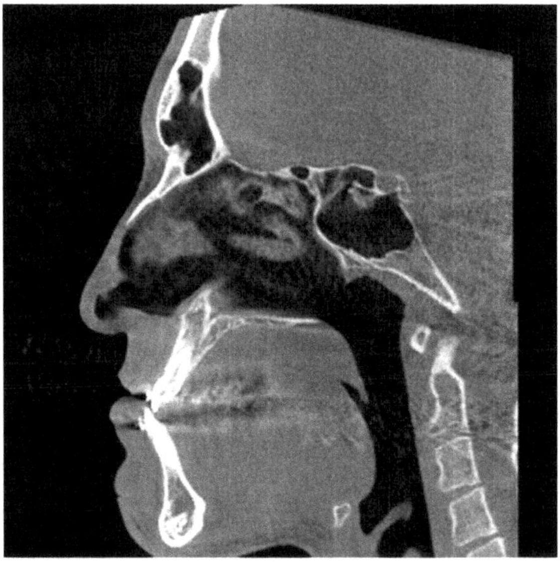

Fig. 4 A computed tomography (CT) image of the patient with maxillary deformity

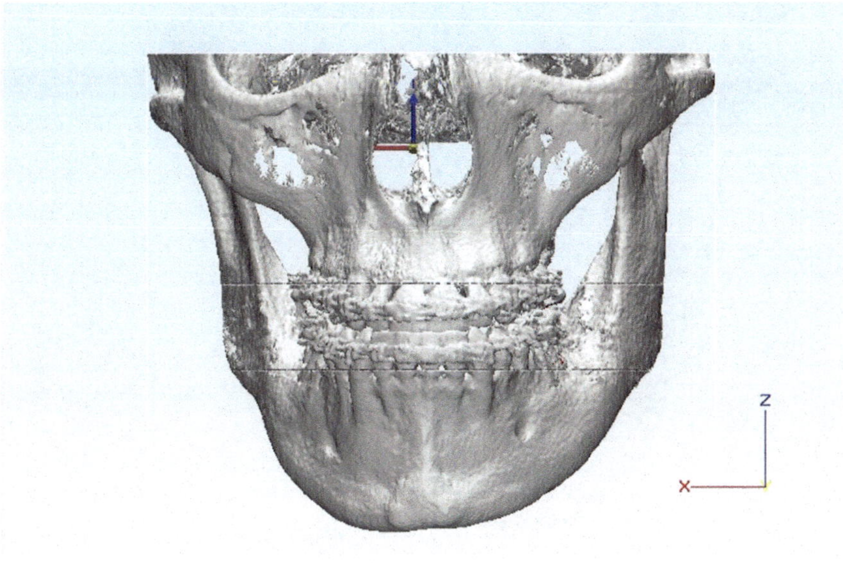

Fig. 5 Three-dimensional (3D) maxillofacial/mandible model of the patient obtained by converting two-dimensional CT data into the 3D model

Then, the Le Fort I osteotomy was simulated on the 3D virtual model. The maxilla was horizontally cut, detached from the skull, and advanced 5 mm along the antero-posterior axis by means of tools available in the software (Fig. 6). The magnitude of advancement was specified by measuring the distances between the reference points.

The maxilla was then fixed to its new position, and osteosynthesis miniplates were designed. Major design requirements were providing (*i*) a sufficient match between the surfaces of the zygomatic and paranasal buttresses and the miniplates surface and (*ii*) a proper connection site between the maxilla and skull (Fig. 7). The process was then modified according to the surgeon's demands and directions.

Osteosynthesis miniplates are generally manufactured by using Ti6AL4V and pure titanium materials [31]. Ti6Al4V alloy and pure titanium are protected by the passive oxide layer formed on the surface. This stabilized and adherent passive oxide film preserves the dental implants against pitting and intergranular corrosions and provides biocompatibility. In addition to those properties, the mechanical properties of the titanium alloys ensure the necessary toughness and stiffness in order to sustain appropriate conditions, especially for bone-implant interactions.

In the case study, the design process was conducted employing an open-source and non-commercial software, namely 3D Slicer due to the economic advantages and easy modification tools on models. Furthermore, the programs such as Autodesk Meshmixer (Autodesk, Inc., San Rafael, CA, USA) and FreeCad (https://www.freecadweb.org/) can also be used to manipulate the solid body models of maxillofacial segments and to design the miniplates attached to those models. The CAD-based methodology enables easier modifications of models and miniplates than the classical techniques.

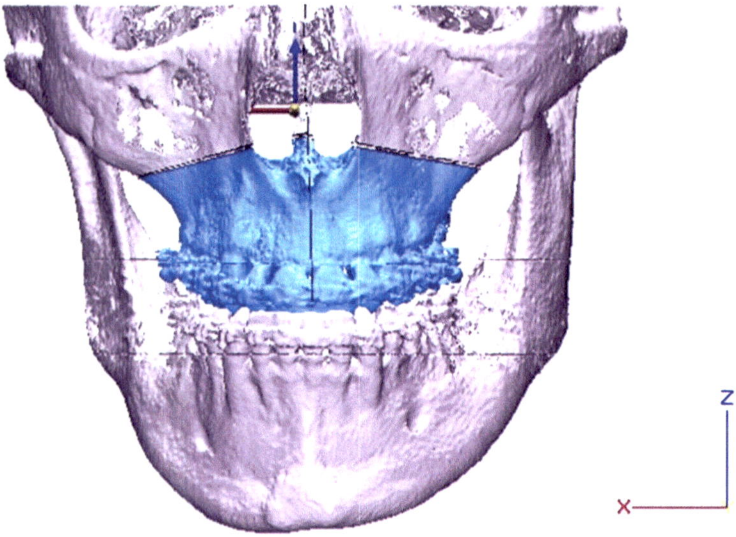

Fig. 6 Three-dimensional models of the maxillofacial and mandibular structures formed in the preoperative planning created from computed tomography data

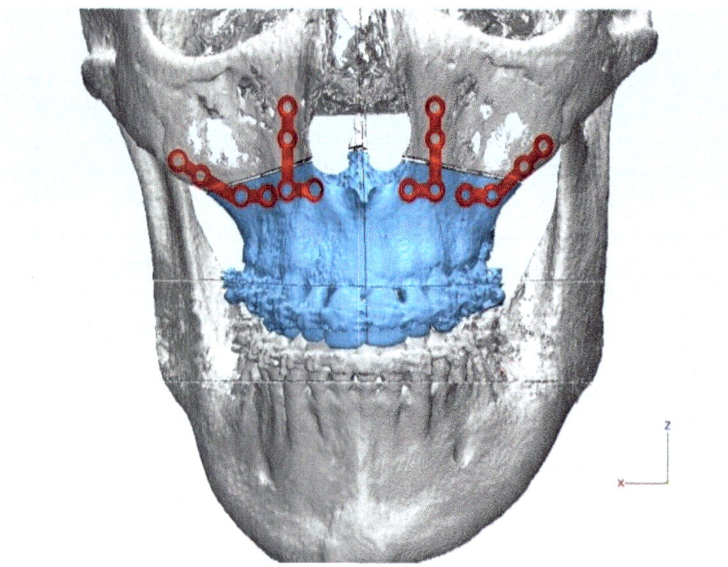

Fig. 7 Simulation of the Le Fort I osteotomy and design of the customized miniplates

By means of this novel approach, the unique and patient-specific miniplates could be designed. In addition, production of miniplates is easier and more applicable than the traditional implants by benefiting from the additive manufacturing process such as 3D printing or laser sintering technology.

4.1 Results and Discussion

Virtual planning of orthognathic surgery provides various useful outcomes. Imaging of the maxillofacial region enables a better diagnosis of the dentoskeletal deformities. Preoperative planning in orthognathic surgery utilizing the improvement of CT technology is a novel and promising method to simulate the bone-implant interaction and surgical interventions before the real operations [30–34]. Various software packages are used for converting CT scans to 3D surface model. In the case study, 3D Slicer was employed to obtain the solid body model of the dentoskeletal structure. The model provided acceptable accuracy and compatibility according to surgeon's feedbacks. The program, Autodesk Meshmixer, which was used to obtain the reconstructed model of the jaw bone, enabled us to reposition the maxilla and to simulate the osteotomies on the surface models. By taking these steps, the osteotomy line was determined and the maxilla was transferred to its desired final position. Unlike from the traditional method, the surgical procedure does not include a cast model on the semi-adjustable articulator. The presented approach avoids time consumption which was caused by the diagnosis of dentoskeletal deformities by cephalograms to prepare the dental cast model of teeth [35, 36].

All through the process from the CT scans to the design of miniplates, a variety of software can be used. Regarding the virtual planning process presented in Fig. 2, several commercial and open-source programs are required to obtain subject-specific miniplate design. In the first step, namely display and segmentation step, a series of software including Mimics (Materialise, NV Leuven, Belgium), 3D Slicer, and 3D-Doctor (Able Software Corp, Lexington, MA) can be alternatively used to process CT data to obtain surface bodies by applying necessary thresholding and masking procedures. In the reconstruction of the surfaces, Rhinoceros 3D NURBS modeling program (Rhinoceros 3D modeling for Windows, v 3.0, Robert McNeel & Assoc., USA), Autodesk Meshmixer, and Geomagics studio software (Raindrop Geomagic, Research Triangle Park, NC, USA) present useful facilities to perform surface modifications such as filling hole and smoothness tasks. CAD operations on solid body models can be carried out using Solidworks (SolidWorks Corporation, Concord, MA, USA), Catia (Dassault Systèmes, France), Solidedge (Siemens, Germany), Autodesk 123D Design (Autodesk Inc., San Rafael, California, USA), or Autodesk inventor (Autodesk Inc., San Rafael, California, USA). Drawing the osteotomy line, and sectioning and movement of the maxillary bone can be implemented by employing a CAD software or 3-matic (Materialise, NV Leuven, Belgium). Design of the miniplates can be implemented by utilizing CAD software as well. By using the presented approach, cost of the surgery would be decreased by approximately 50%, i.e., the cost of surgery would be reduced from 2000 to 1000 € in the Turkish local market. Virtual design of the miniplates also reduces the hardware usage since the pre-bending process is eliminated. In a 90-minute surgery, nearly 12% of the operation duration is allocated for the miniplate adaptation by applying externally bending moment. This process is eliminated by employing the computer-aided miniplate design. Traditionally, 4 four-hole miniplates and 16 screws are required for an ideal fixation of the Le Fort I down fracture and it approximately costs 2000–2200 €. On the other hand, production of the custom miniplate system costs around 1400–1600 € depending on the volume of the instruments. Thus, a one-third reduction of the cost would be achieved by producing custom-made miniplates. Furthermore, repeated attempts of pre-bending may reduce the mechanical endurance of the miniplates. Inflexible characteristics of the standard could increase the material failure risk. Even if any crack does not appear over the material during the surgery, miniplate fracture might occur in the primer and recovery periods of the maxillofacial regions.

Production of the dental instruments by means of additive manufacturing technologies such as 3D printing will be the next step in the progression of the subject-specific computerized treatment planning and decision-making. Materials with different functional, mechanical, and biocompatible characteristics, which are compatible with 3D printing technology, are continuously explored and used for the orthognathic surgery [37].

5 Conclusion

Consequently, the presented systematic protocol for the subject-specific miniplate design would ensure cost, labor, and time efficiency. Feedback from the surgeon enlightened us that the method presented remarkable advantages and promising performance. By using this approach, clinicians or surgeons could perform any scenarios on the solid body models without collecting information from the traditional cast models. Thus, complex surgical operations could be planned and simulated by utilizing various open-source and non-commercial computer programs. The preoperative virtual planning and design of the miniplates in orthognathic surgery would provide optimal functional and aesthetic outcomes, patient satisfaction, and accurate translation of the surgical plan.

References

1. Neville BW, Damm DD, Allen CM, Bouquot J (2009) Oral and maxillofacial pathology, 3rd edn. WB Saunders Co., Philadelphia
2. Ruhl CM, Bellian KT, Van Meter BH, Hoard MA, Pham CD, Edlich RF (1994) Diagnosis, complications, and treatment of dentoskeletal malocclusion. Am J Emerg Med 12(1):98–104
3. Huang CS, Hsu SS, Chen YR (2014) Systematic review of the surgery-first approach in orthognathic surgery. Biomed J 37(4):184–190

4. Kraeima J, Jansma J, Schepers RH (2016) Splintless surgery: does patient-specific CAD-CAM osteosynthesis improve accuracy of Le Fort I osteotomy? Br J Oral Maxillofac Surg 54:1085–1089
5. Drommer RB (1986) The history of the "Le Fort I osteotomy". J Maxillofac Surg 14:119
6. Kim SG, Park SS (2007) Incidence of complications and problems related to orthognathic surgery. J Oral Maxillofac Surg 65:2438–2444
7. Hammoudeh JA, Howell LK, Boutros S, Scott MA, Urata MM (2015) Current status of surgical planning for orthognathic surgery: traditional methods versus 3D surgical planning. Plast Reconstr Surg Glob Open 3(2):e307
8. Hausamen JE (2001) The scientific development of maxillofacial surgery in the 20th century and an outlook into the future. J Craniomaxillofac Surg 29:2–21
9. Zinser MJ, Mischkowski RA, Dreiseidler T et al (2013) Computer-assisted orthognathic surgery: waferless maxillary positioning, versatility, and accuracy of an image-guided visualisation display. Br J Oral Maxillofac Surg 51:827–833
10. Vale F, Scherzberg J, Cavaleiro J, Sanz D, Caramelo F, Malo L, Marcelino JP (2016) 3D virtual planning in orthognathic surgery and CAD/CAM surgical splints generation in one patient with craniofacial microsomia: a case report. Dental Press J Orthod 21:89–100
11. Aboul-Hosn Centenero S, Hernández-Alfaro F (2012) 3D planning in orthognathic surgery: CAD/CAM surgical splints and prediction of the soft and hard tissues results—our experience in 16 cases. J Craniomaxillofac Surg 40:162–168
12. Nilsson J, Richards RG, Thor A, Kamer L (2016) Virtual bite registration using intraoral digital scanning, CT and CBCT: in vitro evaluation of a new method and its implication for orthognathic surgery. J Craniomaxillofac Surg 44:1194–1200
13. Swennen GR, Mollemans W, Schutyser F (2009) Three-dimensional treatment planning of orthognathic surgery in the era of virtual imaging. J Oral Maxillofac Surg 67(10):2080–2092
14. Bootvong K, Liu Z, McGrath C, Hagg U, Wong RW, Bendeus M, Yeung S (2010) Virtual model analysis as an alternative approach to plaster model analysis: reliability and validity. Eur J Orthod 32:589–595
15. Stevens DR, Flores-Mir C, Nebbe B, Raboud DW, Heo G, Major PW (2006) Validity, reliability, and reproducibility of plaster vs digital study models: comparison of peer assessment rating and Bolton analysis and their constituent measurements. Am J Orthod Dentofacial Orthop 129:794–803
16. Hatamleh MM, Bhamrah G, Ryba F, Mack G, Huppa C (2016) Simultaneous computer-aided design/computer-aided manufacture bimaxillary orthognathic surgery and mandibular reconstruction using selective-laser sintered titanium implant. J Craniofac Surg 27:1810–1814
17. McCormick SU, Drew SJ (2011) Virtual model surgery for efficient planning and surgical performance. J Oral Maxillofac Surg 69:638–644
18. Stokbro K, Aagaard E, Torkov P, Bell RB, Thygesen T (2014) Virtual planning in orthognathic surgery. Int J Oral Maxillofac Surg 43:957–965
19. Cansiz E, Arslan YZ, Turan F, Atalay B (2014) Computer-assisted design of patient-specific sagittal split osteotomy guide and soft tissue retractor. J Med Biol Eng 34:363–367
20. Cansiz E, Turan F, Arslan YZ (2016) Computer-aided design and manufacturing of a novel maxillofacial surgery instrument: application in the sagittal split osteotomy. J Med Devices 10:44505-1–44505-4
21. Zinser MJ, Mischkowski RA, Sailer HF, Zöller JE (2012) Computer-assisted orthognathic surgery: feasibility study using multiple CAD/CAM surgical splints. Oral Surg Oral Med Oral Pathol Oral Radiol Endod 113:673–687
22. Hernandez-Alfaro F, Guijarro-Martinez R (2013) New protocol for three-dimensional surgical planning and CAD/CAM splint generation in orthognathic surgery: an in vitro and in vivo study. Int J Oral Maxillofac Surg 42:1547–1556
23. Choi JY, Song KG, Baek SH (2009) Virtual model surgery and wafer fabrication for orthognathic surgery. Int J Oral Maxillofac Surg 38:1306–1323
24. Cansiz E, Dogru SC, Arslan YZ (2015) Comparative evaluation of the mechanical properties of resorbable and titanium miniplates used for fixation of mandibular condyle fractures. J Mech Med Biol 15:1–8

25. Dogru SC, Cansiz E, Arslan YZ (2018) A review of finite element applications in oral and maxillofacial biomechanics. J Mech Med Biol 18:1–26
26. Mišic D, Manic M, Vitkovic N, Korunovic N (2015) Toward an integrated information system for the design, manufacturing and application of customized implants. Facta Univ, Ser: Mech Eng 13(3):307–323
27. Manic M, Stamenkovic Z, Mitkovic M, Stojkovic M, Shephard DE (2015) Design of 3D model of customized anatomically adjusted implants. Facta Univ, Ser: Mech Eng 13(3):269–282
28. Ristic M, Manic M, Misic D, Kosanovic M, Mitkovic M (2017) Implants material selection using expert system. Facta Univ, Ser Mech Eng 15:133–144
29. Fernandes CR, Fernandes BL (2015) Evidence of the semi-solid formation in the medical grade TI6AL4V alloy using induction heating. Facta Univ, Ser: Mech Eng 13(3):229–239
30. Fedorov A, Beichel R, Kalpathy-Cramer J, Finet J, Fillion-Robin JC, Pujol S et al (2012) 3D Slicer as an image computing platform for the quantitative imaging network. J Magn Reson Imaging 30:1323–1341
31. Neumann A (2009) Biomaterials for craniofacial reconstruction. Laryngorhinootologie 88(Suppl. 1):S48–S63
32. Hatamleh MM, Turner C, Bhamrah G, Mack G, Osher J (2016) Improved virtual planning for bimaxillary orthognathic surgery. J Craniofac Surg 27:568–573
33. Swennen GR, Mommaerts MY, Abeloos J, De Clercq C, Lamoral P, Neyt N (2009) A cone-beam CT based technique to augment the 3D virtual skull model with a detailed dental surface. Int J Oral Maxillofac Surg 38:48–57
34. Agrawal JM, Agrawal MS, Nanjannavar LG, Parushetti AD (2013) CBCT in orthodontics: the wave of future. J Contemp Dent Pract 14:153–157
35. Cevidanes LH, Tucker S, Styner M (2010) Three-dimensional surgical simulation. Am J Orthod Dentofacial Orthop 138:361–371
36. Philippe B (2013) Custom-made prefabricated titanium miniplates in Lefort I osteotomies: principles, procedure and clinical insights. Int J Oral Maxillofac Surg 42:1001–1006
37. Lin HH, Lonic D, Lo LJ (2018) 3D printing in orthognathic surgery—a literature review. J Formos Med Assoc 117(7):547–558

Faruk Ortes working as Research Assistant in the Department of Mechanical Engineering, Istanbul University-Cerrahpasa, Turkey. He is working in the area of muscle mechanics, computational biomechanics, applied mechanics, finite element analysis, and design and production of surgical instruments. He has authored more than 5 research papers in various journals and authored 5 chapters. He is a reviewer of various international journals.

Dr. Erol Cansiz working as Assistant Professor in Faculty of Dentistry, Oral and Maxillofacial Surgery, Istanbul University, Turkey. He is working in the area of orthognatic surgery, tumor surgery, maxillofacial reconstruction, temporomandibular joint surgery, and facial plastic surgery. He has authored more than 25 research papers in various journals and 5 chapters. He is a reviewer of various international journals.

Dr. Yunus Ziya Arslan working as Associate Professor in the Department of Mechanical Engineering, Istanbul University-Cerrahpasa, Turkey. He received the B.Sc. degree in 2002, the M.Sc. degree in 2005, and the Ph.D. degree in 2009 all from the Mechanical Engineering Department, Istanbul University, Turkey. He spent six months at the University of Calgary as a Visiting Researcher and one year at the Karlsruhe Institute of Technology as a Visiting Professor. He is working in the area of musculoskeletal biomechanics, computational modeling of human movement, computer-aided analysis of movement disorders, dental biomechanics, and optimization. He has 6 patents in his name and authored more than 30 research papers in various journals and authored 7 chapters. He is a reviewer of various international journals.

Additive Manufacturing: Current Concepts, Methods, and Applications in Oral Health Care

Jagat Bhushan and Vishakha Grover

Abstract Additive manufacturing is an emerging technique that has almost revolutionized the material science and mechanical engineering in recent years. The technique started to help engineers to actualize the concepts of their minds by fabricating diverse three-dimensional objects in a layer-by-layer fashion. Multitude of processes such as stereolithography (SL), polyjet, fused deposition modeling (FDM), laminated object manufacturing (LOM), 3D printing (3DP), prometal, selective laser sintering (SLS), laser engineered net shaping (LENS), and electron beam melting (EBM) is considered as a part of additive manufacturing. Recently, additive manufacturing has received a lot of attention in the healthcare sector for fabricating devices for the purpose to restore, support, and repair defected human body parts. Particularly, the field of dentistry rather extensively deals with such restorative work in order to improve the functional and aesthetic demands of today's oral healthcare sector. This chapter shall present a focussed update on the current concepts, methods and diverse applications in the field of oral health care, based on contemporary published literature relevant to the subject. Further, it shall provide the future scope of research to address the challenges, for the widespread clinical utilization of additive manufacturing technologies in routine clinical dental care.

Keywords Additive manufacturing · Methods · Applications · Materials · Biologic · Oral · Digital · Dentistry

1 Additive Manufacturing: Current Concepts, Methods, and Applications in Oral Health Care

Additive manufacturing is an emerging science in the field of product development that comprises 3D printing and related technologies so as to develop a physical product from the digital image data automatically [1]. As defined by the American Society for Testing and Materials (ASTM), "additive manufacturing is the process of joining

J. Bhushan · V. Grover (✉)
Dr. Harvansh Singh Judge Institute of Dental Sciences, Panjab University, Chandigarh, India
e-mail: vishakha_grover@rediffmail.com

C. Prakash et al. (eds.), *Biomanufacturing*,
https://doi.org/10.1007/978-3-030-13951-3_5

materials to make objects from 3D model data, usually layer upon layer, as opposed to subtractive manufacturing methodologies" [2]. The technique is an outcome of three different domains, viz. technology, materials, and creativity, working together and bringing a new diverse ocean of opportunities for manufacturing process [1]. The additive manufacturing process involves a layer-by-layer addition of material under computer control to create a three-dimensional object composed of multiple layers. The prime basis of this innovative method is that the three-dimensional digital image is virtually split up into sections or slices and the attached printing device uses this geometric data consecutively to form different layers in the same sequence until the final desired product is achieved. The first step is the creation of a virtual design of the object. Scanner may be used to scan any object to produce a 3D model. The 3D model is sliced, and then, it is ready to feed into the 3D printer of compatible brand and type. This transfer of information is done via USB, SD, or Wi-Fi. The versatility of the technique is its ability to virtually to produce objects of any geometric shape, design, and complexity [3–5].

2 Additive Manufacturing: General Aspects

At the outset, technique came into being as an answer to create models or prototype parts by engineers as rapid prototyping aiming at lowering the cost, manpower, and time requisites for product development. The possibility of developing three-dimensional shapes of almost any complexity rapidly expanded the use of additive manufacturing to diverse fields [6–9]. Prototyping entails only for the production of models, but succinct advances and development have now turned prototyping to rapid manufacturing of three-dimensional objects as functional parts for direct applications. Three technologies, viz. computer-aided design (CAD), computer-aided manufacturing (CAM), and computer numerical control (CNC), put together have virtually realized the concept of rapid manufacturing [7, 10–12]. Charles Hull of the University of Colorado, USA, has been regarded as the "founder" of modern 3D printing as he has pioneered the idea of hardening the surface coatings to make a three-dimensional object from the liquid base by using ultraviolet (UV) light. In 1986, Hull filed his first patent related to this process and called it stereolithography (SL) [13]. This was probably the first technique emerged for three-dimensional printing and has been most widely employed, so much so that it is used synonymously with rapid prototyping. The basic process involves a CAD software model data, which is translated to an Standard Tessellation Language (STL) file to analyze the information contained in each layer as a slice. The STL file was also created by 3D Systems Inc. in 1987 and later formed the basis for almost all additive manufacturing process. The thickness of each layer, as well as the resolution of the final product, depends on the equipment used in the AM process. Photopolymerization remains the underlying process for printing in this technique, where an ultraviolet light acts as an energy source to initiate and mediate the polymerization reaction producing a solidified polymer from a monomer in a liquid state [5, 10, 11]. Microstereolithography is capable of developing a much higher resolution in the product, with a layer thickness of less than 10 μm and is the improvised version of the same basic technique [14]. Driesche et al. presented methods based on 3D printing to make microfluidic chip holders with reliable fluidic and electric connections. Such silicon/glass microfluidic chips fabricated are required for various biochemical investigations and diagnostic setups, where we need to have a very precise, transparent, and biocompatible sample handing in critical clinical situations [15]. In another variant of stereolithography, multiple or a variety of materials can be used for printing a single product; this process is called multiple material stereolithography [12].

Since then, there has been an exponential rise in the technical innovations and advancements in the existing techniques for additive manufacturing. In 2004, the Society of Manufacturing Engineers did a classification of the various technologies such as stereolithography (SL), fused deposition modeling (FDM), laminated object manufacturing (LOM), 3D printing (3DP), selective laser sintering (SLS) based upon a criterion identifying the processes into liquid base, solid based, and powder based. [11], but owing to the continuous search for improvised materials and equipment, four additional significant technologies, viz. electron beam melting(EBM), prometal, laminated engineered net shaping (LENS), and polyjet were considered in 2012 [12]. These techniques emerged to curtail the time, cost, and manpower have actually revolutionized the manufacturing process and market in diverse fields. Major industry sector transformed by the application for three-dimensional printing processes was aerospace, wherein there is an inherent need to make the parts lighter and less cumbersome. As the materials used are quite expensive, there is a rising and essential urge to reduce the wastefulness of the manufacturing process to make it more cost effective. Both these considerations were well taken care of with additive manufacturing, significantly changing the face of the aerospace industry. Even, in architectural model making for buildings, the process has made a tremendous impact by providing much faster pace for complex model production with enhanced resolutions. These techniques are transforming the practice of medicine and dentistry by enabling the rapid prototypes printing and very high-quality bone transplants and models of the damaged bone of the patients for analysis and nowadays even the functionalized parts to restore the lost areas or organs of the body to limit disability. The details regarding the contemporary status of additive manufacturing technologies, specific techniques from a biologic standpoint, materials customized for functional body parts and associated information are provided in the subsequent section. Apart from these, the applied arena of additive manufacturing is vast including art, creative work, fuel cell engineering [12, 16]. In a nutshell, now is the time for infusing these technologies in industry and market for our day-to-day consumables, so as to significantly reduce the time and cost burden for the daily needs.

3 Contemporary Status of Additive Manufacturing in Oral Health Care

Health care is one of the most prime affected sectors with the evolution of additive manufacturing and associated methods of 3D printing. Major medical applications of additive manufacturing which have been documented in past are preoperative models, inert models, biomaterial printing, splints, guides or support devices, external fixators, and tools, instruments, and parts for medical devices [16, 17]. The upbeat of additive manufacturing technology in this sector largely leans on its ability to create orthotics, prosthetics, and personalized implants, thus limiting much of disability associated with missing body parts. This is all the more relevant for the field of dentistry or oral health care as the essence of dentistry has been talked of as an art and science focussed on recreating customized lost parts of teeth, alveolar bone, and other oral structures. Dental therapeutic procedures lead to the restoration of the natural oral anatomy and morphology for optimal patient function, aesthetics, and comfort [13]. Dentistry has been digitized with the concept of CAD/CAM techniques, which have successfully served quite a time period for computer-based design and planning restorations, but have been a lot wasteful affair in terms of time and cost. Three-dimensional printing of crowns, bridges, stone models, and various orthodontic appliances which are realized combining latest technique are in terms of oral scanning, 3D printing, and CAD/CAM design. Dental prosthetics and restorative work had been executed in dental laboratories till now with laborious manual work, and a great deal of time, effort, and cost. However, the advent of high-precision printing equipments, advancements in materials used for dental applications, has actuated the small size, high complexity, and high-end customization of products for utilization in oral cavity. RepRap, a desktop 3D printer, is known as a self-replicating machine, as it is able to print its own plastic parts required for printing [2, 18]. Alongside these developments, the number of materials has significantly expanded. Furthermore, most recently, we are witnessing a surge in the enhancement of biologic behavior of 3D printed parts for functionalization of these components in unison with live vital oral cavity structures. In addition, quality controls and regulatory adoption of numerous additive manufacturing technologies and materials are enabling this field to design, validate, and add value return to the market for next-generation medical technology.

With the quantum leap in advancements in materials, computer software, and printing devices, the face of healthcare devices has been literally transformed in the recent past. Highly sophisticated printing devices and production equipment are available today with optimal printing heads, integrated recycling, and feedstock supply, throughput devices and coordinated post-processing. With a paradigm shift in IT technologies, equipment with abilities to monitor, record, detect, and report abnormal functioning of steps of manufacturing and provide real-time feedback of the working equipment. Such smart devices help to reduce the wastage of material, cut down on time, minimize error, and enhances reliability and efficiency of the manufacturing process. Similarly, huge advances in the printable materials range for dental applications as well as improvements in mechanical and biologic properties of the materials have been witnessed. Keeping the pace with both these aspects, computer software has also evolved over the time to consistently perform upgraded and support the changes in parallel aspects to optimize the total process for greater applicability. With the precision and ability to design complex lattice structures, randomized osseointegration, mimicking the biologic behavior of vital tissues of humans and creating geometric arrangements that match or complement existing tissue structures, now the industry is well equipped to revolutionize the field of oral health care [19, 20].

4 Materials for Additive Manufacturing for Dental Applications

As a first and foremost point for 3D printed parts for dental use, the primary requisite for the material will be that it should be an inert, stable, nontoxic, biocompatible and capable of amenable for functional integration based on the application intended on the manufactured part [16]. In some clinical situations, the printed part acts as just physical replacement of the lost dental structure, e.g., inlay or only tooth surface restorations, whereas in other situations, it may have to functionally integrate itself as in an existing tissue complex which is performing a defined function, e.g., dental implants or when 3D printed part or scaffold is used for bone or tissue regeneration for treating periodontal bone loss. In situations of the first kind, it is the mechanical properties of the material which are the basis of the selection of the material. However, in clinical situations of the second kind, the biologic behavior and properties along with the physical characteristics assume great significance. So, the choice of material is guided by the purpose, the area being replaced and properties of that material fulfilling the targeted functional requirement. Apart from these, when 3D printed parts are used as dental tools, e.g., implant guides or as display models for education purposes, mechanical properties such as strength, compressive and shear, hardness, brittleness, color, surface characteristics bear prime influence for the selection of material to be used. For some of the dental uses such as for printing individual impression trays, orthodontic models, restorations, crowns, and bridges materials have been well studied and certified. The material may obtain a flexural strength of more than 80 MPa [21]. A variety of materials have been utilized in the dental applications of additive manufacturing techniques.

Metals such as titanium and its alloys, steel alloys, aluminum alloys, nickel alloys, and some cobalt-based and magnesium alloys are widely used as they have been extensively researched and optimized for additive manufacturing techniques [22]. Ti [23] and Ti6Al4V [24] worked on with additive manufacture for implants have offered a very cost-effective products owing to inherent strength and lesser wastage for complex designed structures. Various kinds of steel alloys have been preferred for high strength and hardness conditions, such as for tools or molding applications

[25–28]. Few Al alloys have been tried in comparison to Ti alloys as they were low cost and easily machinable materials [29]. For biomedical and dental uses, Co–Cr alloys [30] have been studied and utilized for widest applications. Other materials such as Mg alloys [31] (for biomedical resorbable applications), Au [32] and Cu [33] have also been evaluated for similar uses. Polymers are a diverse class of materials, which are very conveniently amenable to different 3D printing processes. These are extensively in practice for various applications from the 3D printing industry. These materials allow to customize geometry with high accuracy and can be worked at much less cost compared to the conventional methods used for the same purpose. Polymers at the outset used for 3D printers were mostly worked out for conceptual prototypes due to minimal strength and limited functional behaviors, but with continued improvisation in material technologies, lots of advanced materials are available with better properties and workability. Photopolymer-based systems are well known for their accuracy of reproduction of details, good resolution layers, and fine precision. Further developments using new resin materials have given rise to new materials with improved strength and temperature resistance. Thermoplastic polymers (e.g., acrylonitrile–butadiene–styrene copolymers (ABS) [34], polycarbonate (PC) [35], and polylactic acid (PLA) [36]) could be processed by various 3D printing methods. The salient advantages offered by these materials are fine surface detailing, accuracy, improved strength, temperature resistance, and high precision. A PLA and a bioactive CaP glass-based 3D printed constructs have been developed for tissue engineering applications where these may act as 3D porous biocompatible scaffolds for harboring cells for regeneration [37]. Apart from this, polymeric cellular materials, materials based on hydrogels for drug delivery, nanocomposites are also been fabricated via 3D printing.

Another category of material for dental usage is ceramics, which have been utilized as biomaterials and for tissue engineering [38]. Accuracy, as well as post-processing of sintered ceramic parts for achieving the final form, had been a cause of concern for the use of these materials as it demanded time and cost. With the continued developments, 3D printing of porous ceramics or lattices has brought in applications in diverse areas. The tissue engineering scaffolds used now a days are processed much faster and are easily printable when compared against the earlier methods of manufacture of such constructs [38, 39]. Figure 1 shows the major categories of materials used for additive manufacturing.

5 Applications of Additive Manufacturing in the Clinical Dentistry

Digital technology and 3D printing have significantly improved the quality and precision of dental restorative and prosthetic work. It has further enhanced the predictability and rate of success in dental implantology using custom surgical guides, etc. In areas such as oral surgery, the planning of various surgical procedures includ-

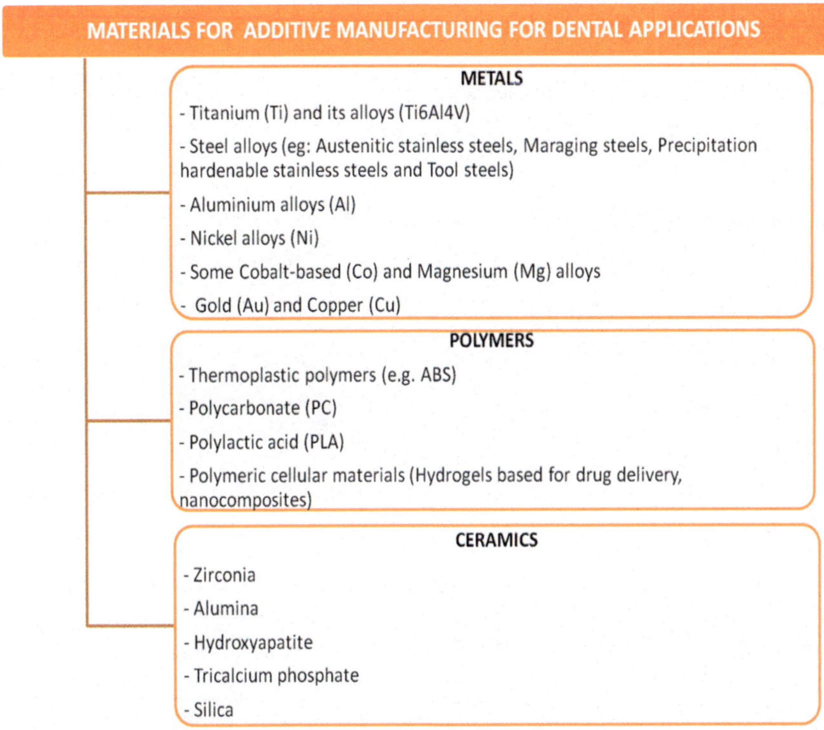

Fig. 1 Major categories of materials used for additive manufacturing for dental applications

ing implants by making surgical guides has increased the precision of surgery with minimal complications and improved healing has been realized. The likelihood of inadvertent damage to the adjacent vital structures such as inferior alveolar nerve and maxillary sinus, etc., is reduced. The technique has also provided display or education models to create jaw bone and conduct and demonstrate various blocks to augment bone defects, etc., and for learning modules for the students for any specific therapeutic procedures [40]. Contemporary 3D printers can print biological tissues such as bone as per the requisites of the bone defect in a particular patient and can act as simulating scaffolds to vital living tissues for better ingress of osteoblasts, i.e., bone-forming cells for bone regeneration. For the same purpose, novel 3D printed alginate–peptide hybrid scaffolds can also be used. These kinds of scaffolds are documented to provide better and a compatible stable environment for the growth of stem cells [41]. Composite scaffolds by combining calcium phosphate (CaP) and calcium sulfate ($CaSO4$) powders have been printed with additive manufacturing techniques and are utilized as bone augmentation material [42]. These innovations are becoming a part of clinical periodontal regenerative procedures now and reflect a promising potential for enhancing regeneration of lost periodontal tissues.

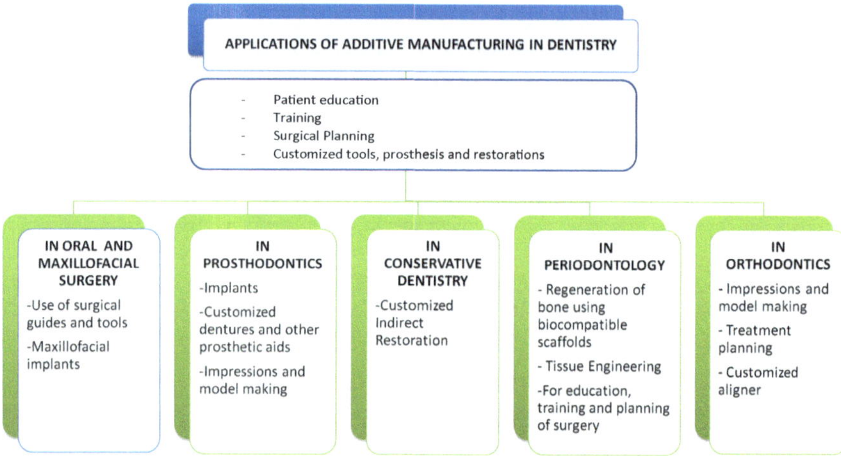

Fig. 2 Applications of additive manufacturing in the various fields of dentistry

Most of the dental laboratories work is now digitized by combining oral scanning, 3D printing, and CAD/CAM design. Contemporary restorative work from prosthodontics as well as endodontics, such as crowns, bridges, stone models, and various orthodontic appliances, is provided by these new generation manufacturing technologies. Moreover, these additive manufacturing techniques have enabled the development of maxillofacial prosthesis with or without functional integration, depending upon the clinical situation to replace the missing, damaged, or lost parts from the oral cavity and face of the patient. Maxillofacial reconstruction for cleft palates, 3D printed mechanical solutions such as obturators, or replacement of parts of the orbital floor, zygoma, maxilla, or mandible due to traumatic loss may sure minimize the disability and morbidity associated for such patients and improved the quality of life for them. Figure 2 shows the various fields of dentistry in which additive manufacturing have been used.

6 Biological Analysis of Additive Manufacturing Processes Part

Dental laboratories have witnessed a significant change in the recent past; additive manufacturing has largely replaced the CAD/CAM milling (subtractive manufacture techniques) for making dental prosthesis. This change over has curtailed most of the issues associated with milling. A prime advantage of the newer technique is to recreate the fine surface characteristics such as undercuts, pores, and complex internal geometric configurations which make a salient feature of dental restorations. The ultimate success of a dental prosthetic is fairly dependent on the recreation of

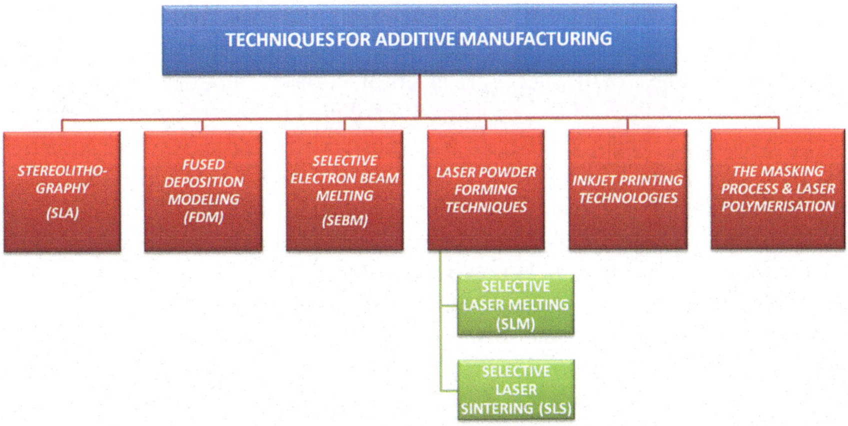

Fig. 3 Techniques of additive manufacturing used for dental application

natural anatomy of the tooth or any other structure which it is replacing. Another added benefit is that multiple parts can be manufactured together, which was a major limitation associated with subtractive manufacturing since only one part could be machined at a time. Digital scanning, combined with 3D printing, also promises to reduce some of the anxiety associated with the dental experience; e.g., traditional impression making with steel or plastic trays loaded with impression materials was anyhow bit uncomfortable procedure, more so for people like young children, elderly, and patients with a high gag reflex. Now, however, it is possible to obtain a digital impression of the patient's mouth for any further restorative work up with a contactless 3D scan of a patient's mouth. Dental clinics with in-house 3D printers are well-positioned to provide restorative solutions to their patients in minutes or hours rather than days or weeks. Digital files and hard models printed on-demand may help dental professionals meet compliance requirements more so in medico-legal cases, where a lot of decision making and treatment depend on the documented records [43]. There are multiple techniques for additive manufacturing as mentioned in Fig. 3 and is discussed in the preceding section, and a continuous increase in the list of new methods is observed day by day. Some of the processes have been specifically utilized in dentistry keeping in view the typical biologic properties which allow these 3D printed parts to be functionally integrated into the oral cavity.

6.1 Stereolithography (SLA)

This process is UV light-based layered polymerization of the synthetics to develop three printed products. Though it is the oldest technique, which led to the evolution of diverse variant techniques for contemporary additive manufacturing methods, it is the best known for its high precision. In the real sense, the process involves a layer-by-layer polymerization from a light-curing plastic bath to form 3D parts. This process is carried out via a support plate lowered in and moved by the corresponding incremental thickness of the successive layers of uniform size. Consecutive polymerization of layers at the surface is continued until the last layer of the desired object has been immersed and polymerized. An almost identical process where successive layers are polymerized by a plate at the bottom of the resin bath forms the basis of top-down printers to improve the printing speed and component height [2]. Commercially available stereolithography currently works with precisions of 25–50 μm (for example, Varseo, BEGO). However, stereolithography in research purposes displays a higher precision up to 5 μm, which is directed at a cost of slowing the speed of construction enormously. For this technique, apart from the physical parameters, the technical machine implementation is an important limiting factor [13]. Earliest applications for medical sciences from additive manufacturing technologies were the development of physical models of anatomic details of human organs and tissues. Even in dentistry, many physical models of teeth, associated oral tissues for display in the dental operatory or clinics have come forth from these techniques. Such models have served an excellent patient education and motivation materials for enhancing the awareness of patients. Customized SLA models for the purpose of diagnosis, planning, and precise execution of surgical procedures are being used in medicine and dentistry nowadays. Further, reconstructive procedures in major maxillofacial surgeries also use replacement of missing/lost part customized implants, e.g., in cranioplasties, orbital floor reconstructions [44–49]. In the early work, the focus was on developing the template models for titanium implants [49] or the surface of the implant was machined [44, 47, 50]. SLA is extensively used to produce surgical implant guides for the better accuracy of placement of dental implants. Its use is further extended for dental prosthetic and restorative work, e.g., crowns and bridges and resin models [2].

6.2 Fused Deposition Modeling (FDM)

Fused filament fabrication involves the printing of the object either by using a thermoplastic wire applied via a heated delivery apparatus or material is taken from a reservoir through a syringe. This technique has been immensely utilized in modeling, prototyping, and product development. S. Scott Crump is considered as the pioneer of this technique, which came into being in the late 1980s and was commercialized in 1990. In this method, the object is produced layer-by-layer deposition of material from nozzle via a wire, which sets immediately as comes out from the nozzle [2]. Most of the materials are workable with this technique, e.g., acrylonitrile butadiene styrene (ABS) polymer, polycarbonates, polycaprolactone, and polyphenylsulfones. Waxes are commonly used as a transient support material during the process and are solubilized at the end of the process by mechanical agitation. The plotter system based on the technique utilizes a reservoir of the material and is capable of utilizing multiple materials. It has been extensively used in the construction of printed constructs for biomedical field applications, e.g., scaffolds for tissue engineering and organ printing. As the system is capable of utilizing a diverse variety of materials, multiple applications like porous bone scaffolds from hydroxyapatite and tricalcium phosphates, vehicles for drug delivery from bioresorbable polymers, and many cell carrier molecules like gelatin and chitosan have been materialized with fused deposition modeling. The instrument has the resolution to create very fine microstructural patterns for biologic activities such as cellular ingress and proliferation [51]. There are reports documented for the creation of vital human body parts such as blood vessels [52], bone [53], and soft tissue [54].

6.3 Selective Electron Beam Melting (SEBM)

Selective electron beam melting (SEBM) works by utilizing electron beam for melting the material (metal powder) in a high vacuum environment, and it forms very fine net shape metal objects layer by layer [2]. The use of electrons rather than light creates a high energy beam, which reflects in the final product in terms of high density, strength, and minimal voids. This salient advantage of this method to obtain very fine mesh-like structures with high strength has popularized it in the field of orthopedics and maxillofacial surgery for the construction of dental implants personalized to the need of the patient. The biocompatible materials like titanium, Ti-6Al-4V, and Co/Cr are utilized and help obtain much closer confirmation of mechanical properties of the implant to natural bone specifically elastic modulus, which has a significant bearing on the stress shielding and thus the longevity of the implant system [55, 56]. Further, the porous structure permits the ingrowth of blood vessels and bone-forming cells, thus helping better osseointegration of the fixture. The accuracy range of the technique is 0.3–0.4, and the surface roughness is in range of 25 μm, which suffices for orthopedic and maxillofacial reconstruction implants but would not be able to provide sufficient resolution for small dental prosthetic works such as crown and bridge frameworks because of the small size of the article.

6.4 Laser Powder Forming Techniques

Laser-based additive manufacturing, such as selective laser melting (SLM) and selective laser sintering (SLS), is based on the fusion of melted powder particles under a high power laser utilizing mirrors [2]. A layer-by-layer deposition is carried out by lowering the powder bed by a single-layer thickness until the part is completed. This technology is being extensively utilized globally to construct intricate geometric structures owing to its ability of transformation from direct digital CAD data. Today, it immensely stands for producing end use functional parts in the market, though it initiated for only for prototyping of models. In industry dealing with polymers and ceramics, the technique is largely known as selective laser sintering, whereas for metals, SLM or DMLS (direct metal laser sintering) is the term used. Compared to other methods of additive manufacturing, a wide range of polymers is workable with this method such as polyamides [57], ultra-high molecular weight polyethylene [58], polycaprolactone [59, 60], and composites such as mixtures of hydroxyapatite and polyethylene and polyamides [61]. A range of metal powders like steel, titanium, titanium alloys, and Co/Cr alloys can be utilized. The scope of this technique in biomedical applications including the uses in dentistry is vast including artificial prosthesis, implants, drug delivery, biologic scaffolds for tissue regeneration, etc. The physical process impacting the material may vary from partial to full melting or liquid-phase sintering. The technology has achieved great success for the construction of multitude of implants such as bone analogs [62], orthopedic [63], and oral implants [64]. Dental prosthetic restorative work such as crowns and bridges [65] and partial denture frameworks [66] have also been successfully constructed and applied. Cutting-edge research is going on at this moment to extend the technique applicability involving pure ceramic powder and make porous apatite scaffolds for bone and tissue regeneration. Some preliminary work has been carried out to produce a bridge to replace missing teeth in a newer restorative material Zirconia, using this technique [67].

6.5 *Inkjet Printing Technologies*

Inkjet printing works by propelling isolated small droplets of material toward a substrate like an "ink." The material may be a resin, metal, ceramic or a cell suspension which is used as similar to an "ink" being sprayed [2, 68, 69]. The material is pushed through a small hole under the differential effect of physical variables, e.g., pressure, heat, and vibration. Generally, product is developed from the droplets of material sprayed to form a layer and then subsequent layer-by layer addition. The liquid droplets must change their state to solid as they are being deposited on the substrate when printing a pattern or object. Drying, heat transfer, UV light, or chemicals are the different methods adopted for the conversion of liquid to solid phase. Another variant technique similar to SLS/SLM exists, where the binder is printed by an inkjet method to hold or fuse or bind the particles together instead of using a laser. This method is exclusive to print multicolored objects of any hue across the range of the color spectrum. The polyjet range of printers from Objet, a commercially available inkjet printing company, is exploring a wide range of dental applications e.g. dental models, guides for placement of orthodontic brackets, surgical guides for implant placement, mouth protective equipment for sportspersons, sleep apnea appliances, etc. The best part of the technique is that it is capable of printing an object using two distinctive materials with unique properties. Such a technique offers the possibility of producing models for oral cavity reflecting the true natural hard and soft anatomic regional variation for better comprehension and use for dental consumers. Many experiments with different materials, e.g., porous calcium polyphosphate, are being worked for forming different tissue engineering constructs and even for collared soft tissue prosthesis [2, 70].

6.6 The Masking Process and Laser Polymerization

This masking process works by simultaneous polymerization of the entire cutting layer of the object with a high powered UV lamp. The underlying principle of the process is via a digital micromirror device (DMD), in which a multitude of mirrors does the job together and achieve layer-by-layer polymerization. This process is widely used in dental applications as a very high degree of accuracy is attained, even for very small-sized printed objects. Laser polymerization is a similar technique, but surface polymerization is performed by a single-laser, point-by-point contact, instead of a UV lamp. On the same lines, this process also offers a high degree of precision and permits the development of very fine articles of complex geometry with a great deal of accuracy such as tissue engineering scaffolds for regeneration of oral tissues [13]. Apart from the diameter of laser and precision of UV light lamp, the natural oscillation through the motors and the size of the construction chamber has a bearing on the final printed outcomes. The light source or laser determine the width of the contours in the X–Y (length and breadth) layer be polymerized, whereas layer thicknesses in the Z (height) direction is largely dependent on the wettability of the preceding, polymerized layer. For dental applications, the viscosity and the surface tension of the plastic material used is a very important determinant for the final printed model [12, 13]

6.7 Emerging Paradigm of 3D Printing and Its Impact on Dentistry

Since 3D printed components such as models, tools, restorations, implants, and prosthesis are becoming an integral part of contemporary dentistry, oral, and maxillofacial surgery, it is an opportune time to lead toward a new era of possibilities. Along with the change in the clinical practice of dentistry owing to newer techniques such as additive manufacture, at par changes are there in the market dynamics of dental materials, dental laboratories, and dental equipment. Many dental companies like 3M, Mitsui Chemicals, Dentsply, Sirona, Bego Medical, Align Technology, Glidewell, Straumann, and 3Shape have actively worked in this field and assured the protection their intellectual property rights by filing patents in this direction. The patent filing started in 1984, and since, then they are rising only, with significant gaps in two years, i.e., 2008 and 2011. The year 2008 had been a cut point in this regard, as most work earlier than that was focussed on to rapid prototyping largely for printing dental molds and other prosthetics. After this, a few patents expired on stereolithography and companies started to use the inventions freely for marketing and commercialization. After that time period, the concept for complete digitization of dental products emerged and a resurge of the field in terms of mass manufacturing in the dental industry came into being. After 2011, many patents were filed regarding materials, printers, and scanners used in this technology. The USA followed by China and Germany has involved in the research pertaining to technology with over **91% of** patents filed from China only. Approximately 462 patent families are there which are related to 3D printing in the dental industry and are owned by 196 different companies/universities.

USA, Germany, China, and France have are the prime markets for this technology. It is evident that 3M has been a lead brand in protecting its technology globally, while the main focus of Align Technology is in the US market. Align Technology filed their first patent US6976627 related to mass customizations under the brand name Invisalign, wherein both aligners and retainers are produced from 3D printers. Dentsply Sirona also filed patent WO2016187155 A1 in 2016 regarding hastening the pace of 3D printing by improvising on the method and eliminating extra steps of irradiation during layer-by-layer 3D printing. Mass manufacturing for end use products through 3D printing is ready to take on the market, with companies continually working on to minimizing the time of product development and to improve on the physical characteristics such as accuracy, precision, and strength of the of the final products to ensure the optimal solutions. Additive manufacturing processes are well incorporated in dentistry for a wide variety of applications such as dental implant drill guides to frameworks and copings. With the 2017 FDA approval of a material used to 3D print denture bases, now 3D-printed full denture solutions can also be offered to the patients. Today, 3D printing of crowns enables to provide restorations with resolutions up to 10 microns and smoother surfaces with minimal post-processing. Using newer techniques such as selective laser sintering, electron beam melting (EBM), or direct metal laser melting (DMLM) machines, crowns from metals such as cobalt-chrome and other alloys have also been realized. In-house production of 3D-printed braces reduces the number of visits required during a lengthy period that often spans 18 to 24 months or more. Using 3D printing in dentistry, it is possible for clinics to fabricate night guards on-site in less than an hour. Similarly, it is possible to quickly print clear aligners and acrylic retainers using AM processes. Another big step in dentistry is to tackle serious reconstructive challenges by using 3D printed scaffolds for tissue regeneration. Mandibular reconstruction and lower arch implant rehabilitation may be revolutionized with the use of these kinds of 3D constructs to support tissue regeneration [43]. It seems 3D printing may emerge as a promising substitute for conventional manufacturing in the healthcare industry for various applications [19].

7 Future and Challenges

Additive manufacture has given ways to tackle and minimize the disability associated with many medical conditions, particularly in the fields of orthopedics with the use of

3D printed orthotics and prosthetic parts for replacing the lost parts of the body. Many splints, braces, immobilization devices, and provisionals are also manufactured by same technology for the postoperative or post-surgical healing phase. So far, dental applications from the technique have not expanded to their fullest scope, because yet there is exists only limited knowledge about the materials more so regarding their biologic behavior, a lack of knowledge in CAD/CAM and factors like a high cost for product development. Nevertheless, engineering principles and methods are changing through functional integration, and improvised methods and materials are on way. The technique was conceived, developed, and nurtured in the domain of engineering and has been extensively explored for mechanical applications like the development of models, rapid prototyping. But now, with the exponential rise in knowledge of technology, and convergence of diverse fields, many applications based on 3D printing are being offered as biologic solutions as therapeutic measures in oral health care. Periodontal regeneration which was being tackled by filling bone grafts as mechanical fillers in the area of bone loss is nowadays targeted with tissue regeneration on customized biologic scaffolds which deliver bone formative cells at the bone defect. Biomedical applications have unique necessities as mostly biologic situations are unique and of high complexity, impacted by individual variations. New biomedical implants, engineered tissues and organs, and controlled drug delivery systems customized by additive manufacture should allow flexibility for developing extremely complex shapes by engineering novel materials such as semi-crystalline polymeric composites [71, 72]. Additive manufacturing has been utilized to enhance the accuracy and precision of placement and minimize the complications and repetitive interventions [73]. Since the products act to replace vital body parts and work in unison with the existing tissue structures, these are to be functionally integrated into the human body. The technique is more speedy, cost, and time-effective compared to conventional manufacturing specifically, for lower production volumes as in case of therapeutic requisites in case of patients. [74]. The sharing of digital records is much easier to reproduce the same design. For example, a project initiative by the National Institutes of Health (NIH), viz. 3D print exchange for sharing additive manufacture files freely has been taken in this direction [39, 75]. With the unique capabilities of the technique, come unique challenges also to utilize the benefits of the technique to the most, so some regulatory issues regarding the use of these products need to be addressed. The products for biologic uses need FDA approval [76]. Class I devices, which have a relatively less rigorous approval process, are the main focus of the biomedical industry. However, among the development of Class II and III devices, some of Class II implants are on way [39, 77]. FDA has issued a statement providing guidance to manufacturers on technical aspects of additive manufacturing regarding submissions to the agency. The FDA has, so far, approved over 100 3D printed medical devices, including knee replacements, custom skull/facial reconstructions, and even a 3D printed drug, called Spritam, which is used to treat seizures. The FDA has categorized this technical guidance as a "leap-frog" guidance, explaining that it is just a mechanism to share initial thoughts in order to bridge where we are today with the innovations of tomorrow [78]. Newer biomaterials are being developed and extensively researched to check for their properties and biologic behaviors as now, mass manufacturing of 3D printed bioconstruction has become the major goal of this technology. The mechanical properties of novel materials are yet to be studied and researched, to explore their full potential for application [79]. So, this aspect needs a specific focus to delineate materials for use in dentistry. Future trends shall focus on the development of customized biologic solutions for personalized patient care. The innovations in information technology to couple CT scan data with design analyses tools with AM technologies will allow for manufacturing patient-specific implants in a very efficient, precise, time, and cost-effective way [74]. Individual patient parameters and case characteristics shall guide for making drugs and drug delivery systems personalized to the case to case situations [80, 81]. Additive manufacturing could optimize all requisite biologic properties such as cellular inflow, growth, blood vessel growth, and vascularity with novel designs. Composite implants such as metallic implants coated with biologic coatings to enhance osseointegration can be developed with greater predictability for regeneration. Biofabrication is the next-generation additive manufacturing, specifically, when we focus on the health care and dental applications. Repair of diseased/lost or injured organs and tissues (anatomic and functional) would become possible at the site of the injury itself with biologic scaffolds implantation in the future [82]. Studies for manufacturing AM artificial organs including vascularization, innervation, and accomplishing the multifunctionality provided by each organ are ongoing [81]. These organs can be further connected with electronic devices known as cyborg organs e.g bionic ear has been developed. It has been functionalized by an inductive coil, which is used to collect electromagnetic signals for hearing [39, 83].

8 Conclusion

Additive manufacturing has led to the emergence of a new era in design and manufacturing, which is impacting the clinical practice of contemporary dentistry in a huge way. The technology enables the dental laboratories to offer rapid, cost-effective, and very precise treatment solutions for complex clinical situations. Additive manufacturing accelerates product development, offers design freedom, optimizes part structures, and allows for a high degree of functional integration. Moreover, it allows the dentist to provide a very efficient, conservative, and customized clinical care to the patient. Overall, such technical changes lead to a better and optimized patient care and cost-effectiveness also for a high-quality patient care at the end of the clinician. With the continued improvements in the speed, reliability, and accuracy of the hardware, materials, and the computer technologies, additive manufacturing is heading to give a tough competition to traditional manufacturing in creating end-user products for dental health care.

References

1. Brown C (2015) Additive manufacturing: it's a positive thing. Inside Dent Technol 6(3):50–51
2. Noort RV (2012) The future of dental devices is digital. Dent Mater 28:3–12
3. Jain R, Supriya Bindra S, Gupta K (2016) Recent trends of 3-D printing in dentistry—a review. Ann Prosthodont Restrorat Dent 2(4):101–104
4. Dawood A, Marti Marti B, Sauret-Jackson V, Darwood A (2015) 3D printing in dentistry. Br Dent J 219(11):521–525
5. Liu Q, Leu MC, Schmitt SM (2006) Rapid prototyping in dentistry: technology and application. Int J Adv Manuf Technol 29:317–335
6. Ashley S (1991) Rapid prototyping systems. Mech Eng 113(4):34
7. Noorani R (2006) Rapid prototyping—principles and applications. Wiley, Hoboken
8. Flowers J, Moniz M (2002) Rapid prototyping in technology education. Technol Teach 62(3):7
9. Chua CK, Chou SM, Lin SC, Eu KH, Lew KH (1998) Rapid prototyping assisted surgery planning. Int J Adv Manuf Technol 14(9):624–630
10. Cooper K (2001) Rapid prototyping technology. Marcel Dekker, New York
11. Kruth PP (1991) Material incress manufacturing by rapid prototyping techniques. CIRP Ann Manuf Technol 40(2):603–614
12. Wong KV, Herndez A (2012) A review of additive manufacturing. ISRN Mech Eng 2012:1–10. https://doi.org/10.5402/2012/208760 (Article ID 208760)
13. See CV, Meindorfer M (2016) 3D printing: additive processes in dentistry. Laboratory, May 2016
14. Halloran JW, Tomeckova V, Gentry S et al (2011) Photopolymerization of powder suspensions for shaping ceramics. J Eur Ceram Soc 31(14):2613–2619
15. Driesche SVD, Lucklum F, Bunge M, Vellekoop MJ (2018) 3D printing solutions for microfluidic chip-to-world connections. Micromachines 9(2):71
16. Emilia M, Marek M, Łukasz Z, Sonia S, Patryk K, Dariusz M (2014) 3D printing technologies in rehabilitation engineering (Technologiedruku 3D w inżynieriirehabilitacyjnej). J Health Sci 4(12):78–83. ISSN 1429-9623/2300-665X
17. Dawood A, Marti B, Sauret-Jackson V, Darwood A (2015) 3D printing in dentistry. Br Dent J 219:521–529
18. http://reprap.org/wiki/MainPage
19. https://3dprinting.com/dental/3d-printing-and-its-impact-on-the-dental-world-a-radical-shift-in-paradigm-and-future-possibilities/
20. https://www.medicaldesignbriefs.com/component/content/article/mdb/features/articles/26718
21. Zaharia C, Gabor AG, Gavrilovici A, Stan AT, Idorasi L, Sinescu C, Negruțiu M (2017) Digital dentistry—3D printing applications. J Interdisc Med 2(1):50–53
22. Herzog D, Seyda V, Wycisk E, Emmelmann C (2016) Additive manufacturing of metals. Acta Mater 117:371–392
23. Attar H, Calin M, Zhang L, Scudino S, Eckert J (2014) Manufacture by selective laser melting and mechanical behavior of commercially pure titanium. Mater Sci Eng A 593:170–177
24. Vaithilingam J, Kilsby S, Goodridge RD, Christie SD, Edmondson S, Hague RJ (2015) Functionalisation of Ti6Al4V components fabricated using selective laser melting with a bioactive compound. Mater Sci Eng C 46:52–61
25. Carlton HD, Haboub A, Gallegos GF, Parkinson DY, MacDowell AA (2016) Damage evolution and failure mechanisms in additively manufactured stainless steel. Mater Sci Eng A 651:406–414
26. Casalino G, Campanelli S, Contuzzi N, Ludovico A (2015) Experimental investigation and statistical optimisation of the selective laser melting process of a maraging steel. Optic Laser Technol 65:151–158
27. Murr LE, Martinez E, Hernandez J, Collins S, Amato KN, Gaytan SM, Shindo PW (2012) Microstructures and properties of 17-4 PH stainless steel fabricated by selective laser melting. J Mater Res Technol 1(3):167–177

28. Mazumder J, Choi J, Nagarathnam K, Koch J, Hetzner D (1997) The direct metal deposition of H13 tool steel for 3-D components. J Occup Med (JOM) 49(5):55–60
29. Brice C, Shenoy R, Kral M, Buchannan K (2015) Precipitation behaviour of aluminium alloy 2139 fabricated using additive manufacturing. Mater Sci Eng A 648:9–14
30. Demir AG, Previtali B (2017) Additive manufacturing of cardiovascular Co–Cr stents by selective laser melting. Mater Des 119:338–350
31. Gieseke M, Noelke C, Kaierle S, Wesling V, Haferkamp H (2013) Selective laser melting of magnesium and magnesium alloys. In: Magnesium technology. Springer, Cham, pp 65–68
32. Khan M, Dickens P (2010) Selective laser melting (SLM) of pure gold. Gold Bull 43(2):114–121
33. Ramirez D, Murr L, Martinez E, Hernandez D, Martinez J, Machado B, Medina F, Frigola P, Wicker R (2011) Novel precipitate–microstructural architecture developed in the fabrication of solid copper components by additive manufacturing using electron beam melting. Acta Mater 59(10):4088–4099
34. Postiglione G, Natale G, Griffini G, Levi M, Turri S (2015) Conductive 3D microstructures by direct 3D printing of polymer/carbon nanotube nanocomposites via liquid deposition modeling. Compos Appl Sci Manuf 76:110–114
35. Yang J-u, Cho JH, Yoo MJ (2017) Selective metallization on copper aluminate composite via laser direct structuring technology. Compos B Eng 110:361–367
36. Zhuang Y, Song W, Ning G, Sun X, Sun Z, Xu G, Zhang B, Chen Y, Tao S (2017) 3D–printing of materials with anisotropic heat distribution using conductive polylactic acid composites. Mater Des 126:135–140
37. Serra T, Planell JA, Navarro M (2013) High-resolution PLA-based composite scaffolds via 3-D printing technology. Acta Biomater 9(3):5521–5530
38. Wen Y, Xun S, Haoye M, Baichuan S, Peng C, Xuejian L, Kaihong Z, Xuan Y, Jiang P, Shibi L (2017) 3D printed porous ceramic scaffolds for bone tissue engineering: a review. Biomater Sci 5(9):1690–1698
39. Ngoa TD, Kashania A, Imbalzanoa G, Nguyena KTQ, Huib D (2018) Additive manufacturing (3D printing): a review of materials, methods, applications and challenges. Compos Part B 143:172–196
40. Lal K, White GS, Morea DN, Wright RF (2006) Use of stereolithographic templates for surgical and prosthodontic implant planning and placement. Part I. The concept. J Prosthodont 15:51–58
41. Zhou Z, Buchanan F, Mitchell C, Dunne N (2014) Printability of calcium phosphate: calcium sulfate powders for the application of tissue engineered bone scaffolds using the 3D printing technique. Mater Sci Eng C Mater Biol Appl 38:1–10
42. Sykes LM, Parrott AM, Owen CP, Snaddon DR (2004) Applications of rapid prototyping technology in maxillofacial prosthetics. Int J Prosthodont 17:454–459
43. https://www.ge.com/additive/additive-manufacturing/industries/dentistry
44. vanRoekel NB (1992) Electircal discharge machining in dentistry. Int J Prosthodont 5:114–121
45. Winder J, Bibb R (2005) Medcial rapid prototyping technologies: state of the art and current limitations for applications in oral and maxillofacial surgery. J Oral Maxillofac Surg 63:1006–1015
46. Bibb R, Brown R (2000) The application of computer aided product development techniques in medical modeling topic: rehabilitation and prostheses. Biomed Sci Instrum 36:319–324
47. Joffe J, Harris M, Kahugu F et al (1999) A prospective study of computer-aided design and manufacture of titanium plate for cranioplasty and its clinical outcome. Br J Neurosurg 13:576–580
48. Winder J, Cook RS, Gray J et al (1999) Medical rapid prototyping and 3D CT in the manufacture of custom made cranial titanium plates. J Med Eng Technol 23:26–28
49. Bartlett P, Carter L, Russell JL (2009) The leeds method for titanium cranioplasty construction. Brit J Oral Maxillofac Surg 47:238–240
50. Eufinger H, Wehmoller M (1998) Individual prefabricated titanium implants in reconstructive craniofacial surgery: clinical and technical aspects of the first 22 cases. Plast Reconstruct Surg 102:300–308

51. Zein I, Hutmacher DW, Tan KC, Teoh SH (2002) Fused deposition modeling of novel scaffold architectures for tissue engineering. Biomaterials 23:1169–1185
52. Centola M, Rainer A, Spadaccio C et al (2010) Combining electrospinning and fused deposition modeling for the fabrication of a hybrid vascular graft. Biofabrication 2(1):014102
53. Chen M, Le DQ, Baatrup A et al (2011) Self assembly composite matrix in a hierarchical 3D scaffold for bone tissue engineering. Acta Biomater 7(5):2244–2255
54. Wiggenhauser PS, Muller DF, Melchels FP et al (2011) Engineering of vascularized adipose constructs. Cell Tissue Res 347(3):747–757
55. Ponader S, von Wilmowsky C, Widenmayer M et al (2010) In vivo performance of selective electron beam-melted Ti-6Al-4V structures. J Biomed Mater Res 92A:56–62
56. Murr LE, Amato KN, Li SJ et al (2011) Microstructure and mechanical properties of open-cellular biomaterials prototypes for total knee replacement implants fabricated by electron beam melting. J Mech Behav Biomed Mater 4:1396–1411
57. Mueller AA, Paysan P, Schumacher R et al (2011) Missing facial parts computed by a morphable model and transferred directly to a polyamide laser-sintered prosthesis: an innovation study. Brit J Oral Maxillofac Surg 49(8):e67–e71
58. Rimmell JT, Marquis PM (2000) Selective laser sintering of ultra high molecular weight polyethylene for clinical application. J Biomed Mat Res 53:414–420
59. Sudarmadji N, Tan JY, Leong KF et al (2011) Investigation of the mechanical properties and porosity relationships in selective laser-sintered polyhedral for functionally graded scaffolds. Acta Biomater 7:530–537
60. Leong KF, Wiria FR, Chua CK, Li SH (2007) Characterization of a poly-epsilon-caprolactone polymeric drug delivery device built by selective laser sintering. Biomed Mater Eng 17:147–157
61. Zhang Y, Hao L, Savalani MM et al (2008) Characterization and dynamic mechanical analysis of selective laser sintered hydroxyapatite-filled polymeric composites. J Biomed Mater Res A 86:607–616
62. Fukuda A, Takemoto M, Saito T et al (2011) Osteoinduction of porous Ti implants with a channel structure fabricated by selective laser melting. Acta Biomater 7:2327–2336
63. Mullen L, Stamp RC, Brooks WK et al (2009) Selective laser melting: a regular unit cell approach for the manufacture of porous, titanium, bone in-growth constructs, suitable for orthopedic applications. J Biomed Mater Res Appl Biomater 89B:325–334
64. Traini T, Mangano C, Sammons RL et al (2008) Direct laser metal sintering as a new approach to fabrication of an isoelastic functionally graded material for manufacture of porous titanium dental implants. Dental Mater 24:1525–1533
65. Tara MA, Eschbach S, Bohlsen F, Kern M (2011) Clinical outcome of metal–ceramic crowns fabricated with laser-sintering technology. Int J Prosthodont 24:46–48
66. Williams RJ, Bibb R, Eggbeer D, Collis J (2006) Use of CAD/CAM technology to fabricate a removable partial denture framework. J Prosthet Dent 96:96–99
67. Xiao K, Dalgarno KW, Wood DJ, Goodridge RD, Ohtsuki C (2008) Indirect selective laser sintering of apatite–wollostonite glass–ceramic. Proc Inst Mech Eng H 222:1107–1114
68. Silva NRFA, Witek L, Coelho PG, Thompson VP, Rekow ED, Smay J (2011) Additive CAD/CAM process for dental prostheses. J Prosthont 20:93–96
69. Schuurman W, Khristov V, Pot MW et al (2011) Bioprinting of hybrid tissue constructs with tailorable mechanical properties. Biofabrication 3(2):021001 [Epub 2011 May 20]
70. Shanjani Y, Amritha De Croos N, Pilliar R et al (2010) Solid freeform fabrication and characterization of porous calcium polyphosphate structures for tissue engineering purposes. J Biomed Mater Res Appl Biomater 93B:510–519
71. Ventola CL (2014) Medical applications for 3D printing: current and projected uses. Pharm Ther 39(10):704
72. Garcia-Gonzalez D, Garzon-Hernandez S, Arias A (2018) A new constitutive model for polymeric matrices: application to biomedical materials. Compos B Eng 139:117–129
73. Jardini AL, Larosa MA, MacielFilho R, de CarvalhoZavaglia CA, Bernardes LF, Lambert CS, Calderoni DR, Kharmandayan P (2014) Cranial reconstruction: 3D biomodel and custom-built implant created using additive manufacturing. J Cranio Maxillofac Surg 42(8):1877–1884

74. Banks J (2013) Adding value in additive manufacturing: researchers in the United Kingdom and Europe look to 3D printing for customization. IEEE Pulse 4(6):22–26
75. NIH 3D print exchange, a collection of biomedical 3D printable files and 3D printing resources supported by the National Institutes of Health (NIH). https://www.ncbi.nlm.nih.gov/pubmed/
76. Morrison RJ, Kashlan KN, Flanangan CL, Wright JK, Green GE, Hollister SJ, Weatherwax KJ (2015) Regulatory considerations in the design and manufacturing of implantable 3d-printed medical devices. Clin Transl Sci 8(5):594–600
77. Di Prima M, Coburn J, Hwang D, Kelly J, Khairuzzaman A, Ricles L (2015) Additively manufactured medical products—the FDA perspective. 3D Print Med 2(1):1
78. Additive manufacturing in medical devices: insights from J&J and the FDA, Generis.html
79. Campbell T, Williams C, Ivanova O, Garrett B (2011) Could 3D printing change the world? technologies, potential, and implications of additive manufacturing strategic for resight initiative, October 2011
80. Ursan ID, Chiu L, Pierce A (2013) Three-dimensional drug printing: a structured review. J Am Pharmaceut Assoc 53(2):136–144
81. Wang X, Xu S, Zhou S, Xu W, Leary M, Choong P, Qian M, Brandt M, Xie YM (2016) Topological design and additive manufacturing of porous metals for bone scaffolds and orthopaedic implants: a review. Biomaterials 83:127–141
82. Di Bella C, Duchi S, O'Connell CD, Blanchard R, Augustine C, Yue Z, Thompson F, Richards C, Beirne S, Onofrillo C (2017) In-situ handheld 3D Bioprinting for cartilage regeneration. J Tissue Eng Regen Med. Ahead of print (in press)
83. Mannoor MS, Jiang Z, James T, Kong YL, Malatesta KA, Soboyejo WO, Verma N, Gracias DH, McAlpine MC (2013) 3D printed bionic ears. Nano Lett 13(6):2634–2639

Material Processing of PLA-HAp-CS-Based Thermoplastic Composite Through Fused Deposition Modeling for Biomedical Applications

Nishant Ranjan, Rupinder Singh and IPS Ahuja

Abstract Hydroxyapatite (HAp) is a commonly used biomaterial (due to its excellent bioactivity, biocompatibility, and osteoconductive properties) having potential applications in clinical dentistry, tissue engineering, orthopedic, and maxillofacial surgery. It is structurally and chemically similar to the enamel, dentin, and bone, which is one of the main reasons to use it in bone/tissue regeneration. The usage of virgin HAp is limited due to less strength and brittleness. In the last two decades, lot of development has taken place in bioactive polymers–ceramics composites (bioanalogue), especially in the field of biomedical application for bone analogue design, mechanical and biological performances as per clinical and biological needs. In this chapter, procedure for reinforcement of HAp and chitosan (CS) in biocompatible and biodegradable polymer polylactic acid (PLA) has been outlined for fabrication of biocompatible feedstock filament by using twin-screw extruder (TSE). Since PLA is one of the most commonly used thermoplastic for fused deposition modeling (FDM) of scaffolds, the PLA-HAp-CS composite can be gainfully employed in various biomedical applications. Finally, the functional prototypes have been printed on commercial FDM printer and mechanical properties have been optimized (by using biocompatible feedstock filament prepared with TSE). Also, the results have been supported by SEM and optical photomicrographs for better understanding.

Keywords PLA · HAp · CS · TSE · FDM · Reinforcement · Biomedical application · Thermoplastic composite · Feedstock filament

N. Ranjan · R. Singh (✉)
Department of Production Engineering, Guru Nanak Dev Engineering College, Ludhiana, India
e-mail: rupindersingh78@yahoo.com

N. Ranjan · I. Ahuja
Department of Mechanical Engineering, Punjabi University, Patiala, India

C. Prakash et al. (eds.), *Biomanufacturing*,
https://doi.org/10.1007/978-3-030-13951-3_6

1 Brief of Material Processing, Composites of Materials, Reinforcement of Materials, and Functional Prototypes for Biomedical Applications

In the last two decades, rapid-prototyping-(RP)-based technologies have been commercially established for making/printing of functional prototypes [1]. Usually, RP-based technologies develop framework from a 3D computer-aided design model. Many AM technologies such as stereolithography (SLA), selective laser sintering (SLS), silicon molding (SM), FDM etc., are available commercially. Each technology has its own merits and demerits. Usually, the selection is based upon end-user applications [2, 3]. The FDM-based 3D printing innovation uses industrial grade thermoplastic materials [4]. Therefore, many times the parts prepared by using FDM are not suitable from a thermal, mechanical, chemical, and synthetic quality viewpoint. FDM based 3D printer, utilizes, base up approach in which part is printed layer-by-layer by heating the thermoplastic filament [5].

The new age biocompatible/biodegradable polymer indicates superb blood similarity and biocompatibility and endorsed for biomedical utilization [6]. Biomaterial-based implants/inserts used for bone recovery must have some properties such as biodegradability, biocompatibility, bio-practical, and non-lethal attributes [7]. PLA-based materials have got high adaptability for tissue-specific applications. For instance, electrospun nanofibers and microsphere-sintered frameworks are more reasonable for bone recovery; comparably, nanofibrous/miniaturized scale scaffolds appropriate for tendon recovery [8, 9]. Natural bone is actually an inorganic/organic composite mainly made up of nanostructure HAp and collagen fibers. It is important to synthesize nano-composites of inorganic/organic in order to have good biocompatibility, high bioactivity, and great bonding properties. HAp nanoparticles and HAp/CS nano-composite with a homogeneous microstructure were prepared and characterized in previous studies. It has been concluded that nanostructure of HAp/CS composite has the best biomedical properties [10]. Bioceramics have the best bioactivity and biocompatibility properties, along with high corrosion resistance. Bioceramics have excellent osseointegration and osseo-conduction properties, which promote the healing of damaged and fractured bones. The discovery of HAp and bioglass has been a revolutionary invention in the biomedical field. But, the bioceramics have exhibited lower mechanical/tensile properties, which restrict their application in load-bearing scaffolds/inserts/implants [11].

HAp particles can be reinforced with high-density polyethylene (HDPE) to fabricate composites for use in biomedical inter-mashing. Some studies highlighted that TSE produces a homogeneous proportion of HAp particles in a HDPE grid [12, 13]. The HAp-based biomedical composites are used for restorative reasons. In the field of orthopedic/dental applications, HAp-reinforced biopolymers have got more consideration, because of their extremely appropriate biocompatibility and biodegradability properties. CS (mostly deacetylate type of chitin), like glycosaminoglycan, has numerous attractive properties as tissue engineering biomedical implants/inserts/scaffolds [14]. HAp has a major inorganic component of bone. It has been used extensively for biomedical implant applications and bone regeneration due to its biocompatibility, biodegradability, and osteoconductive properties. Pure HAp has very limited use due to its brittleness. Since the natural bone is a composite mainly consisted of nano-sized, needlelike HAp crystals and collagen fibers, efforts have been made to modify HAp by polymers, such as PLA, CS, and polyethylene. [15].

Regular polymers and their derivatives have been progressively utilized as an alternative option to engineered polymers in light of their biodegradability and organic movement. CS, a characteristic biodegradable polymer, is a low acetyl substituted type of chitin. In the recent past much consideration has been paid to CS reinforced composite materials due to extraordinary properties (like biodegradability, non-toxicity, antibacterial impact and biocompatibility) [16, 17]. CS-HAp composite materials have turned out to be promising for the treatment of bone imperfections. The study establishes the high capability of CS-HAp composite frameworks/inserts/implants created by co-precipitation technique as a filling material for orthopedic and stomatology. The permeable CS-HAp materials have indicated great osteoconductive properties while in vivo trials [18]. Different types of orthopedic joints have been prepared with welding methods. There were no inclusions and porosity detected inside the joining area [19, 20].

2 Research Gap and Challenges

The reported literature reveals that some studies have been reported on improving the biocompatibility of scaffolds/ functional prototypes (by using biocompatible polymers with different fillers and reinforcements (organic/inorganic) like HAp, and Catechin (Cat). But very less has been reported for use of PLA-matrix-based scaffolds (which has good biocompatibility and biodegradability) comprising of CS and HAp for Trauma/cancer patients. The use of such functional prototypes (scaffold) in the human body will help in the development/regeneration of bone/tissue in a sustainable manner. The scaffold is used to fill the gap in the fracture site of the bone produced due to trauma or surgery. The scaffold must possess adequate porosity at the level of more than 50% with interconnected pores to facilitate bone growth. It should also have the strength to withstand the compressive load, biocompatibility, and biodegradability. The literature also reveals that some researchers have worked in the field of welding/joining of scaffolds/biomedical implants with different techniques (like pin joint, laser welding, and screw joint).

But very less has been reported on the welding/joining of the cracked/damaged biocompatible scaffolds/implants (which was inserted in the body) by using friction welding through rapid tooling. The study may be conducted to perform online repair of orthopedic implants/scaffolds (prepared as a functional prototype from biocompatible feedstock filament on FDM) with portable rapid tooling assisted with friction welding. In this chapter, systematic procedure to prepare a spool of PLA-HAp-CS composite by using TSE (with different compositions/proportions of the reinforcements in PLA) has been reported. Thereafter developed feedstock, filaments can be deployed on commercial FDM setup for the fabrication of scaffold/functional prototype. The composite-material-based functional prototype will provide maximum strength to the scaffold by mixing varying proportions of the ingredients at optimized process parameters of the FDM process. Biodegradability study of the specimens may be conducted to assess the degradation time of scaffold. Multi-objective optimization with respect to compressive strength and biodegradability may be performed to arrive at optimized process factors of the FDM process. A systematic analysis needs to be performed to confirm that material degradation rate matches with the rate of regeneration of bone/tissue growth (which may be performed on the prepared scaffolds/biomedical implants).

3 Development/Fabrication of Feedstock Filaments of Polymeric Composites

3.1 Materials

An injection grade of PLA procured from a commercial manufacturer (Nature Works, USA). The density, melting temperature, glass transition temperature, and melt flow rate of the filament are 1.24 g/cm^3, 173-178 °C, 55–60 °C, 14 gm/10 min, respectively. Chitosan (CS, degree of deacetylation >90%), HAp (colorless & brittle) were procured from local manufacturer.

3.2 Preparation of Materials

PLA pellets were pre-dried at 80 °C in a vacuum oven overnight (8–10 h.), while HAp powder and CS were dried at 40 °C for 4 h. [21]. In the pilot experimentation, an effort has been made to prepare the feedstock filament of biocompatible/biodegradable polymers (PLA). The first step was to check the possibility of preparation of PLA filament with biocompatible fillers (HAp and CS). In this experimentation, PLA wire was prepared on the TSE by taking the random composition of HAp and CS by weight. Further according to MFI tester suitable four compositions (other than 100% PLA) were selected for drawing feedstock filament in the specific range of TSE temperature, rotational speed, and dead weight. After drawing feedstock filament, tensile test was performed and the best two samples were selected for further experimentation. Table 1 shows the selected process parameters and their level for the final experimentation.

Table 1 A parameter selected for the experimentation on TSE

Factors	Level 1	Level 2	Level 3
Temperature (°C)	170	180	190
Revolution (rpm)	100	140	180
Dead weight (Kg)	8	10	12

3.3 Extrusion Process

Extrusion is a process in which materials (polymers and metals) are pushed through a suitable/fixed cross-sectional area/profile/die to produce a desirable shape of products according to nozzle/output shape of the extruder. In other words, extrusion is a mechanical process in which polymers/metals are melted and extruded through the fixed shape. The single-screw extrusion process is widely used in common polymer processing, such as filmmaking, blow molding, and injection molding. On the other hand, a TSE is mainly used for compounding of various fibers, nano-size fillers, and polymer blending prior to final molding. Main functions of screw type extruders are:

- Grinder for granules, fibers, and fine particles,
- Pressure riser for highly viscous fluids;
- Conveyor for particles, granules, and viscous fluids;
- Squeezer for particles, granules, and viscous fluids;
- Heat-exchanger for particles and viscous fluids;
- Melter for organic and inorganic particles;
- Crystallizer for organic and inorganic fluids;
- Dryer for organic and inorganic particles and fluids;
- Continuous pyrolizer for organic materials;
- Mixture for granules and highly viscous fluids.

For the development of biomedical scaffolds/implants, the mixing of the polymer with different ceramics is must to be uniformly distributed. TSE is better than single-screw extruder for mixing of polymers with ceramics due to the interpenetration of the twin screws and the broad diversity of specialized screw designs. Figure 1 shows the different types of the extruder.

The twin-screw extrusion process is generally used to deliver bio-sourced plastics, fabrication of feedstock filament, and reinforcement of polymers with fillers or ceramics. Figure 2 demonstrates the 3D view of TSE. There is an extensive variety of barrel and screw plan, different process capacities and screw profiles to be set up as indicated by requirements of the process. TSE is competent to guarantee blending, shearing, cooling, warming, compacting, molding, and so forth with an abnormal state of adaptability. The fundamental points of interest of TSE (intermeshing co-turning) are their blending capacity that gives the amazing qualities to expelled items. In the twin-screw extrusion process, the crude materials could be solids (granules and powders form) or fluids.

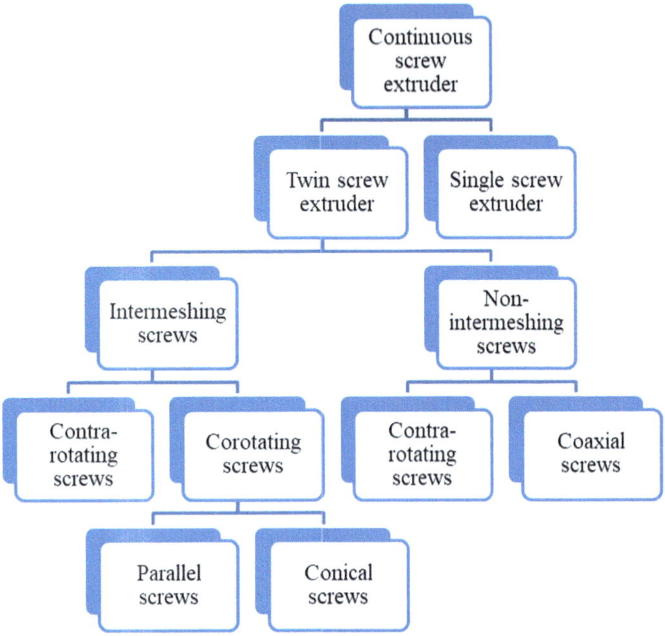

Fig. 1 Different types of extruder [3]

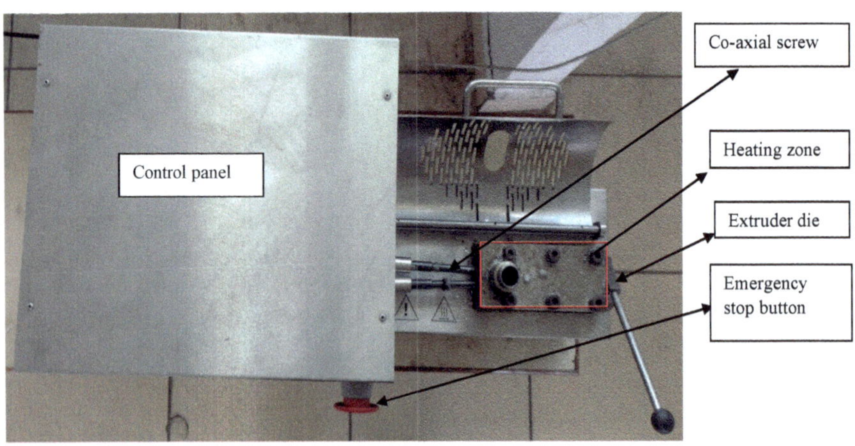

Fig. 2 Pictorial view of TSE

Table 2 DOE for TSE and output of tensile strength

Exp. no.	Input parameters			Output parameters		
	Temperature (°C)	Revolution (rpm)	Dead weight (Kg)	Strength at peak (MPa)	Peak load (N)	Strength at break (MPa)
1	170	100	8	3.14	6.40	2.83
2	170	140	10	3.64	7.80	3.28
3	170	180	12	4.12	9.00	3.71
4	180	100	10	3.62	7.70	3.22
5	180	140	12	4.25	9.20	3.80
6	180	180	8	3.10	6.30	2.73
7	190	100	12	4.85	10.10	4.38
8	190	140	8	3.76	7.90	3.38
9	190	180	10	4.54	9.50	4.10

3.4 *Preparation of Feedstock Filament (PLA-HAp-CS)*

Ranjan et al. [22] have made an attempt to prepare the feedstock filament of the biocompatible/biodegradable grade of PLA. The first step was to check the possibility of preparation of feedstock filament of PLA reinforced with HAp and CS with varying input parameters. Taking a random selection of different composition/proportions of polymer materials and fillers first step is to check the flow ability, followed by thermal, tensile, and microscopic testing. Suitable composition/proportion of PLA-HAp-CS for the fabrication of feedstock filament on TSE and also for making scaffolds on FDM is to be selected. For this study commercial TSE (make: HAAKE Mini CTW, Germany) has been used [22]. In this experimentation for material processing, PLA-HAp-CS are mixed in different compositions/proportions as per reported literature. Basic investigations like flow continuity, melt flow index (MFI), thermal analysis, tensile testing, and scanning electron microscope (SEM) were performed on these compositions/proportions for determining the best material composition for processing on TSE and FDM. Based upon pilot study, PLA-HAp-CS as 91-8-1 (weight percentage) has been selected for the fabrication of feedstock filament and development of biomedical scaffolds. After the selection of composition/proportion of PLA-HAp-CS composite, next step was to optimize the TSE for material processing. For the optimization of TSE Taguchi L9 (3^3) OA (Table 1) has been used with input factor; temperature of barrel, the revolution speed of screw and dead weight. Based upon Table 1 designs of experimentation (DOE) have been prepared (see Table 2). Table 2 shows that DOE for optimization of TSE setting and output of tensile strength of feedstock filaments.

According to Table 2 at first 27 (9×3) samples have been prepared for minimizing the environmental and human error, and the tensile test has been done on commercially available UTT. Figure 3 shows the best setting of input parameters for

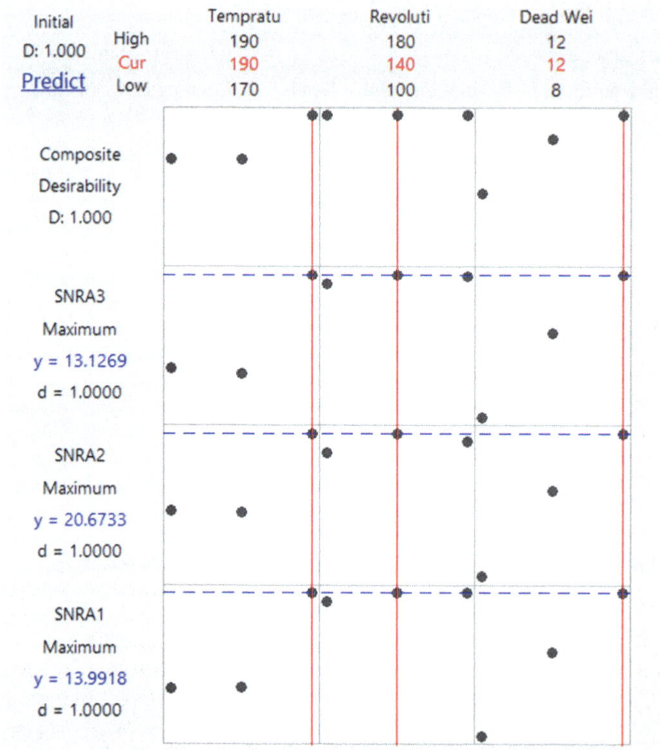

Fig. 3 Multi-factor optimization plot for TSE best setting [22]

the development of feedstock filaments of material composition 91-8-1 (PLA-HAp-CS), which are 190 °C barrel temperature, 140 rpm of screw speed and 12 kg dead weight.

4 Development/Fabrication of Framework for Optimization of FDM

FDM constructs conception models, frameworks, and end-utilization components in the standard. The most skilled 3D printing innovation uses industrial grade thermoplastic materials [23, 24].

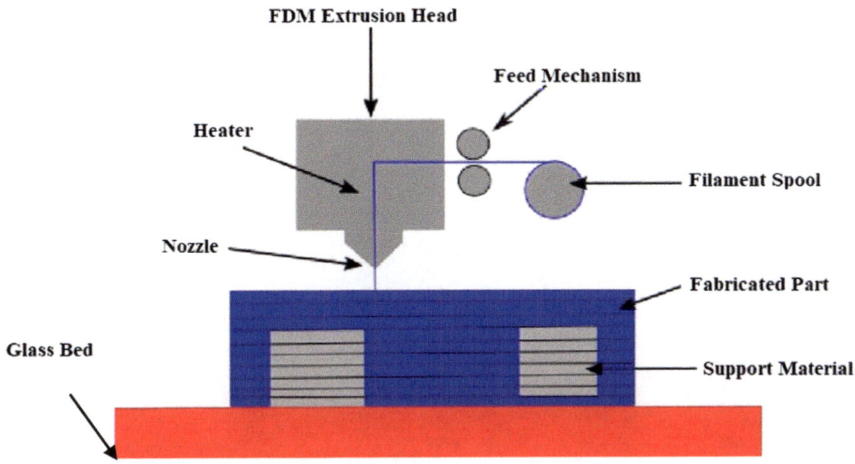

Fig. 4 Schematic diagram of FDM

4.1 *FDM: A 3D Printing Process*

FDM, refer to Fig. 4, utilizes base up approach in which part is worked as a layer-by-layer strategy by the procedure of warming and ousting the thermoplastic filament [25]. The procedure, for the most part, comprises of following three stages:

- **Pre-fabrication**: Under this stage, the product consequently cuts the three-dimensional CAD document and positions it. In addition, a way is likewise figured under this phase to extrude the thermoplastic material and some other supportive material.
- **Fabrication**: Under this stage, the thermoplastic material is warmed to the semi-fluid stage and begins depositing material as additional fine beads along extrusion path. Printer naturally deposits removable material going acting as framework wherever there is need of any support.
- **Post-fabrication**: Lastly, under this stage, the support material is split away by the client or dissolves up it in a blend of water and cleanser.

4.2 *Preparation of Biomedical Scaffolds*

The biomedical scaffolds must have good tensile and flexural strength. The fabricated feedstock filaments are ready for printing of biomedical scaffolds but before fabrication of scaffolds on open-source FDM final part strength needs to be optimized. So, as per ASTM standards, tensile and flexural samples were prepared on open-source FDM. Fabricated parts were tested on UTT and results were optimized by using Minitab-17 software. Further based upon tensile and flexural samples multi-factor

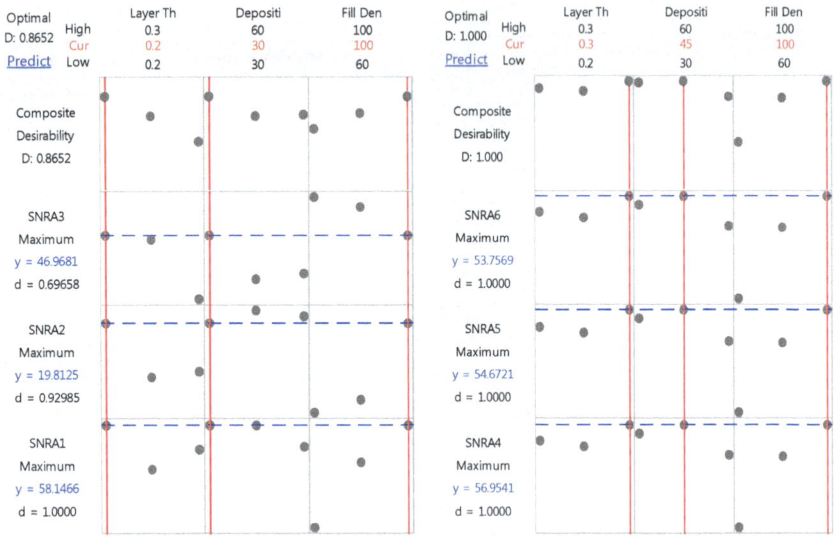

(a) Optimization plot of open-source FDM with the best setting for fabrication of a tensile specimen

(b) Optimization plot of open-source FDM with the best setting for fabrication of a flexural specimen

Fig. 5 Optimization plot of open-source FDM for tensile and flexural specimens [22]

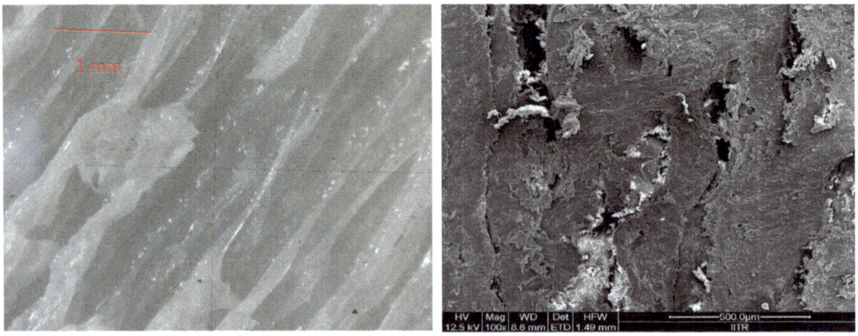

(a) ×100 microphotographs of tensile specimen on optical microscope

(b) ×100 microphotographs of tensile specimen on SEM

Fig. 6 Photomicrographs of tensile specimen on microscope and SEM

optimization was performed for determining the best setting of factors. The results of the study suggest that layer thickness as 0.2 and 0.3 mm, respectively, deposition angle of material as 30 °C or 45 °C with 100% infill density are the best settings, refer Fig. 5.

Finally, at best settings, fabricated parts were subjected to photomicrographic analysis by using optical microscope and SEM, refer Figs. 6 and 7.

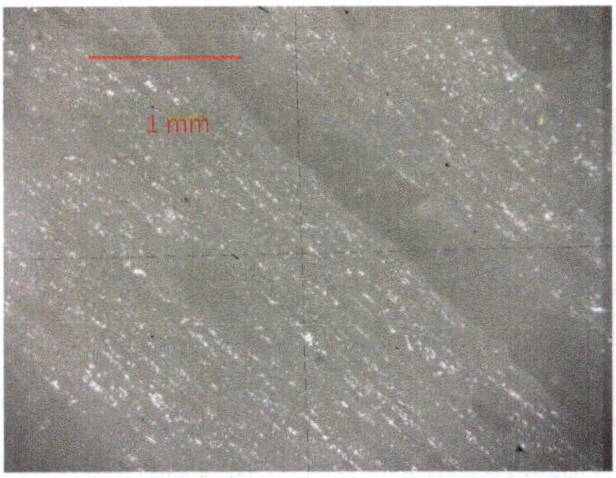

(a) ×100 microphotographs of the flexural specimen on optical microscope

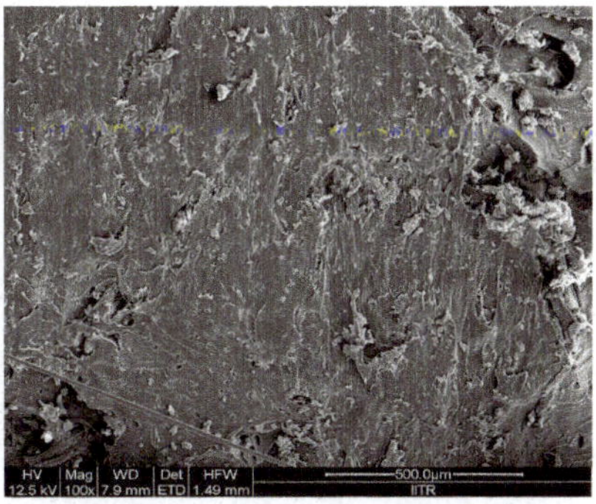

(b) ×100 microphotographs of flexural specimen on SEM

Fig. 7 Photomicrographs of flexural specimen on microscope and SEM

4.3 SEM and Optical Microphotographs Analysis

Photomicrographs were taken on Mitutoyo Tool maker Microscope and SEM (Model no. LEO 435Variable Pressure, Make: North Billerica, Massachusetts, United States) with 100X magnification on flat surface and fractured surface as shown in Figs. 6 and 7.

As observed from Figs. 6 and 7, at proposed material processing conditions, the HAp and CS are uniformly distributed in PLA. In biomedical applications, revealing the capability of supporting cell adhesion and cell proliferation, in vitro and in vivo analysis needs to be performed. Further, in order to analyze the biocompatibility of specimen, under in vitro and in vivo test, one has to make sure the bonding of fibroblast cells to the manufactured structures. Further studies may be conducted by using mouse embryonic fibroblasts cells (NIH-3T3) in serum-containing medium for in vitro and in vivo analysis.

5 Summary

Following observations have been noted from our study for material processing of PLA-HAp-CS-based thermoplastic composite through fused deposition modeling for biomedical applications:

- For ascertaining the best material composition of PLA-HAp-CS, thermal, rheological and mechanical analyses were performed on eight different compositions/proportions. It has been established that 91-8-1 (percentage by weight) is the best composition/proportion according to different tests performed for the preparation of feedstock filament prepared on TSE printing of scaffolds/functional prototypes via FDM in biomedical applications.

- For development of the biocompatible feedstock filament of PLA with reinforcement of HAp, and CS by using TSE, the best setting of TSE are: barrel temperature—190 °C, TSE screw rotational speed—140 rpm, and dead weight—12 kg.
- For fabrication of biomedical scaffolds/implants/functional prototypes, open-source FDM is parametrically optimized according to tensile and flexural samples and found that best setting is layer thickness—0.2 mm or 0.3 mm, deposition angle—30 °C or 45 °C, and infill density—100%.
- In optical and scanning electron microscopes, it was found that mixing capability of TSE is very good which is highly preferred for the fabrication of biomedical scaffolds/functional prototype/implants because in biomedical functional prototypes mixing must be good and equally distributed.

Acknowledgements The authors are highly thankful to SERB under AISTDF Secretariat (File No. IMRC/AISTDF/R&D/P-10/2017, Dated 01-02-2018) for financial support.

References

1. Pham DT, Gault RS (1998) A comparison of rapid prototyping technologies. Int J Mach Tools Manuf 38(10–11):1257–1287
2. Ngo TD, Kashani A, Imbalzano G, Nguyen KT, Hui D (2018) Additive manufacturing (3D printing): a review of materials, methods, applications and challenges. Compos Part B Eng 143:172–196
3. Sharma R, Singh R, Penna R, Fraternali F (2018) Investigations for mechanical properties of Hap, PVC and PP based 3D porous structures obtained through biocompatible FDM filaments. Compos B Eng 132:237–243
4. Chohan JS, Singh R, Boparai KS (2016) Parametric optimization of fused deposition modeling and vapour smoothing processes for surface finishing of biomedical implant replicas. Measurement 94:602–613
5. Singh R (2010) Three dimensional printing for casting applications: a state of art review and future perspectives. In: Advanced materials research, vol. 83. Trans Tech Publications, pp 342–349
6. Tanaka M, Hayashi T, Morita S (2013) The roles of water molecules at the biointerface of medical polymers. Polym J 45(7):701
7. Keogh MB, O'Brien FJ, Daly JS (2010) A novel collagen scaffold supports human osteogenesis applications for bone tissue engineering. Cell Tissue Res 340(1):169–177
8. Narayanan G, Vernekar VN, Kuyinu EL, Laurencin CT (2016) Poly (lactic acid)-based biomaterials for orthopaedic regenerative engineering. Adv Drug Deliv Rev 107:247–276
9. Peter SJ, Miller MJ, Yasko AW, Yaszemski MJ, Mikos AG (1998) Polymer concepts in tissue engineering. J Biomed Mater Res Part A 43(4):422–427
10. Chen F, Wang ZC, Lin CJ (2002) Preparation and characterization of nano-sized hydroxyapatite particles and hydroxyapatite/chitosan nano-composite for use in biomedical materials. Mater Lett 57(4):858–861
11. Ibrahim MZ, Sarhan AA, Yusuf F, Hamdi M (2017) Biomedical materials and techniques to improve the tribological, mechanical and biomedical properties of orthopedic implants—a review article. J Alloy Compd 714:636–667
12. Wang M, Porter D, Bonfield W (1994) Processing, characterization, and evaluation of hydroxyapatite reinforced polyethylene. Br Ceram Trans 93:91–95
13. Hench LL, Wilson J (1993) An introduction to bio-ceramics. World Scientific, Singapore, pp 43–56

14. Mi FL, Shyu SS, Wu YB, Lee ST, Shyong JY, Huang RN (2001) Fabrication and characterization of a sponge-like asymmetric chitosan membrane as a wound dressing. Biomaterials 22(2):165–173
15. Wang M, Bonfield W (2001) Chemically coupled hydroxyapatite–polyethylene composites: structure and properties. Biomaterials 22(11):1311–1320
16. Mi FL, Tan YC, Liang HF, Sung HW (2002) In vivo biocompatibility and degradability of a novel injectable-chitosan-based implant. Biomaterials 23(1):181–191
17. Zhang Y, Venugopal JR, El-Turki A, Ramakrishna S, Su B, Lim CT (2008) Electrospun biomimetic nano-composite nano-fibers of hydroxyapatite/chitosan for bone tissue engineering. Biomaterials 29(32):4314–4322
18. Danhier F, Ansorena E, Silva JM, Coco R, Le Breton A, Préat V (2012) PLGA-based nanoparticles: an overview of biomedical applications. J Controlled Release 161(2):505–522
19. Ferraris S, Spriano S, Lorenzon G (2016) Intraoral welding of titanium dental implants: characterization of the joints. J Mater Process Technol 235:85–91
20. Mai R, Lauer G, Pilling E, Jung R, Leonhardt H, Proff P, Stadlinger B, Pradel W, Eckelt U, Fanghänel J, Gedrange T (2007) Bone welding—a histological evaluation in the jaw. Ann Anat Anat Anz 189(4):350–355
21. Singh R, Ranjan N (2017) Experimental investigations for preparation of biocompatible feedstock filament of fused deposition modeling (FDM) using twin screw extrusion process. J Thermoplast Compos Mater https://doi.org/10.1177/0892705717738297
22. Ranjan N, Singh R, Ahuja IPS, Singh JP (2018) Fabrication of PLA-HAp-CS based biocompatible and biodegradable feedstock filament using twin screw extrusion. Springer International Publishing AG, part of Springer Nature (in press) https://doi.org/10.1007/978-3-319-91713-9_11
23. Singh R, Kumar R, Ranjan N, Penna R, Fraternali F (2018) On the recyclability of polyamide for sustainable composite structures in civil engineering. Compos Struct 184:704–713
24. Singh R, Kumar R, Ranjan N (2018) Sustainability of recycled ABS and PA6 by banana fiber reinforcement: thermal, mechanical and morphological properties. J Inst Eng (India) Ser C, 1–10
25. Singh R, Fraternali F, Bonazzi G, Hashmi MS, Kumar R, Ranjan N (2018) Investigations for development of feed stock filament of fused deposition modeling from recycled polyamide. Reference module in materials science and materials engineering. https://doi.org/10.1016/B978-0-12-803581-8.10391-1

Mr. Nishan Ranjan is a Ph.D. research scholar at Punjabi University, Patiala, and is working in the area of additive manufacturing and feedstock developments. He has authored more than five research papers.

Prof. Rupinder Singh is a Professor in Department of Production Engineering, Guru Nanak Dev Engineering College, Ludhiana, India. He has received Ph.D. in Mechanical Engineering from Thapar Institute of Engineering & Technology, Patiala, India. His area of research is non-traditional machining, additive manufacturing and development of porous biomaterials using 3D printing and rapid prototyping techniques. He has more than 18 years of teaching and research experience. He has contributed extensively to the world in additive manufacturing literature with publications appearing in Journal of Manufacturing Processes, Composite Part: B, Rapid Prototyping Journal, Journal of Mechanical Science and Technology, Measurement, International Journal of Advance Manufacturing Technology, and Journal of Cleaner Production. He has authored 17 books and ~3 book chapters. He has received more than 3 crores research grants from various funding agencies such as DST-SERB, AICTE, CSIR, DAE, and IE. He is working with Prof. Seeram Ramakrishna, NUS Nanoscience & Nanotechnology Initiative, Prof. Fernando Fraternali, Full Professor of Structural Mechanics, Department of Civil Engineering, University of Salerno.

Dr. IPS Ahuja is working as Professor in Mechanical Engineering Department of Punjabi University, Patiala. His research areas are optimization, machining, industrial engineering, and manufacturing. He has authored more than 150 research papers. He served as reviewer for various journals. He guided more than 10 Ph.D. theses and 20 M.Tech. theses.

Neurosurgical Bone Grinding

Atul Babbar, Vivek Jain and Dheeraj Gupta

Abstract Brain cancer is one of the major causes of death worldwide which occurs due to the tumors present in the brain. If these tumors not diagnosed at right time, then it will lead to loss of life. Neurosurgery, radiotherapy, and chemotherapy are the treatments which are used to remove tumors from the brain. Neurosurgery is often used as a treatment to diagnose this life-threatening disease. In surgical treatment especially in neurosurgery and orthopedics, bone grinding is commonly used. Bone grinding is an operation in which some part of the bone is removed to gain clearer operative access to the bones. Heat generated during bone grinding results in a rise in temperature which may cause harmful effects like osteonecrosis, blood coagulation, and optic nerve damage. This is the major concern for experts working in this area. Microstructure and thermophysical properties of the bone significantly affect the response of bone toward machining. The outcomes of research work done by experts are explained with their experimental setups. Different mathematical models are being explained by their key equations. Computational models and the role of automation in surgical operations are highlighted to reduce human involvement in such operations.

Keywords Grinding · Burr · Temperature · Necrosis

1 Introduction of Neurosurgical Grinding

A brain tumor is a protuberance inside the brain due to the unrestrained division and growth of brain cells. Brain tumor which instigates inside the brain is called primary brain tumor and if originated in other parts of the body and then spread to the brain is recognized as metastasis or secondary brain tumor. Diagnosis of brain tumors includes three different treatments (surgery, radiography, and chemotherapy) which can be used individually or a combination of two, depending upon the severity

A. Babbar · V. Jain (✉) · D. Gupta
Mechanical Engineering Department, Thapar Institute of Engineering and Technology,
Patiala 147003, India
e-mail: vivek.jain@thapar.edu

C. Prakash et al. (eds.), *Biomanufacturing*,
https://doi.org/10.1007/978-3-030-13951-3_7

of tumor present in the brain. Surgery is the first treatment which is used to remove a tumor in which incisions are made and a part of the skull is removed by bone grinding. In bone grinding, a ball end miniature grinding wheel (a diamond bur), rotating at 60,000 rpm, is used by neurosurgeons to reach the target regions [1]. When grinding wheel comes in contact with the bone surface, then a lot of heat is generated; thus, temperature increases to an unsafe level. If this temperature goes beyond threshold value, then it may result in severe consequence like thermal necrosis or osteonecrosis in the bone [2]. It was observed that increase in temperature can cause different sorts of heat-related damage. For example, when the temperature rises beyond a basic estimation of 50 °C then it causes osteonecrosis, contingent upon the sort of nerve, the basic temperature for beginning the heat damage could begin at 43 °C, coagulation of blood in the carotid corridor may happen due to rise in temperature [3]. Bone encompassing the cranial and optic nerves needs widespread grinding of bone to expose and resect tumor sores. Heat evolution is a major worry for specialists working in this field as heat produced can cause visual impairment and loss of muscle (facial) control and optic nerve damage [4–6]. We can reduce the risk of thermal damage by limiting the temperature to safeguard level. A few neurosurgeons utilized miniature grinding wheel to reduce the risk of trauma and exact movement control [7, 8]. Grinding wheel with little coarseness helps to feel a specific level of resistive power in bone grinding. To enhance the view of tumor present in the skull base, neurosurgeons often use the spherical grinding wheel (diameter = 3–4 mm) made from diamond grits [9]. Cooling plays an important role in deteriorating rise in temperature, and saline irrigation is the basic mode implemented for such purpose [7–12]. Amid surgical bone grinding, the grinding bur and restricted surgical space both contrive in skull base to confine the viability of saline for cooling. So, there is a need for the development of advanced cutting-edge technologies which can serve the purpose of reducing temperature evolved during the grinding of bone (Fig. 1).

A diamond bur was used for grinding the synthetic bone. The forces induced in the bone burring tools at high-speed were studied. Statistical analysis was employed along with the experimental methods to establish the correlations between burring motion and feedback of forces induced in the surgical tool. It was observed that there was no noteworthy correlation exists between the feed rate and forces. During experimentation, higher efficiency and lower burring force were achieved at a tilt angle 30° [7, 9, 13].

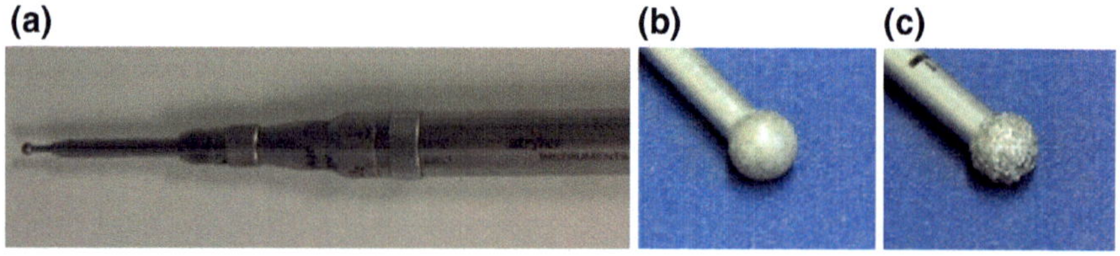

Fig. 1 **a** Neurosurgical grinding device, **b** diamond wheel specification 711 FEPA, and **c** 76 FEPA diamond grit [1]

Reducing heat evolution is the key to hinder the rise in temperature during bone machining [14–17]. To overcome this issue, a new grinding wheel coated with titanium dioxide (TiO_2) was developed. The size of TiO_2 particles was in nanometer range with diameter 180 nm. A significant reduction in temperature was observed with TiO_2 coating [12]. The temperature generated during bone machining can be reduced to a great extent by using irrigation. For example, few researchers (Morris et al. [18]; Gupta et al. [19]; Shih et al. [13]) used spray irrigation system to avoid localized heating during drilling of bone. Tai et al. [9] used cryogenic mist to decrease the initial temperature of the bone prior to the grinding. Experimental outcomes revealed that cryogenic mist not only helped in decreasing the temperature but also increased the visibility of surgical area. In a similar manner, nanoparticles were mixed in the mist and their effect on the temperature was studied. A positive correlation was observed between nanoparticle size and surface temperature [14]. Noticeable reduction in heat generation was achieved by using coolant which demonstrates its effectiveness against temperature rise [15–17, 20–30].

Heat flux produced during bone grinding was approximated by utilizing a widely implemented method known as inverse heat transfer method (IHTM) [31–47]. One of the major benefits of this method is that it gives us temporal changes in heat flux distribution with respect to varying time [42]. Heat flux distribution can also be calculated by combining experimental data with IHTM method [48]. Furthermore, a technique was developed in which feedback of electrical motor was used to determine temperature rise and heat flux was estimated with the help of IHTM under different environmental conditions. This technique showed its capabilities in estimating possible temperature rise with less than 20% error in temperature estimation, and outcomes revealed that with 0.4–0.5 W grinding power, the temperature can raise beyond 50 °C in surrounding nerves [4]. The performance of grinding process depends upon the number of independent and dependent factors for example the volume of material removal from the workpiece, the velocity of the workpiece, the angular velocity of the wheel, depth of cut, wheel diameter, uncut chip thickness, and specific energy consumption [49–54]. In view of this, the next section is dedicated to the mechanics of grinding, in which different parameters involved during the grinding operation are being deliberated.

2 Bone Grinding

2.1 Mechanics of Grinding

Grinding is a process in which material removal takes place by means of hard abrasive particles which may or may not be bonded to a tool. Generally, grinding operation is used to achieve a high-quality surface with minimum surface and subsurface damage. It is often applied as a last finishing operation. Material removal from the workpiece surface occurs in the form of microchips. The chips produced may be continuous, discontinuous, and chip with built up edge (BUE), and it is completely dependent on nature of workpiece and tool material. The amount of material removed during the grinding process is comparatively small in comparison to other machining processes. Commonly used abrasive materials are silicon carbide (SiO_2), aluminum oxide (Al_2O_3), cubic boron nitride (CBN), boron carbide (B_4C), and diamond. The performance of these abrasive grits is entirely dependent upon certain physical and geometrical properties such as friability, shape, size, hardness, toughness, and fracture resistance. Friability is related to self-sharpening ability and is dependent on the shape and size of the abrasive grit. Hardness enables the grits to resist penetration during scratching of the workpiece surface. More the difference in relative hardness of grit and workpiece material, more efficient is the grinding process. Fracture resistance helps in maintaining the sharp edges of the grit and resist attrition [55–57]. The parameters of the grinding process are articulated in different ways. The ratio of the volume of material removed during grinding to the volume of wheel material is defined as grinding ratio (G), which is given below:

$$G = \frac{\text{volume of material removed}}{\text{volume of wheel material}} \tag{2.1}$$

Grinding ratio, G, depends upon the number of factors for example workpiece material, cooling conditions, speed, nature of wheel, feed rate, and depth of grinding, etc. The approximate mathematical relation used to represent the undeformed length (l) of the workpiece is written below, where 'D' represents the diameter of the wheel and depth of cut is denoted by d.

$$l \approx (Dd)^{\frac{1}{2}} \tag{2.2}$$

A few assumptions were made during analysis which is as follows:

1. The workpiece width was taken to unity.
2. The velocity of the workpiece (v) was considered very small in comparison to the velocity of circumferential abrasive grain adhered on the wheel surface (V), and corresponding material removed per unit time was vd and VC, respectively, where C is the number of cutting edges per unit area of wheel surface.
3. The width of the chip removed during grinding was taken as constant. Assuming $v \ll V$, the volume of the chip is given by

$$V_{\text{chip}} = \frac{\omega t l}{2} \tag{2.3}$$

where 't' is the depth of cut by abrasive grain and 'ω' is the angular velocity of the grinding wheel. Furthermore, the subsequent relationship has been used to represent the volume of material removed from the workpiece.

$$vd = VC.V_{\text{chip}} = \frac{VC\omega t l}{2} \tag{2.4}$$

The schematic diagram of the surface grinding process has been shown below (refer to Fig. 2). Next, the thickness (t) of the undeformed portion of the workpiece in contact with the wheel is given by:

$$t = \frac{2v^{\frac{1}{2}} X^{\frac{1}{2}} d^{\frac{1}{4}}}{V^{\frac{1}{2}} C^{\frac{1}{2}} \omega^{\frac{1}{2}} D^{\frac{1}{4}}} \tag{2.5}$$

where 'X' is average chip thickness. The accuracy achieved on the surface after grinding is dependent on the relative grit force which in turn depends upon different significant parameters and is given by:

$$\text{Relative grit force} \propto \frac{vd^{\frac{1}{2}}}{VCD^{\frac{1}{2}}} \tag{2.6}$$

Furthermore, stresses are induced in the workpiece due to temperature generated during grinding, and these stresses can dimensionally distort the workpiece surface. The temperature generated in the vicinity of heat-affected zone, ΔT, depends upon specific energy consumption by chip (u) and is written as:

$$\Delta T \propto ud \tag{2.7}$$

$$\Delta T \propto \frac{d}{t} \propto d^{\frac{3}{4}} V^{\frac{1}{2}} C^{\frac{1}{4}} D^{\frac{1}{4}} V^{\frac{1}{2}} \tag{2.8}$$

Nevertheless, the temperature can rise beyond 1000 °C at the localized region of the wheel-work interface, but the material may melt depending on the nature of different materials and shorter contact time of wheel with the workpiece.

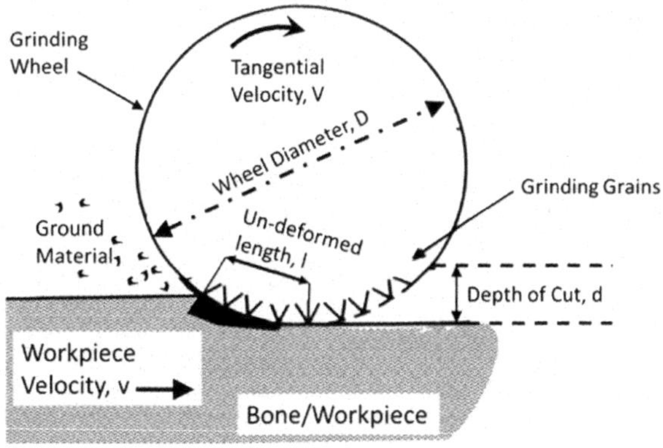

Fig. 2 Surface grinding operation [58]

2.2 Process parameters

There are different factors which significantly affect the performance of bone grinding operation, for instance, the rotational speed of bur, feed rate, the diameter of the tool, depth of cut, contact angle, and type of cooling. A lot of research has been carried out to determine their impact on the temperature rise during bone grinding. Bone drilling has been investigated a lot, while a few research articles are available for grinding of bone. The work done by different researchers in the field of bone grinding has been summarized in Table 1.

2.3 Computational Modeling

Mathematical modeling and simulation using finite element analysis (FEM) allow us an in situ examination of physical phenomena. Due to the complex nature of bone, numerical or analytical approach fits appropriate for designing bone grinding process. The selection of optimized parameters, tool design, and workpiece material is crucial to achieving wanted outcomes. So, this section is devoted to reviewing heat transfer models, key equations, and validation results. To minimize the damage caused by osteonecrosis during bone grinding, accurate temperature prediction is required. The basic mathematical equation that rules heat transfer creates a strong base for thermal FEMs.

$$\rho C_{\mathrm{p}}\left(\frac{\partial T}{\partial t}\right)_{(x,y,z)} = k'\left[\left(\frac{\partial T}{\partial x}\right)_{(y,z,t)} + \left(\frac{\partial T}{\partial y}\right)_{(x,z,t)} + \left(\frac{\partial T}{\partial z}\right)_{(x,y,t)}\right] \tag{3.1}$$

where k' is the constant of thermal conductivity and C_{p} is the specific heat at constant pressure. The models developed by the different researchers have been accentuated below with their validation results (refer Table 2).

The versatile feature of FEM is that it converts the complex problem into a simple one by dividing the problem into smaller parts. In the above figure, we can see that the grinding wheel is decomposed into several elemental grinding wheels (EGWs) and moves with the same rotational speed. Moreover, heat flux (q) is also constant for all elemental grinding wheels (Fig. 3).

The power consumed by the grinding wheel is partially transferred to the bone, and percentage of energy transferred is referred to as partition ration (ε). In the light of the above-mentioned literature, we can say that thermal FEM-based computational modeling is an excellent tool to predict the temperature on the surface as well as the subsurface region of the workpiece. Still, more efforts are needed from engineers and medical practitioners to come to a common platform and exchange their views, so as to develop better, precise, and accurate models. This will form a milestone step in upgrading the present procedure. As the time is passing, the scenario for surgical operations is also changing. Latest innovations in artificial intelligence (AI), novel

Table 1 Literature recap of bone grinding

Authors and year	Title of paper	Input parameters	Output parameters	Findings/conclusion
Tai et al. (2012)	Prediction of bone grinding temperature in skull base neurosurgery [1]	• Rotational speed = 60,000 rpm • Shank diameter of tool = 4 mm • Diamond abrasive size = 711 FEPA • Feed rate = 20 mm/min • Depth of cut = 0.5 mm • Workpiece—bovine cortical bone • Tool—diamond bur	Temperature	To compute the heat flux (q) and thermal conductivity (k) of the bone, inverse heat transfer method was used Peak temperature surpasses 50 °C at optic nerve at 10s
Tai et al. (2013)	Temperature prediction in high-speed bone grinding using motor PWM signals [4]	• Rotational speed = 60,000 rpm • Contact angle α = 30° • Feed rate = 20–60 mm/min • Depth of cut = 0.5 mm • Workpiece—bovine femur bone • Tool—diamond round bur	Temperature	The proposed method proved to be feasible for temperature prediction with less than 20% error
	Mist cooling in neurosurgical bone grinding [9]	• Rotational speed = 60,000 rpm • Feed rate = 20 mm/min • Depth of cut = 0.5 mm • Workpiece—bovine cortical bone • Tool—spherical bur	Temperature	The rise in temperature decreases on using cryogenic saline mist The temperature achieved was less than 43 °C
	Thermal model to investigate the temperature in bone grinding for skull base neurosurgery [3]	• Rotational speed = 60,000 rpm • Feed rate = 20 mm/min • Cutting depth = 0.4 mm • Workpiece—bovine cortical bone • Tool—spherical bur		The maximum temperature at tool–work interface was over 200 °C, while at the edge of the slot was 100 °C The thermal injury prevailed up to 3 mm in the transverse direction and 3 mm below ground surface in dry grinding conditions

(continued)

Table 1 (continued)

Authors and year	Title of paper	Input parameters	Output parameters	Findings/conclusion
Enomoto et al. (2014)	A new surgical grinding wheel for suppressing grinding heat generation in bone resection [12]	• Diameter of diamond bur = 5 mm • Rotational speed = 5000 rpm • Feed rate = 1 mm/min • Coolant flow rate = 180 ml/h. • Coatings on the wheel—SiO_2 and TiO_2 • Workpiece—bovine femur bone • Tool—abrasive coated wheel	Temperature	The significant reduction in temperature was observed by using the TiO_2 coating on the wheel surface
Tai et al. (2016)	An inverse method to reconstruct the heat flux produced by bone grinding tools [45]	• Diameter of bur = 4 mm • Rotational speed = 60,000 rpm • Contact angle = 30° • Feed rate = 20 mm/min • Bone density = 2050 kg/m^3 • Specific heat = 516 J/(kgk) • Thermal conductivity = 0.5 W/(mk) • Initial temperature = 20 °C • Workpiece—bovine femur bone • Tool—spherical diamond bur	Heat flux	Experimental data incorporated with IHTM to estimate heat flux distribution SFSM and SQP combined to calculate temporal and spatial variables simultaneously
Avinash danda (2016)	Characterization of forces in high-speed bone cutting and grinding for haptics rendering [7]	• Rotational speed = 60,000 rpm • Depth of cut = 0.5 mm • Feed rate = 1.6 mm/s and 3 mm/sec. • Burring angle = 30°, 60° and 90° • Workpiece—synthetic bone (made using the 3D printer)	Force	Higher efficiency and lower burring force were obtained at 30° burring angle

(continued)

Table 1 (continued)

Authors and year	Title of paper	Input parameters	Output parameters	Findings/conclusion
Jia et al. (2017)	Research on microscale skull grinding temperature field under different cooling conditions [59]	• Rotational speed = 60,000 rpm • Feed rate = 120 mm/min • Depth of cut = 0.01 mm • Drip flow rate = 50 ml/h • Mist flow rate = 10 ml/h • NJMC flow rate = 10 ml/h • MQL pressure = 0.54 MPa • Nozzle diameter = 15 mm • Nozzle angle = 15° • Workpiece—bovine femur bone • Tool—diamond coated bur	Temperature	It was found that NJMC showed minimum temperature (27.5 °C) followed by mist cooling, drip cooling, and dry grinding consecutively The peak temperature obtained was 41.6 °C in case of dry grinding, which was 10.1% higher than drip cooling, 29.3% higher than mist cooling, and 37% higher than NJMC

TiO_2 Titanium dioxide, *IHTM* Inverse heat transfer method, *SFSM* Sequential function specification method, *SQP* Sequential quadratic programming

Table 2 Different analytical models for bone grinding

Author and year [references]	Assumptions	Proposed analytical model	Validation results	Conclusions/remarks
Tai et al. (2013) [1]	Each EGW is moving at the same speed Heat partition ratio, ε, is constant Heat flux is constant	Inverse heat transfer model gives $q = \varepsilon q_{\text{total}} = \varepsilon \frac{F_t v_s}{b_w l_c}$; Objective function is $Min\ f(k, h) = \sum_{i=1}^{l} \sum_{j=1}^{j} T_{PIJ} - T_{ij}$	The discrepancy in the validation results found to be less than 6%	The proposed IHTM proved to be accurate and robust in accordance with the validation results
Zhang et al. 2013 [3]	Each EGW is moving with the same speed Heat partition ratio, ε, is constant Heat flux is constant Each EGW perform grinding with its own tangential speed and cutting force	FETM incorporated with IHTM and gives $q = \varepsilon \frac{F_t v_s}{b_w l_c}$; $a_i = r_i - r_{i-1} - b_E \text{Sin}\alpha$; $q_i = \varepsilon \frac{(ka_i)(\omega r_i)}{b_E l}$; $q_i = (\varepsilon k \omega)\left(\frac{a_i r_i}{b_E l_b}\right) = q_m R_h$; Objective function is $F(q_m, k) = \sum_{i=1}^{l} \sum_{j=1}^{j} \left(T_{ij}^{\text{exp}} - T_{ij}^{\text{FETM}}\right)^2$	The discrepancy in results found to be less than 4% FETM used in conjunction with IHTM gives 10 times faster computational power	The proposed model proved to be reliable and free from sensitivity toward noise factors
Wang et al. 2015 [45]	The material is inhomogeneous The geometrical variables are constant and are not a direct function of time The coordinating system is moving in nature	IHTM coupled with SFSM and SQP, using two geometrical variables k_x and k_y. The coordinates conversion is given by $x' = x + vt$; $y' = -y + \left(\frac{l_y}{2} + \frac{l_b}{2}\right)$; $q(x'y't) = k_x x' + k_y y' + c(t)$;	To measure the robustness of the model, arbitrary parameters were taken The values hence obtained fall in the range of the normal distribution curve	Results obtained from IHTM were highly influenced by future time step (r) and initial geometrical values of k The proposed model gave consistent results, showing its robustness

(continued)

Table 2 (continued)

Author and year [references]	Assumptions	Proposed analytical model	Validation results	Conclusions/remarks
Jia et al. 2017 [59]	The effect of cutting parameters on the contact state and deformation is neglected The length of the contact arc is constant under microgrinding conditions Tangential force is taken as constant Due to the poor airflow and closed environment, CHTC took as 5×10^{-6} W/mm^2·K The workpiece is a rectangular plane and decomposed into plane grid structures Three surfaces are considered as adiabatic	Four theoretical models are developed in this study which is as follows: Microgrinding model: $r' = \sqrt{r^2 - (r - a_p)^2}$; $h\text{max} = t' V_w$ Heat flux density model: $l_c = \frac{\pi d_c \arcsin(a_p/d_e)^{\frac{1}{2}}}{180}$; $q_w = 0.9R \int_{r-a_p}^{r} F\omega r^{n2} dz / r'^2$ Convective heat transfer coefficient model: $h = 105.6k_{nf}^{0.6}.\rho_{nf}^{0.8}.C_{nf}^{0.9}/\mu^{0.4}$ Intra-workpiece heat conduction model: $T_{t+\Delta t}(i, 1) = \frac{\Delta t}{\rho_w C_w \Delta x.\Delta z}(X, Y, Z)$; $X = k_x[T(i-1, 1) + T(i+1, 1-2T(i, 1))$; $Y = k_z[T(i, 2) - T(i, 1)$; $Z = \Delta z[q_x - G(T_t - T_a)]$	Simulation results were validated by experimentation The model error came out to be less than 6.5%	Theoretical results were consistent with experimental results. Hence, proposed models proved to be effective in determining the rise in temperature

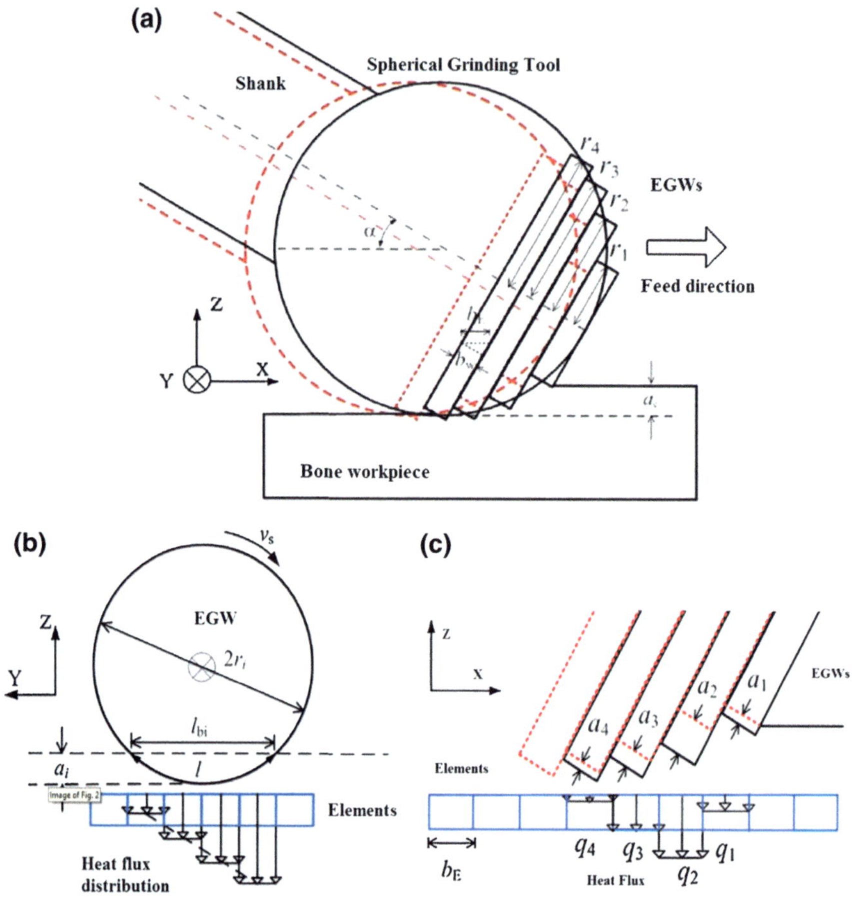

Fig. 3 **a** Schematic of the geometry of grinding wheel, the configuration of EGWs with heat flux distribution in **b** YZ-plane and **c** XZ-plane [3]

methods, and computing power lead us to another important aspect of automation in surgical operations.

2.4 Recent Case Studies

It is clear from the literature discussed in the introduction that temperature should be below the threshold value of temperature at which thermal injury starts. So, coolant plays an important role in clinical operations. Different authors used different types of modes of cooling. Shih et al. [9] used cryogenic mist to keep the temperature under a safe limit. The bone was pre-cooled before grinding due to which initial temperature of the bone decreased significantly. Moreover, the localized rise in the temperature was quite low during the course of grinding. Li et al. [14] used the jet mist cooling method to perform microscale grinding of bone with different particle sizes. The particle size taken for nanoparticle was 30, 50, 70, and 90 nm. Real-time dynamic heat flux model was generated, and temperature obtained for nanofluids was 21.4, 17.6, 16.1, and 8.3% lower than the mist cooling temperature (33.6 °C) for 30, 50, 70, and 90 nm, respectively. A positive correlation was achieved between particle size and the surface temperature of the workpiece. The results acquired from the experiments validate the theoretical results obtained from the model. Danda et al. [7] brought out the concept of using haptic technology for surgical simulation. He performed characterization of the forces during bone grinding. Experiments were conducted to measure these forces under different input factors like tool orientation, feed rate, and paintbrush like motion in different axes. The results revealed that the orientation of the tool highly affects the forces, while feed rate was least significant. In this study, the vibrational frequency was not considered, which is a major limitation.

Jia et al. [59] carried out skull grinding to monitor the temperature distribution under different cooling conditions. They developed different models which are micro-grinding model (μ-grinding), heat flux density model (HFD), convective heat transfer coefficient method (CHTC), and intra-workpiece heat conduction model (IWHC). Experimental conditions were simulated under different cooling conditions like dry grinding, drip cooling, mist cooling, and nanoparticle jet mist cooling (NJMC). Highest temperature peak was obtained in the case of dry grinding in comparison to other modes of cooling. Nanoparticle jet mist cooling gave excellent cooling results and lowest temperature peak (27.5 °C). To check the feasibility of nanoparticles, Wang et al. [60] used the concept of minimum quantity lubrication along with different types of nanoparticles during microscale bone grinding. The temperature was measured against different mass fractions of different nanoparticles. The mass fraction was 10, 8, 6, 4, and 2% of nanofluids, carbon nanotube, Fe_2O_3, SiO_2, and hydroxyapatite, respectively. It was observed that temperature peak was proportional to a mass fraction in certain range while inversely proportional outside that range. Moreover, results showed that micromotion, adsorption layer, and rheological properties were highly dependent on the shape, thermophysical properties, and mass fraction of nanoparticles. Tai et al. [45] used high-speed surgical diamond burr for grinding of the bone (refer Fig. 4). Four K-type thermocouples were placed beneath the path of grinding. To estimate the heat flux distribution, an inverse method coupled with sequential function specification method (SFSM) and sequential quadratic programming (SQP) was used. This method was validated by using numerical tests, and consistent results were obtained which demonstrates its robustness. The effect of the arrangement of sensors and their location will be taken as the future research by authors.

Sullivan et al. [2] used the inverse heat transfer method to inspect the temperature rise during bone grinding. A thermal model was developed, and FEM was used to estimate the distribution of temperature. A spherical diamond bur was used for grinding of bovine bone, and severe thermal damage was observed as the temperature

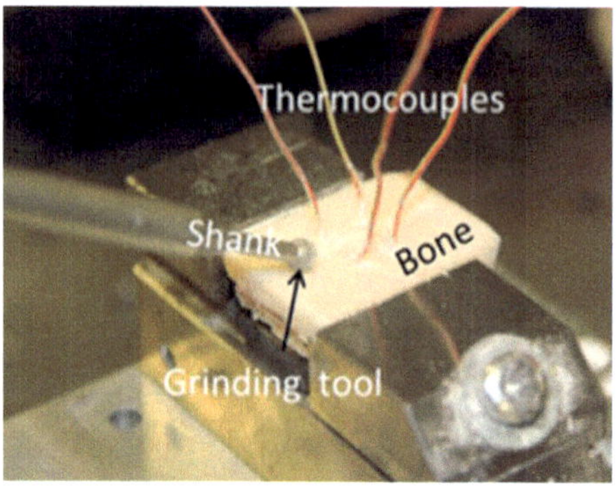

Fig. 4 Close-up view of grinding setup [45]

achieved was more than the threshold value (50 °C), but model developed proved to be efficient for thermal analysis.

3 Future Work Scope

The primary objective of the research is to lessen the involvement of human beings in surgical operations. Human errors are always involved in such operations. So, there is a need to use automation either fully automatic or semiautomatic in surgical operations. The information gathered from the workpiece (bone) is sent to algorithm; then, the plan is executed. Steps involved in the whole process are image data acquisition, processing, matching, referencing, tool tracking, and tool visualization [61–67]. While developing a FEM model, certain factors should be considered for example blood perfusion, direction and depth of grinding, saline irrigation, etc. Moreover, further investigation is required to determine the effect of speed, feed, tilt angle, and type of wheel on temperature produced during grinding of bone. To increase the robustness of models developed, hybrid soft computing techniques like particle swarm optimization, artificial neural network, fuzzy logic, and the evolutionary algorithm can be incorporated with experimental data to optimize and predict the response of bone toward grinding process.

Recent technological advancements enabled us to use robots in the surgical operation theater. In robotic surgery, well-organized integration of advanced computer technology can be seen. The state-of-the-art equipment (ECMO, call Vizio, da Vinci Si surgical system) offers benefits like less tissue damage, reduction in loss of blood, less infection danger, less pain, and rapid recovery of the patient. Robotic arm tracks predetermined reference points to perform machining. Robotic systems are categorized into passive and active systems. The passive system refers that the robot is under the control of surgeon, while active systems provide full autonomy. The worldwide healthcare system is undergoing tremendous changes which have paved the way for surgical simulators. Surgical simulators are playing a prominent role in the medical industry as they provide real-time images, actual controls, and imitate problems that may happen during actual surgery. The concept originally started in 1980 with video games, but in the 1990s with the development of 3-D graphics and motion sensors in 2000s, simulators came to its present versatile form. With the advancement in 3D rapid prototyping, patient-specific anatomy, 3-D printing, and telesurgery, surgical simulator training has become useful for both novice and expert neurosurgeons as skills attained from training may lead to improvement in clinical operations. At present, different surgical simulators are used to for example virtual reality (VR), human cadavers, benchtop models, and live animals. These simulators are used for different treatments like repair of cleft palate, laparoscopic examination of the bile duct, intestinal anastomosis, and skull base neurosurgery. These surgical simulators can be incorporated with potential haptic technology in future. A haptic device acts as a link between user and computer, and a feeling of touch is experienced when a user applies force or vibrations. Haptic is a Greek word which is originated from word 'haptikos' which refers to a sense of touch. At present, a lot of research has been going on to hone the skills of tactile interaction with holograms. If this research is completed successfully, then it will benefit the medical industry as well.

4 Concluding Remarks

The cases of high-grade brain tumors are increasing worldwide, but the survival rate is less than 10% since 1971, and it highlights the investment and research in this field. In UK, only 1.9% of £580 million is spent by the government [68]. Investment alone is not sufficient to pave the way for brain tumor research. Three parameters are very important in leadership, workforce, and resources. Leadership from government authorities is crucial, and they should create a positive environment for the research. It is the responsibility of the government to identify research gaps and should be prioritized according to the needs of this field. Opportunities must be created for the people working in health and life sciences with attractive incentives. Moreover, collaboration between non-governmental organizations, and government organization should be established to endorse a better research environment. Each hospital should make a separate section of biobank to store biological samples of humans. These biobanks can play an eye-catching role in genomics and personalized medicine. Virtual biobanks, tissue banks, and population banks are the types of biobanks. At present catalyzation, the rate is very slow as it can take 15 years to translate experimental and clinical trial discoveries into a new treatment. This has a huge impact on the patient's survival and his quality of life. This process of translation should be accelerated so that new state-of-the-art treatments and drug delivery methods should be given to the patients suffering from malignancies of brain tumors.

References

1. Shih AJ, Tai BL, Zhang L et al (2012) Prediction of bone grinding temperature in skull base neurosurgery. CIRP Ann Manuf Technol 61:307–310. https://doi.org/10.1016/j.cirp.2012.03.078
2. Tai BL, Zhang L, Wang A et al (2013) Neurosurgical bone grinding temperature monitoring. Procedia CIRP 5:226–230. https://doi.org/10.1016/j.procir.2013.01.045
3. Zhang L, Tai BL, Wang G et al (2013) Thermal model to investigate the temperature in bone grinding for skull base neurosurgery. Med Eng Phys 35:1391–1398. https://doi.org/10.1016/j.medengphy.2013.03.023
4. Tai BL, Zhang L, Wang AC et al (2013) Temperature prediction in high speed bone grinding using motor PWM signal. Med Eng Phys 35:1545–1549. https://doi.org/10.1016/j.medengphy.2013.05.011
5. Manić M, Stamenković Z, Mitković M et al (2015) Design of 3D model of customized anatomically adjusted implants. Facta Univ Ser Mech Eng 13:269–282
6. Fernandes CR, Fernandes BL (2003) Evidence of semi-solid formation in medical grade Ti6Al4V alloy using induction heating. Facta Univ Ser Mech Eng 1:1–2
7. Danda A, Kao YT, Kuttolamadom MA, Tai BL (2016) Characterization of forces in high-speed bone cutting and grinding for haptics rendering. ASME 2016 11th Int Manuf Sci Eng Conf MSEC 2016 2:1–8. https://doi.org/10.1115/msec2016-8794
8. Mišić D, Manić M, Vitković N, Korunović N (2015) Toward an integrated information system for the design, manufacturing and application of customized implants. Facta Univ Ser Mech Eng 13:307–323
9. Zhang L, Tai BL, Wang AC, Shih AJ (2013) Mist cooling in neurosurgical bone grinding. CIRP Ann 62:367–370. https://doi.org/10.1016/j.cirp.2013.03.125
10. Liang S, Shih AJ (2016) Analysis of machining and machine tools. Springer, US, Boston, MA
11. Alam K, Khan M, Muhammad R et al (2015) In-vitro experimental analysis and numerical study of temperature in bone drilling. Technol Heal Care 23:775–783. https://doi.org/10.3233/THC-151035
12. Enomoto T, Shigeta H, Sugihara T, Satake U (2014) A new surgical grinding wheel for suppressing grinding heat generation in bone resection. CIRP Ann 63:305–308. https://doi.org/10.1016/j.cirp.2014.03.026
13. Cseke A, Heinemann R (2018) The effects of cutting parameters on cutting forces and heat generation when drilling animal bone and biomechanical test materials. Med Eng Phys 51:24–30. https://doi.org/10.1016/j.medengphy.2017.10.009
14. Yang M, Li C, Zhang Y et al (2018) Microscale bone grinding temperature by dynamic heat flux in nanoparticle jet mist cooling with different particle sizes. Mater Manuf Process 33:58–68. https://doi.org/10.1080/10426914.2016.1244846
15. Strbac GD, Giannis K, Unger E et al (2014) A novel standardized bone model for thermal evaluation of bone osteotomies with various irrigation methods. Clin Oral Implants Res 25:622–631. https://doi.org/10.1111/clr.12090
16. Gehrke SA, Neto HL, Mardegan FEC (2013) Investigation of the effect of movement and irrigation systems on temperature in the conventional drilling of cortical bone. Br J Oral Maxillofac Surg 51:953–957. https://doi.org/10.1016/j.bjoms.2012.10.023
17. Siljander BR, Wang AC, Zhang L et al (2014) Cool mist irrigation improves heat dissipation during surgical bone drilling. J Neurol Surgery, Part B Skull Base 75:243–246. https://doi.org/10.1055/s-0034-1368098
18. Sasaki M, Morris S, Goto T et al (2010) Spray-irrigation system attached to high-speed drills for simultaneous prevention of local heating and preservation of a clear operative field in spinal surgery. Neurol Med Chir (Tokyo) 50:900–904. https://doi.org/10.2176/nmc.50.900
19. Gupta V, Pandey PM, Gupta RK, Mridha AR (2017) Rotary ultrasonic drilling on bone: A novel technique to put an end to thermal injury to bone. Proc Inst Mech Eng Part H J Eng Med 231:189–196. https://doi.org/10.1177/0954411916688500

20. Strbac GD, Unger E, Donner R et al (2014) Thermal effects of a combined irrigation method during implant site drilling. A standardized in vitro study using a bovine rib model. Clin Oral Implants Res 25:665–674. https://doi.org/10.1111/clr.12032
21. Gaspar J, Borrecho G, Oliveira P et al (2013) Osteotomy at low-speed drilling without irrigation versus high-speed drilling with irrigation: an experimental study. Acta Med Port 26:231–236
22. Leite PHAS, Barreto-Filho TA, Oliveira TS et al (2011) Microscopic evaluation of bone fragments obtained by different methods of irrigation and osteotomy. Pesqui Bras Odontopediatria Clin Integr 11:539–545. https://doi.org/10.4034/PBOCI.2011.114.14
23. Trisi P, Berardini M, Falco A et al (2014) Insufficient irrigation induces peri-implant bone resorption: An in vivo histologic analysis in sheep. Clin Oral Implants Res 25:696–701. https://doi.org/10.1111/clr.12127
24. Tawy GF, Rowe PJ, Riches PE (2016) thermal damage done to bone by burring and sawing with and without irrigation in knee arthroplasty. J Arthroplasty 31:1102–1108. https://doi.org/10.1016/j.arth.2015.11.002
25. Isler SC, Cansiz E, Tanyel C et al (2011) The effect of irrigation temperature on bone healing. Int J Med Sci 8:704–708. https://doi.org/10.7150/ijms.8.704
26. Owens BBD, White DW, Wenke JC (2009) Comparison of irrigation solutions and devices in a contaminated musculoskeletal wound survival model. J Bone Jt Surg - Ser A 91:92–98. https://doi.org/10.2106/JBJS.G.01566
27. Bullon B, Bueno EF, Herrero M et al (2015) Effect of irrigation and stainless steel drills on dental implant bed heat generation. J Mater Sci Mater Med 26:1–10. https://doi.org/10.1007/s10856-015-5412-8
28. Gehrke SA, Pazetto MK, de Oliveira S et al (2014) Study of temperature variation in cortical bone during osteotomies with trephine drills. Clin Oral Investig 18:1749–1755. https://doi.org/10.1007/s00784-013-1163-4
29. Sener BC, Dergin G, Gursoy B et al (2009) Effects of irrigation temperature on heat control in vitro at different drilling depths. Clin Oral Implants Res 20:294–298. https://doi.org/10.1111/j.1600-0501.2008.01643.x
30. Draeger RW, Dirschl DR, Dahners LE (2006) Debridement of cancellous bone: A comparison of irrigation methods. J Orthop Trauma 20:692–698. https://doi.org/10.1097/BOT.0b013e31802b41e2
31. Huang CH, Jan LC, Li R, Shih AJ (2007) A three-dimensional inverse problem in estimating the applied heat flux of a titanium drilling—theoretical and experimental studies. Int J Heat Mass Transf 50:3265–3277. https://doi.org/10.1016/j.ijheatmasstransfer.2007.01.031
32. Li X, Zhu W, Wang J, Deng Y (2016) Optimization of bone drilling process based on finite element analysis. Appl Therm Eng 108:211–220. https://doi.org/10.1016/j.applthermaleng.2016.07.125
33. Tai BL, Stephenson DA, Shih AJ (2012) An inverse heat transfer method for determining workpiece temperature in minimum quantity lubrication deep hole drilling. ASME J Manuf Sci Eng 134:021006. https://doi.org/10.1115/1.4005794
34. Tai BL, Jessop AJ, Stephenson DA, Shih AJ (2012) Workpiece Thermal Distortion In Minimum Quantity Lubrication Deep Hole Drilling—finite element modeling and experimental validation. J Manuf Sci Eng 134:011008. https://doi.org/10.1115/1.4005432
35. Tai BL, Stephenson DA, Shih AJ (2012) An inverse heat transfer method for determining workpiece temperature in minimum quantity lubrication deep hole drilling. J Manuf Sci Eng Trans ASME 134. https://doi.org/10.1115/1.4005794
36. Brandao LC, Coelho RT, Lauro CH (2011) Contribution to dynamic characteristics of the cutting temperature in the drilling process considering one dimension heat flow. In: Applied Thermal Engineering. pp 3806–3813
37. Fernandes AP, dos Santos MB, Guimarães G (2015) An analytical transfer function method to solve inverse heat conduction problems. Appl Math Model 39:6897–6914. https://doi.org/10.1016/j.apm.2015.02.012
38. Ristić M, Manić M, Mišić D et al (2017) Implant material selection using expert system. Facta Univ Ser Mech Eng 15:133. https://doi.org/10.22190/FUME160723004R

39. Babbar A, Singh P, Farwaha HS (2017) Regression model and optimization of magnetic abrasive finishing of flat brass plate. Indian J Sci Technol 10:1–7. https://doi.org/10.17485/ijst/2017/v10i31/113860
40. Prakash C, Singh S, Pabla BS, Uddin MS (2018) Synthesis, characterization, corrosion and bioactivity investigation of nano-HA coating deposited on biodegradable Mg-Zn-Mn alloy. Surf Coatings Technol 346:9–18. https://doi.org/10.1016/j.surfcoat.2018.04.035
41. Tai BL, Stephenson DA, Shih AJ (2012) An inverse heat transfer method for determining workpiece temperature in minimum quantity lubrication deep hole drilling. J Manuf Sci Eng 134:021006. https://doi.org/10.1115/1.4005794
42. Tai BL, Palmisano AC, Belmont B et al (2015) Numerical evaluation of sequential bone drilling strategies based on thermal damage. Med Eng Phys 37:855–861. https://doi.org/10.1016/j.medengphy.2015.06.002
43. Orlande HRB (2011) Inverse heat transfer problems. Heat Transf. Eng. 32:715–717
44. Kumar R, Panda SS (2013) Drilling of bone : A comprehensive review. J Clin Orthop Trauma 4:15–30. https://doi.org/10.1016/j.jcot.2013.01.002
45. Wang G, Zhang L, Wang X, Tai BL (2016) An inverse method to reconstruct the heat flux produced by bone grinding tools. Int J Therm Sci 101:85–92. https://doi.org/10.1016/j.ijthermalsci.2015.10.021
46. Mokhtar MM, Fawad H (2012) Bone drilling modelling and simulation techniques. In: ISBEIA 2012 - IEEE Symposium on Business, Engineering and Industrial Applications. pp 357–361
47. Li X, Zhu W, Wang J, Deng Y (2016) Optimization of bone drilling process based on finite element analysis. Appl Therm Eng 108:211–220. https://doi.org/10.1016/j.applthermaleng.2016.07.125
48. Wang Y, Cao M, Zhao X et al (2014) Medical engineering & physics experimental investigations and finite element simulation of cutting heat in vibrational and conventional drilling of cortical bone. Med Eng Phys 36:1408–1415. https://doi.org/10.1016/j.medengphy.2014.04.007
49. Wang W, Shi Y, Yang N, Yuan X (2014) Medical engineering & physics experimental analysis of drilling process in cortical bone. Med Eng Phys 36:261–266. https://doi.org/10.1016/j.medengphy.2013.08.006
50. Alam K, Mitrofanov AV, Silberschmidt VV (2011) Medical engineering & physics experimental investigations of forces and torque in conventional and ultrasonically-assisted drilling of cortical bone. Med Eng Phys 33:234–239. https://doi.org/10.1016/j.medengphy.2010.10.003
51. Lee J, Gozen BA, Ozdoganlar OB (2012) Modeling and experimentation of bone drilling forces. J Biomech 45:1076–1083. https://doi.org/10.1016/j.jbiomech.2011.12.012
52. Pandey RK, Panda SS (2015) Evaluation of delamination in drilling of bone. Med Eng Phys 37:657–664. https://doi.org/10.1016/j.medengphy.2015.04.008
53. Alam K, Hassan E, Bahadur I (2015) Experimental measurements of temperatures in ultrasonically assisted drilling of cortical bone. Biotechnol Biotechnol Equip 29:753–757. https://doi.org/10.1080/13102818.2015.1034176
54. Xu L, Wang C, Jiang M et al (2014) Drilling force and temperature of bone under dry and physiological drilling conditions. Chin J Mech Eng 27:1240–1248. https://doi.org/10.3901/CJME.2014.0912.151
55. Singh G, Jain V, Gupta D (2017) Multi-objective performance investigation of orthopaedic bone drilling using Taguchi membership function. 1–7. https://doi.org/10.1177/0954411917735129
56. Singh G, Jain V, Gupta D, Ghai A (2016) Optimization of process parameters for drilled hole quality characteristics during cortical bone drilling using Taguchi method. J Mech Behav Biomed Mater 62:355–365. https://doi.org/10.1016/j.jmbbm.2016.05.015
57. Singh G, Jain V, Gupta D (2015) Comparative study for surface topography of bone drilling using conventional drilling and loose abrasive machining. 229:225–231. https://doi.org/10.1177/0954411915576945
58. Dahotre NB, Joshi SS (2016) Machining of bone and hard tissues
59. Yang M, Li C, Zhang Y et al (2017) Research on microscale skull grinding temperature field under different cooling conditions. Appl Therm Eng 126:525–537. https://doi.org/10.1016/j.applthermaleng.2017.07.183

60. Yang M, Li C, Zhang Y et al (2017) Experimental research on microscale grinding temperature under different nanoparticle jet minimum quantity cooling. Mater Manuf Process 32:589–597. https://doi.org/10.1080/10426914.2016.1176198
61. Le HM, Do TN, Phee SJ (2016) A survey on actuators-driven surgical robots. Sens Actuators, A Phys 247:323–354
62. Bergeles C, Yang GZ (2014) From passive tool holders to microsurgeons: Safer, smaller, smarter surgical robots. IEEE Trans Biomed Eng 61:1565–1576. https://doi.org/10.1109/TBME.2013.2293815
63. Aston G (2012) Surgical robots: Worth the investment? Hosp Heal Networks 86:38–42
64. Kirkpatrick K (2014) Surgical robots deliver care more precisely. Commun ACM 57:14–16. https://doi.org/10.1145/2632042
65. Payne CJ, Yang G-Z (2014) Hand-held medical robots. Ann Biomed Eng 42:1594–1605. https://doi.org/10.1007/s10439-014-1042-4
66. Zhao J, Feng B, Zheng MH, Xu K (2015) Surgical robots for SPL and NOTES: A review. Minim Invasive Ther Allied Technol 24:8–17. https://doi.org/10.3109/13645706.2014.999687
67. Kapoor A, Li M, Taylor RH (2006) Constrained control for surgical assistant robots. In: Proceedings - IEEE International Conference on Robotics and Automation. pp 231–236
68. Howlader N, Noone AM, Krapcho M, Garshell J, Miller D, Altekruse SF, Kosary CL, Yu M, Ruhl J, Tatalovich Z, Mariotto A, Lewis DR, Chen HS, Feuer EJ, Cronin KA (eds). SEER Cancer Statistics Review, 1975-2011, National Cancer Institute

Micro-machining Performance Assessment of Ti-Based Biomedical Alloy: A Finite Element Case Study

Swastik Pradhan, Kalipada Maity, Sunpreet Singh and Chander Prakash

Abstract Titanium alloy is one of the most abundantly used materials in the field of biomedical application due to its good physical and mechanical characteristics. But the machining of titanium alloy, to achieve the precise shape and dimension as per the tolerance, is extremely difficult, particularly due to the low thermal conductivity and high chemical reactivity properties. Since these materials product and components are to be used inside the human body, hence the machining of titanium alloy is of utmost importance. Therefore, in the present investigation, FEA modeling has been carried out at dry conditions in order to investigate the performance of microgroove-textured cutting tools in three-dimensional machining of titanium alloy grade 5 (Ti-6Al-4 V) using SNMG120408 carbide insert. Various types of microgroove were designed in the rake face of cutting tool using SolidWorks 2012 and Stereolithography file format. The lower cutting forces, temperature, effective stress, and strain are obtained during the FEM simulation of microgroove-textured cutting tool compared to a non-textured cutting tool. Furthermore, variation in the geometry of microgroove on the machinability criteria has been analyzed.

Keywords Microgroove-textured cutting tool · DEFORM-3D · Effective stress · Titanium alloy · Cutting force

1 Introduction

Titanium alloy is the most compatible materials used for replacing any tissue, organ or functional part of the body. The advancement seen on the hard tissue reconstruction and rejoining was also possible due to the involvement of the titanium alloy.

S. Pradhan (✉) · S. Singh · C. Prakash
School of Mechanical Engineering, Lovely Profesional University,
144411 Phagwara, Punjab, India
e-mail: swastik.rock002@gmail.com

K. Maity
Mechanical Engineering Department, National Institute of Technology,
Rourkela 769008, Odisha, India

C. Prakash et al. (eds.), *Biomanufacturing*,
https://doi.org/10.1007/978-3-030-13951-3_8

The elastic modulus of the titanium alloys also had a good agreement with the bone elastic modulus (30 GPa) as compared to the other alloys [1]. The most peculiar characteristics of the material with respect to the mechanical properties are high strength-to-weight ratio, high corrosion resistance, and the ability to withstand moderately high temperature without creeping. Apart from these, the diversity of properties in terms of nontoxic and adaptable nature by the body increases its application toward the medical field including the surgical implements and implants. The hip balls and sockets, use for joint are made up titanium alloy that stay inside the body for up to 20 years, dental implants that remain for over 30 years, total hip replacement as well as knee and shoulders replacements surgeries, trauma and spinal fixation devices, cardiovascular stents, spinal discs and pins for setting bones [2]. Some of the medical–surgical instruments use in image-guided surgery and some of the product uses for the medical field such as wheelchairs and crutches are also manufactured from this high strength to low weight materials [3]. However, a low level of thermal conductivity and high chemical reactivity, the tool wear rate, adhesion, and poor surface finish of the product are often witnessed by many [2, 4–6].

From the literature review, it is evident that titanium alloy is one of the most potential materials in the field of biomedical application, so the emphasis on the bio-manufacturing in order to fabricate complex shapes of the body parts with accurate dimensions. In terms of micro-textured machining, not such adequate work has been carried out in the field of modeling of textured cutting tools using microgroove patterns. Further, it has been observed from the literature reviewed that the cutting forces and temperature get reduced using microgroove-textured cutting tool with the improvement of the machinability criteria. Therefore, some more detailed investigations are required to understand this process and benefits of microgroove patterns on the cutting tool in a more effective way some of the reviews, refer Table 1.

From Table 1, it has been found that the researchers used a wide range of advanced materials, processing techniques, and these combinations. However, very little attempt has been made to simulate the micro-grooving of titanium alloy with tungsten carbide tool by using FEM. Therefore, in this work, an effort has been made to investigate the optimized processing condition, under dry machining condition, using DEFORM-3D commercial software. The simulations have been carried out with the available different microgrooves patterns and cutting speeds at constant feed and depth of cut.

2 Modeling of Textured Geometry on Rake Face of Cutting Tool

The plain cutting tool without any texture was modeled using SolidWorks 2012. The specification of the cutting tool is SNMG120408. Full description of the shape of plain cutting tool used for modeling is given in Table 2.

Table 1 List of the literature review on titanium alloy used for biomedical application

S. No.	Title	Author	Work material	Technique	Remarks
1	Coupling machining and heat treatment to enhance the wear behavior of an Additive Manufactured Ti-6Al-4 V titanium alloy	Stefania Bruschi, et al. [7]	Ti-6Al-4 V	Coupling machining with a subsequent heat treatment EBM	It can be used for the efficient strategy for the improvement of the wear resistance during EBM of titanium alloy
2	Finite element analysis of porous commercially pure titanium for biomedical implant application	Nicolas Soroe et al. [8]	Pure titanium	Finite element analysis	It found that shape of the pore and porosity level is the most significant parameter. Young's modulus predicted from the simulation is accurate with 0.9% errors
3	Review: the potential impact of surface crystalline states of titanium for biomedical applications	Julien Barthes et al. [9]	Titanium alloy	Surface treatment	The surface modification techniques are to ensure the long-term functionality and safety in implantology
4	Antibacterial activity and biofilm inhibition by surface modified titanium alloy medical implants following application of silver, titanium dioxide, and hydroxyapatite nano-coatings	A Besinis et al. [10]	Titanium alloy	Nano-coating	The silver/nHA nano-coating on the implant made of titanium alloy is highly antibacterial for the oral pathogen. The dual-layered composite used is more effective toward the bacterial growth

(continued)

Table 1 (continued)

S. No.	Title	Author	Work material	Technique	Remarks
5	Effect of an amorphous titanium nanotubes coating on the fatigue and corrosion behaviors of the biomedical Ti-6Al-4 V an d Ti-6Al-7Nb alloys	Campanelli et al. [11]	Ti-6Al-4 V and Ti-6Al-7Nb alloys	Surface modification	The electrochemical response of the nano-tubular oxide surface coating reduces the cracks, fatigue, and corrosion resistance
6	Surface modification of bulk titanium substrates for biomedical applications via low-temperature microwave hydrothermal oxidation	Cheng et al. [12]	Titanium	Microwave hydrothermal (MWHT) oxidation process for modification of Ti surfaces	The MWHT process enhanced wetting of the implants without degrading the microscale surface structure of Ti by providing scalable, low-temperature route tailoring nanoscale topographies
7	Behavior of a biocompatible titanium alloy during orthogonal micro-cutting employing green machining techniques	Carlos et al. [1]	Titanium alloy Ti-6Al-7Nb	Green machining techniques	Compared the MQL and HSM machining with dry condition low-speed machining. It is observed that lower surface roughness is obtained through dry machining, whereas lower burr formation was achieved through cooled and lubricated condition

(continued)

Table 1 (continued)

S. No.	Title	Author	Work material	Technique	Remarks
8	Improved fatigue properties with maintaining low Young's modulus achieved in biomedical beta-type titanium alloy by oxygen addition	Liu et al. [13]	Beta titanium alloy Ti-29Nb-13Ta-4.6Zr	Surface treatment	The fatigue limits of the alloy increases monotonically with oxygen increment. At 0.7 concentration of oxygen, the fatigue crack initiation and propagation increases, i.e., 635 MPa with 1100 MPa tensile strength and 76 GPa of Young's modulus
9	Study the formation of porous surface layer for a new biomedical titanium alloy	Mohammed et al. [14]	Ti-Zr-Nb (TZN) alloy	Oxidation treatment	Studied the porous formation in the surface of the biomedical titanium alloy. A stable ultrafine porous film is found after the oxidation treatment
10	Experimental investigations in powder-mixed electric discharge machining of Ti-35Nb-7Ta-5Zr β-titanium alloy	Prakash et al. [15]	β-Ti alloys	Powder-mixed electrical discharge machining	Through PMEDM process the generation carbides and oxides enriched the surface and enhanced the biocompatibility of Titanium alloy

(continued)

Table 1 (continued)

S. No.	Title	Author	Work material	Technique	Remarks
11	Innovative surface treatments of titanium alloys for biomedical applications	Caroline [16]	Titanium alloy	Surface treatment	Studied various types of surface modification process to enhance the life of the implants from wear, corrosion, tribo-corrosion and reduce the toxic effects
12	Polycrystalline diamond coating of additively manufactured titanium for biomedical applications	Rifai et al. [17]	Titanium alloy	Additive manufacturing with selective laser melted titanium (SLM-Ti)	The polycrystalline diamond coatings on the titanium through SLM-Ti enhanced the biological characteristics such as mammalian cell growth, apatite deposition
13	Additive manufacturing of titanium alloys in the biomedical field: processes, properties, and applications	Trevisan et al. [18]	Titanium alloy	Additive manufacturing	Study revealed that the additive manufacturing such as selective laser melting, electron beam melting, and laser metal deposition enhanced the porous complex structure and mechanical properties of titanium alloy used for implants by lowering the processing costs and waste

(continued)

Table 1 (continued)

S. No.	Title	Author	Work material	Technique	Remarks
14	Nano-structured commercially pure titanium for development of miniaturized biomedical implants.	Valiev et al. [19]	Grade 4 titanium alloy	Bioactive coating	It is observed that the chemical etching and deposition of bioactive coating on the surface of nano-structured CP-Ti enriched the surface bioactivity and lower the release of Ti ion
15	Micro-texture at the coated tool face for high-performance cutting	Obikawa et al. [20]	Al6061	Micro-textured cutting tool	As a result, it was observed that parallel type and square-dot type of micro-texture effectively improved the machinability of the aluminum alloy
16	Improving anti-adhesive properties of cutting tool surfaces by nano-/micro-textures	Enomoto et al. [21]	Aluminum	Micro-texture	It was seen that face-milling experiment with micro-texture cutting tool improved the lubricity and anti-adhesive properties in the tool–chips zone during machining of aluminum alloy

(continued)

Table 1 (continued)

S. No.	Title	Author	Work material	Technique	Remarks
17	A study of micro-pool lubricated cutting tool in machining of mild steel	Lei et al. [22]	Titanium	Micro-holes	It was found that formation of micro-holes on the surface of cutting tool with minimum lubricant on the micro-holes avoided the uses of large amount of cutting fluid used during machining of titanium alloy which was environmental friendly
18	Performance of cemented carbide tools with microgrooves in Ti-6Al-4 V titanium alloy cutting	Yang et al. [23]	Titanium alloy	Microgroove cutting tool, cryogenic MQL machining	It was observed that microgroove cutting tool combined with cryogenic minimum quantity lubrication machining reduced the average coefficient of friction on the rake surface of the cutting tool

Table 2 Geometrical specification of cutting tool used for modeling and simulation

Insert number	Description	Value
S	Insert shape	Square
N	Insert clearance angle	0°
M	Tolerance class	0.06
G	Insert features	Cylindrical hole with both side chip breaker
12	Size	12.70 mm
04	Thickness	4.76 mm
08	Corner radius	0.8 mm

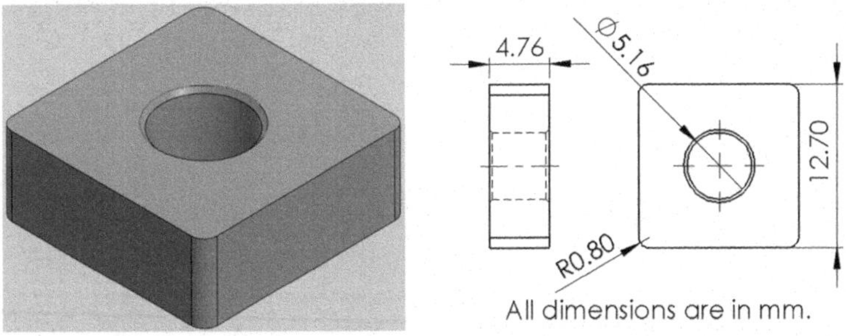

(a) 3D model of SNMG120408 cutting inserts (b) Drafting view of SNMG120408 cutting insert

Fig. 1 Three-dimensional and drafting view of SNMG120408 non-textured cutting insert

The plain or non-textured cutting tool is included for comparison with the other textured tool. The 3D model of the cutting tool geometry files was exported to Stereolithography (STL) file format. This STL file was then imported into DEFORM-3D software. Various types of patterns have been made in the surface of the cutting tool. The 3D model and drafting view of the plain cutting tool are shown in Fig. 1.

The 3D models with various types of patterns designed on the rake surface of the cutting tool used for the simulation are shown in Fig. 2. As shown in Fig. 2, the different types of patterns engraved on the cutting tool are horizontal [24], vertical, grid, inclined at 45° to the *X*-axis [25], inclined at 135° to the *X*-axis of [26] cutting tool.

A new C-shape micro-texture cutting tool was also developed and included for the present investigations. The horizontal pattern had grooves parallel to cutting edge, and the vertical patterns had grooves perpendicular to cutting edge; grid patterns had grooves both parallel and perpendicular to cutting edge. The textured cutting tools of 45° and 135° had grooves 45° and 135° angles inclined to the cutting edge. The C-shape patterns had grooves with respect to the primary cutting edge nose radius and secondary cutting edge of the cutting tool, respectively. The dimensions of the

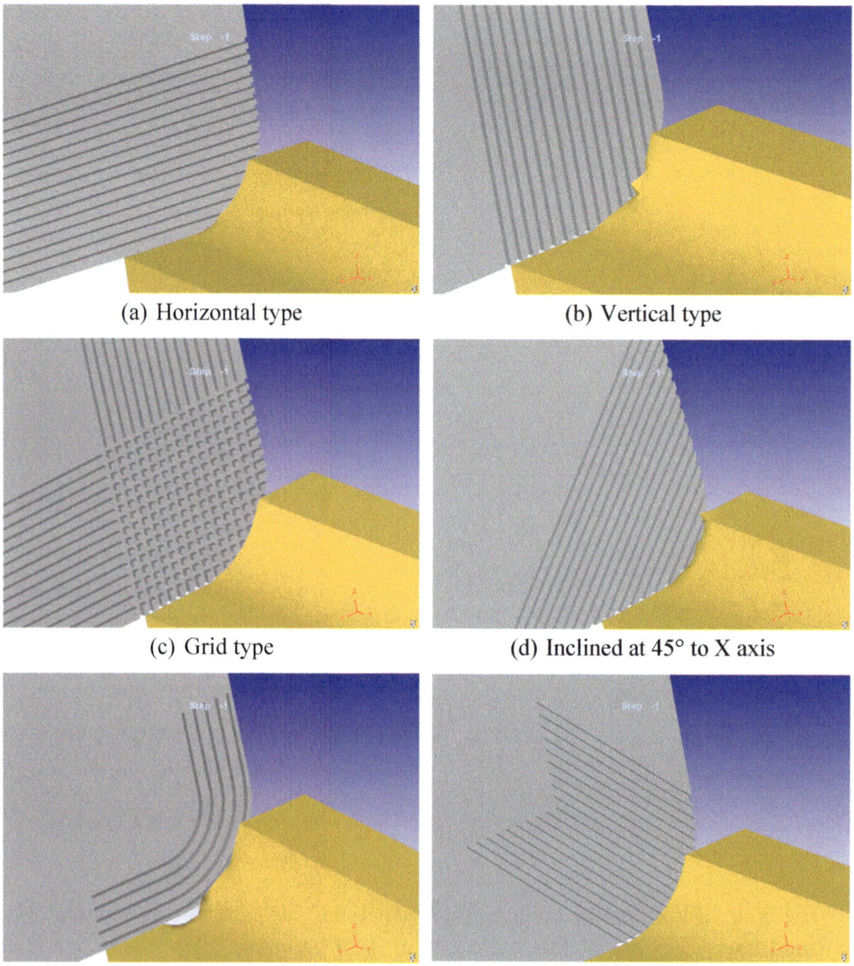

(a) Horizontal type (b) Vertical type

(c) Grid type (d) Inclined at 45° to X axis

(e) C-shaped type (f) Inclined at 135° to X axis

Fig. 2 Various types of microgroove patterns

microgroove width, the distance between the microgroove, and the edge distance of the microgroove for all the modeled patterns cutting tool are shown in Fig. 3.

The boundary conditions for each of the simulations were fixed. The cutting tool was considered as rigid, and the workpiece as plastic. The materials used for cutting tool and workpiece were tungsten carbide and titanium alloy, respectively. Material properties for both workpiece and cutting tool are taken from the material library of DEFORM-3D. The base portion of the workpiece was fixed in *X*-, *Y*-, and *Z*-directions. The horizontal motion of the cutting tool was governed by specifying cutting speed of the cutting tool, whereas movement of the workpiece was restricted. The boundary conditions of the cutting tool and the workpiece are shown in Fig. 4.

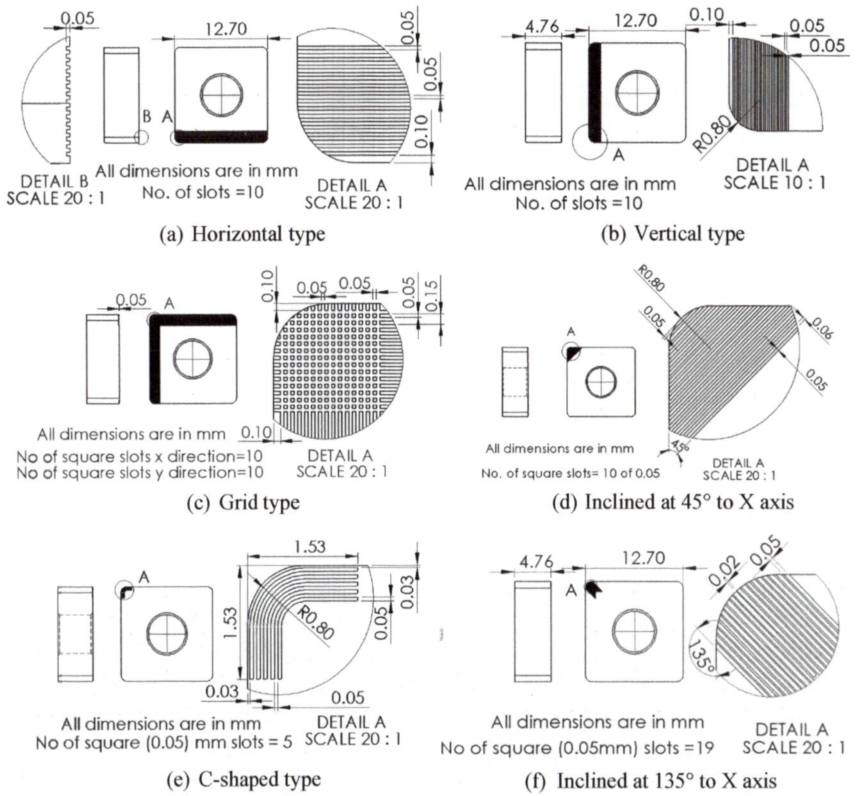

Fig. 3 Detailing and drafting of various types of microgroove patterns

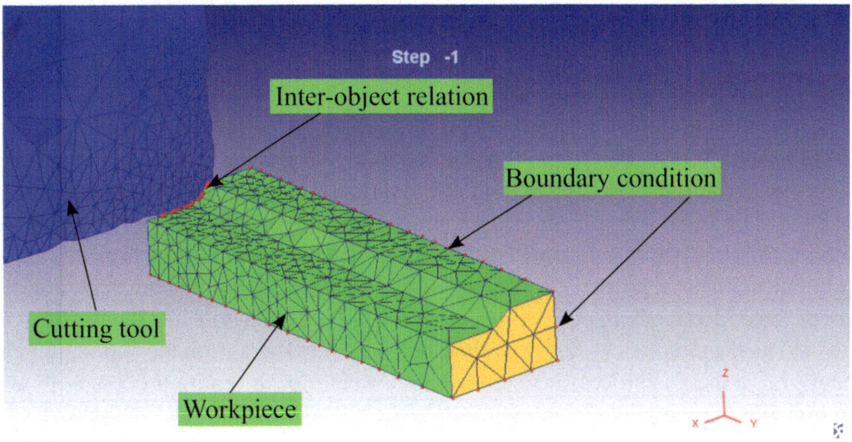

Fig. 4 Boundary conditions of cutting tool and workpiece

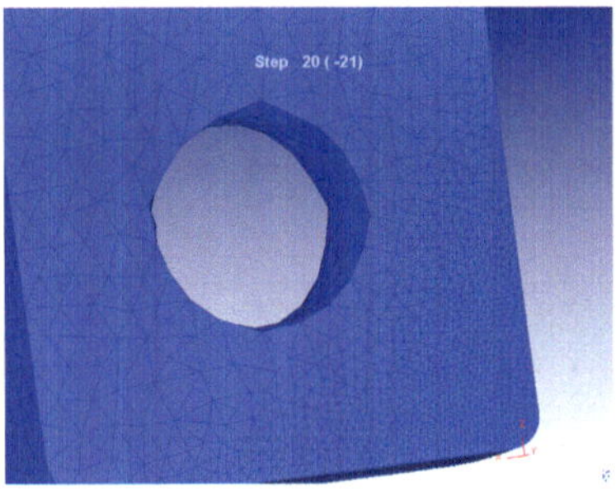

(a) High density meshing on non-textured cutting tool

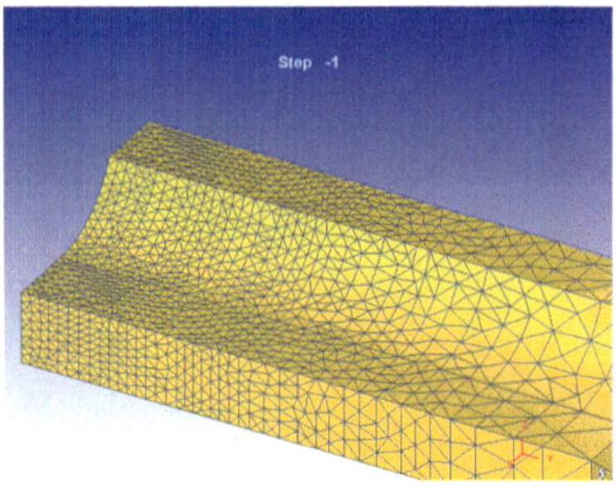

(b) High density meshing on workpiece

Fig. 5 High-density meshing of cutting tool and workpiece

The cutting variables are cutting speed of 65, 112, and 124 mm/min, feed rate of 0.1 mm/rev and depth of cut of 0.5 mm. The environmental cutting temperature was assumed to be 20 °C. The tool–workpiece interface conditions for machining were defined by shear friction factor and heat transfer coefficient. As per the literature, it was assumed and defined in the preprocessor. The meshing of the cutting tool and workpiece has been defined through the mesh window of DEFORM-3D. Tetrahedral mesh was used for meshing both the cutting tool and workpiece. A high-density mesh was defined in the textured part of the cutting tool using a mesh window tab of DEFORM-3D as shown in Fig. 5a for the non-textured cutting tool. Similarly, Fig. 5b shows a high-density meshing in the cutting zone of the workpiece. The element size of the mesh is limited to a minimum of 0.01 mm. In the same way, the high-density meshing for the textured cutting tools has been created as shown in Fig. 6.

2.1 3D Simulation

DEFORM-3D software was used to simulate the 3D machining process of cutting tool and workpiece. The DEFORM-3D software was categorized into three different parts, i.e., preprocessor, postprocessor, and simulator. These helped to formulate the model for simulation, importing of cutting tool and creation of workpiece, mesh properties, material properties, boundary conditions and cutting parameters (cutting speed, feed, depth of cut, and number of steps) of the cutting tool and the workpiece, and to run the simulation, respectively. The results obtained from the simulation are analyzed in the postprocessor section. The flowchart of the entire process is shown in Fig. 7.

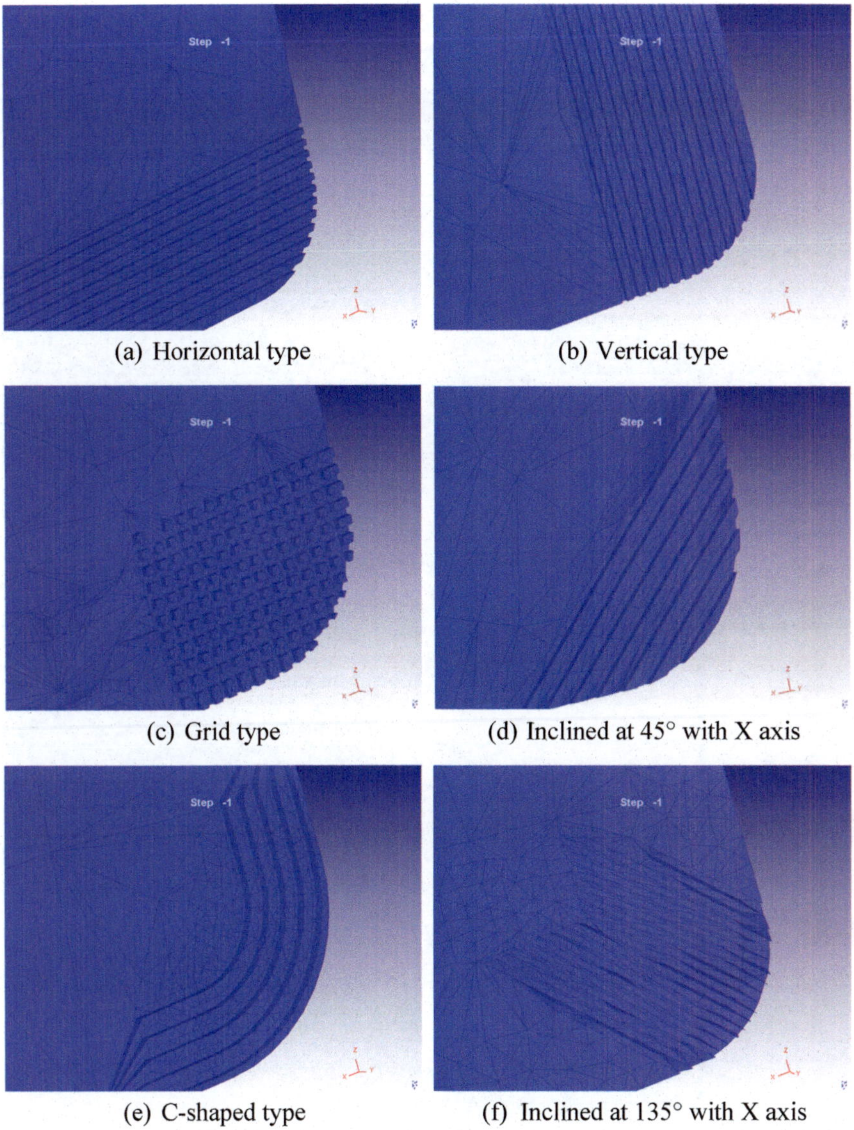

Fig. 6 High-density meshing on various types of microgroove patterns

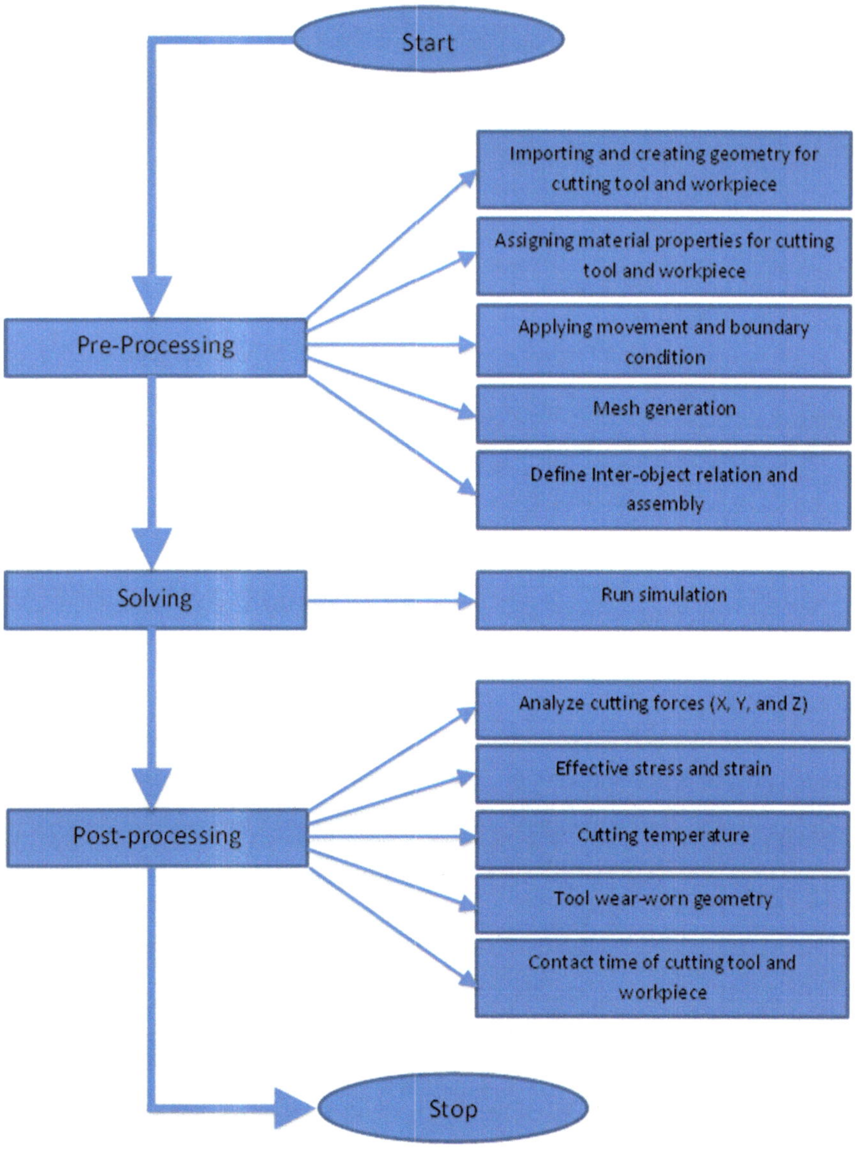

Fig. 7 Flowchart diagram of DEFORM-3D modeling and simulation

3 Results and Discussions

The effects of different types of microgroove geometry on the cutting performance were studied during 3D turning simulations of titanium alloy while using tungsten carbide as the cutting tool. The variations of cutting forces, cutting temperature, effective total strain, and effective stress have been studied with different cutting speeds and microgroove patterns. The tool–chip contact area during the microgroove FEM simulation was less due to the generation of microgroove on the rake surface of the cutting tool. It reduced the friction and thereby the cutting force, and cutting temperature. The variation in the microgroove created different types of contact area and, therefore, impacted the machinability index of the cutting tool [27].

3.1 Effects of Microgroove Geometry and Cutting Speed on Radial Force or Thrust Force

The variations of radial force with respect to time for different cutting speeds (65, 112, and 124 mm/min) are shown in Fig. 8a–c, respectively, with different type of textured cutting tools. The radial cutting force was observed to be least with the textured cutting tool with 45° inclined. The reduction in the radial force for textured cutting tool in comparison with the non-textured tool is 60%.

3.2 Effects of Microgroove Geometry and Cutting Speed on Tangential Force

The variations of the tangential cutting force for the different textured tool with respect to time are shown in Fig. 9a–c for cutting speed of 65, 112, and 124 mm/min, respectively. It has been observed that the tangential cutting force is least at low and high cutting speed using textured cutting tool, whereas the tangential cutting force is minimum for 45° inclined textured cutting tool in the medium cutting speed. The minimum difference of tangential cutting force for the textured and non-textured tool is 26%.

3.3 Effects of Microgroove Geometry and Cutting Speed on Feed Force or Axial Force

Figure 10a–c indicates the variations of feed force with respect to the time for different types of microgroove geometries at cutting speed of 65, 112, and 124 mm/min, respectively. It has been observed that feed force was minimum for the vertical

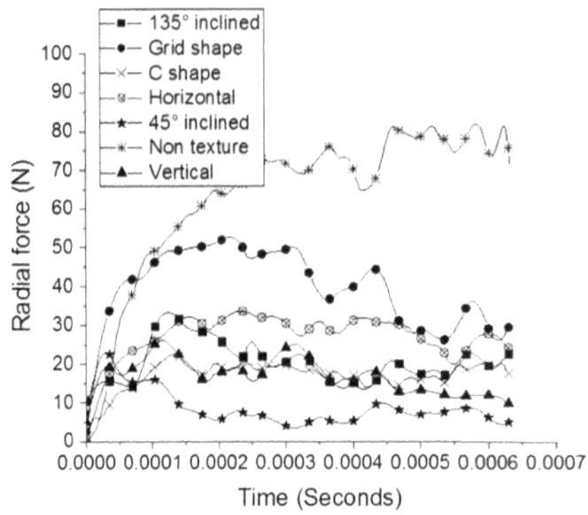

(a) Cutting speed 65 mm/min, feed 0.1 mm/rev, depth of cut 0.5 mm

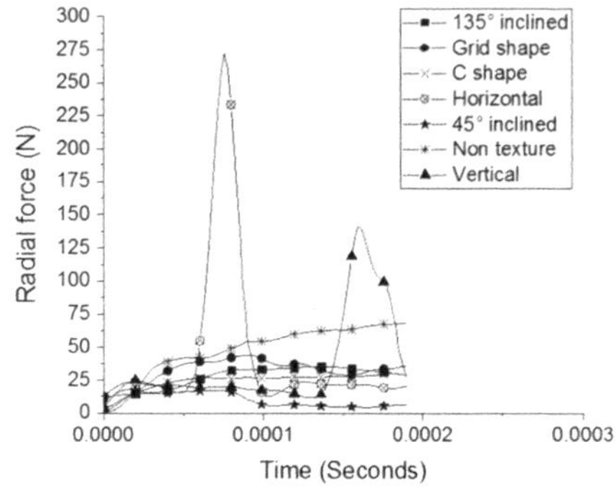

(b) Cutting speed 112mm/min, feed 0.1 mm/rev, depth of cut 0.5 mm

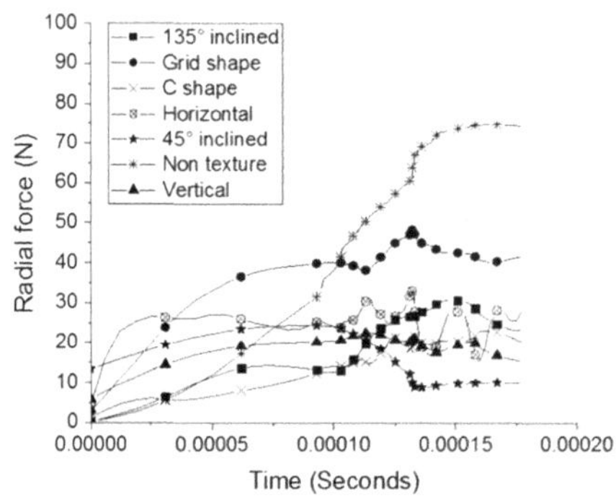

(c) Cutting speed 124 mm/min, feed 0.1 mm/rev, depth of cut 0.5 mm

Fig. 8 Effects of process parameters of different type of microgroove patterns on radial cutting force

microgroove pattern. The minimum difference between the textured and non-textured cutting tool is 70%.

3.4 Effects of Microgroove Geometry and Cutting Speed on Cutting Temperature

The variation in the cutting temperature with respect to time is shown in Fig. 11a–c for 65, 112, and 124 mm/min, respectively. It has been observed that the cutting temperature is highest for the non-textured cutting tool. And, the minimum cutting temperature was obtained for the vertical textured tool.

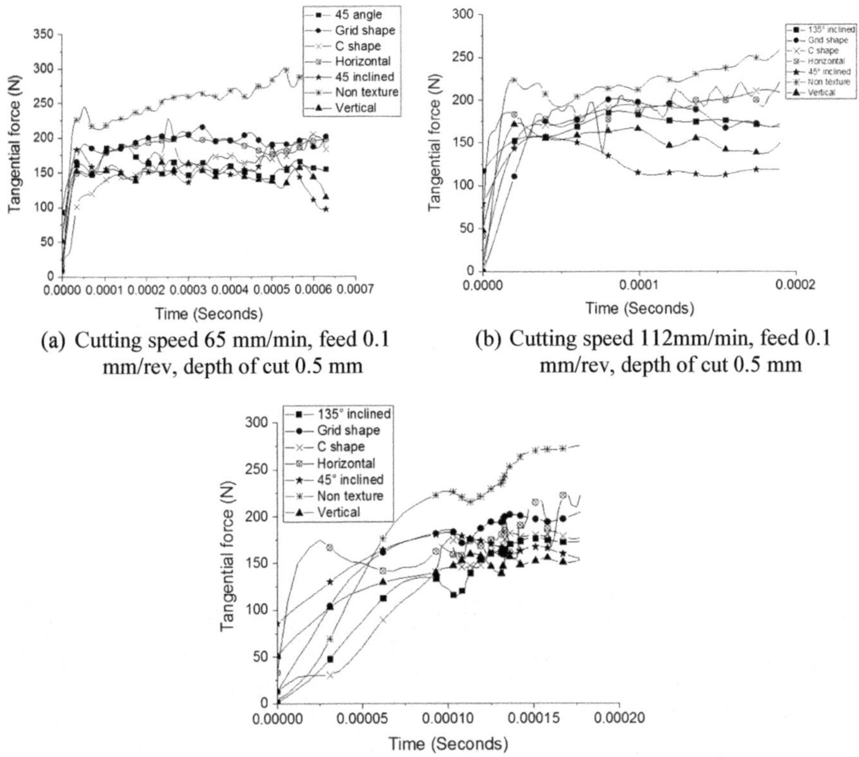

(a) Cutting speed 65 mm/min, feed 0.1 mm/rev, depth of cut 0.5 mm

(b) Cutting speed 112mm/min, feed 0.1 mm/rev, depth of cut 0.5 mm

(c) Cutting speed 124 mm/min, feed 0.1 mm/rev, depth of cut 0.5 mm

Fig. 9 Comparison of tangential cutting force with different types of microgroove patterns

3.5 Effects of Microgroove Geometry and Cutting Speed on Effective Stress

The stress is being resolved in terms of normal and shear stress, and the complete stress state can be resolved with three normal stress components and three shear stress components. The shear stress acting along a defined axis is zero, and the resultant normal stress acting on same axis is termed as principal stress. DEFORM-3D uses the von Mises or effective stress to analyze the characteristic of the stress. The effective stress is defined as follows

$$\overline{\sigma} = \sqrt{\frac{(\sigma_1 - \sigma_2)^2 + (\sigma_2 - \sigma_3)^2 + (\sigma_3 - \sigma_1)^2}{2}} \quad (1)$$

where σ_1,σ_2, and σ_3 are the three principal stresses, and $\overline{\sigma}$ is effective stress [28].

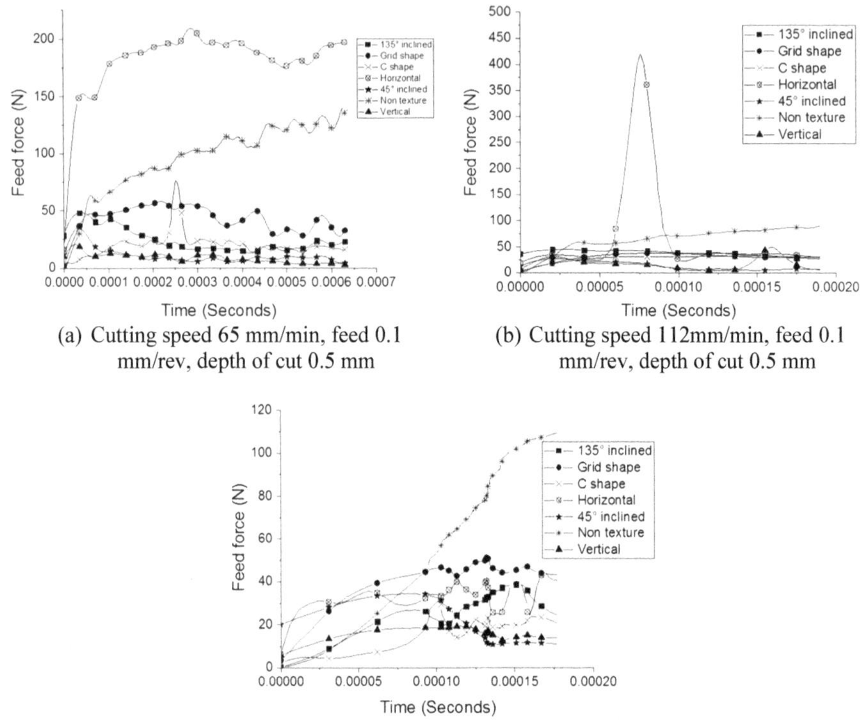

(a) Cutting speed 65 mm/min, feed 0.1 mm/rev, depth of cut 0.5 mm

(b) Cutting speed 112mm/min, feed 0.1 mm/rev, depth of cut 0.5 mm

(c) Cutting speed 124 mm/min, feed 0.1 mm/rev, depth of cut 0.5 mm

Fig. 10 Comparison of feed cutting force with different microgroove patterns in cutting tool

Figure 12 shows the effects of the microgroove cutting tool on effective stress with variation of time at various cutting speed. It is observed that the effective stress of 1900 MPa is nearly same for all textured cutting tool.

The undesirable effect of any machining process led to the creation of stress and strain. The formation of such stress and strain during machining process affects the surface integrity of the machined surface and deteriorates the life of the cutting tool also, leading to the formation of the creep and fatigue lives of the machined components. Hence, a study had been made regarding the effective stress and effective strain developed in the workpiece [29].

3.6 *Effects of Microgroove Geometry and Cutting Speed on Effective Strain*

The DEFORM-3D which determines strains in terms of true strain can be defined as the sum of a large series of arbitrarily small strain increments, i.e., integrating the

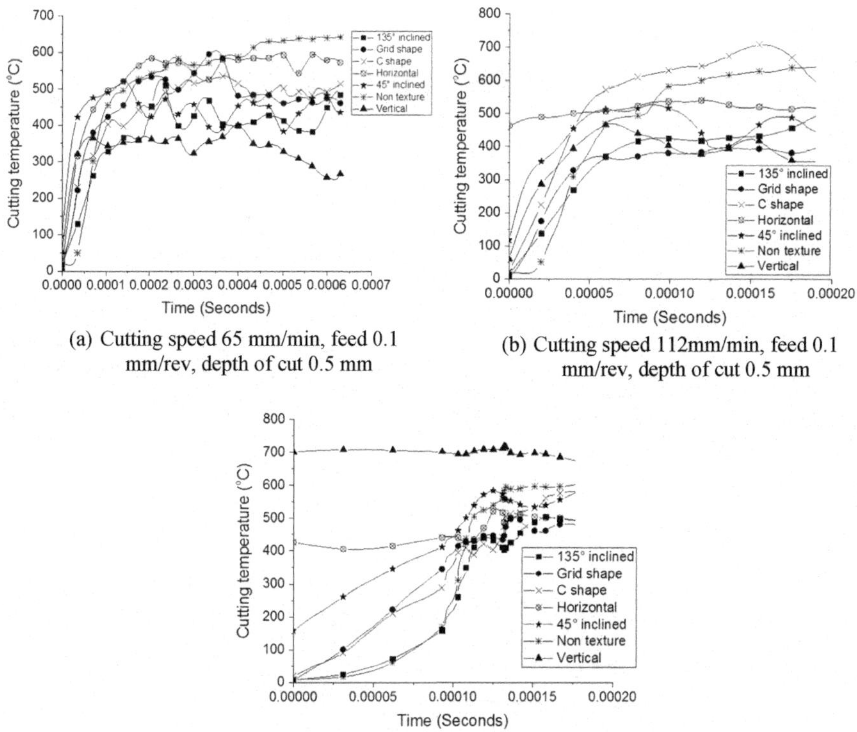

(a) Cutting speed 65 mm/min, feed 0.1 mm/rev, depth of cut 0.5 mm

(b) Cutting speed 112mm/min, feed 0.1 mm/rev, depth of cut 0.5 mm

(c) Cutting speed 124 mm/min, feed 0.1 mm/rev, depth of cut 0.5 mm

Fig. 11 Effects of process parameters of different type of microgroove patterns on cutting temperature

total change in the length that measures the value of true strain. DEFORM-3D uses a value common to metal forming analysis, as shown in Eq. 2:

$$\overline{\varepsilon} = \frac{\sqrt{2}}{3}\sqrt{(\varepsilon_1 - \varepsilon_2)^2 + (\varepsilon_2 - \varepsilon_3)^2 + (\varepsilon_3 - \varepsilon_1)^2} \tag{2}$$

where ε_1,ε_2, and ε_3 are principal strains, and $\overline{\varepsilon}$ is the effective strain.

Figure 13 shows the uneven variation of effective strain with respect to time at various cutting speed for different microgroove geometries. In case of effective strain, it has been observed that the 45° inclined shaped microgrooved patterns cutting tool has the lowest effective strain of 2–4 mm/mm. On the other hand, the non-textured microgrooves patterns cutting tool has the highest effective strain of 6 mm/mm.

The comparisons made for various types of microgroove cutting inserts, and it was found that that the 45° inclined microgrooved cutting tool has given better results than other types of micro-textured cutting inserts. It has been observed that by using 45° inclined microgrooved cutting tool, most of the factors such as cutting forces, temperature, and effective stress and strain reduce. It may be due to the reduction in

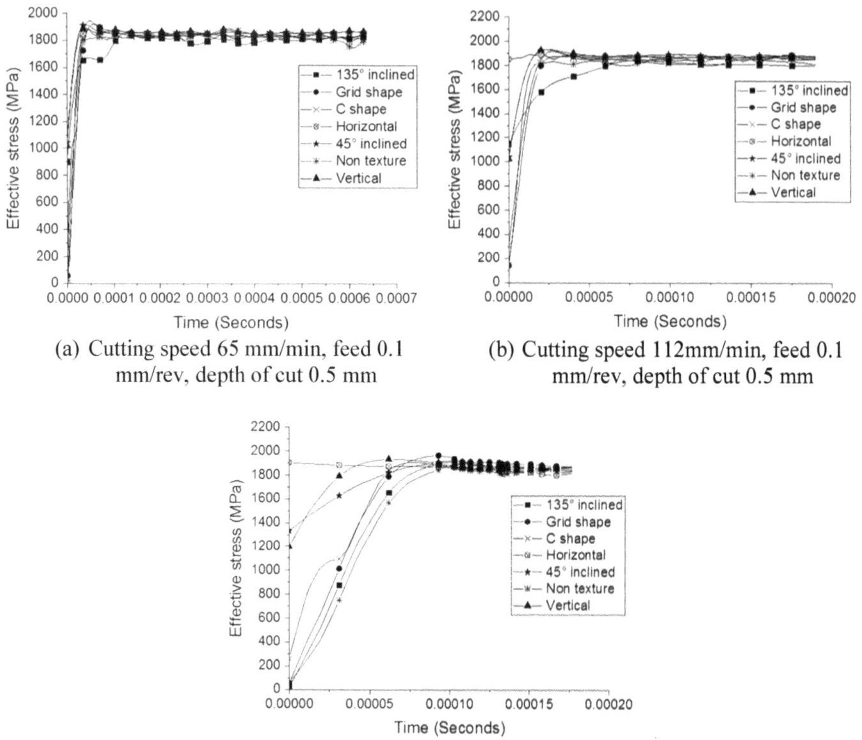

(a) Cutting speed 65 mm/min, feed 0.1 mm/rev, depth of cut 0.5 mm

(b) Cutting speed 112mm/min, feed 0.1 mm/rev, depth of cut 0.5 mm

(c) Cutting speed 124 mm/min, feed 0.1 mm/rev, depth of cut 0.5 mm

Fig. 12 Effects of process parameters of different type of microgroove patterns on effective stress

tool–chip contact area during microgroove FEM simulation. Thus, it reduces friction and thus reduced the cutting force and cutting temperature. So, further 45° inclined microgroove patterned cutting tool was further used in the analysis of minimum distance, contact time, and tool wear-worn geometry.

Figure 14 shows that when the cutting speed increases from 65 to 124 mm/min, the effective stress of the 45° inclined cutting tool increased from 1890 to 1950 MPa. Figure 15 shows that with an increase in cutting speed of 65 to 112 mm/min, the cutting temperature generated during machining operation increased from 553 to 612 °C. However, increasing the cutting speed beyond the high-speed machining range, i.e., 60–100 mm/min [30], the cutting temperature started decreasing at 124 mm/min. The minimum distance and contact time between the workpiece and cutting tool during machining was difficult to analyze during machining operation for microgroove cutting tool, but with the simulation of a machining operation in DEFORM-3D, it can be analyzed. Refer to Fig. 16, it can be observed that the minimum cutting depth of the cutting tool and workpiece nearly remains unchanged with an increase in cutting speed. Further, Fig. 17 shows that the contact time between the cutting tool and workpiece increased from

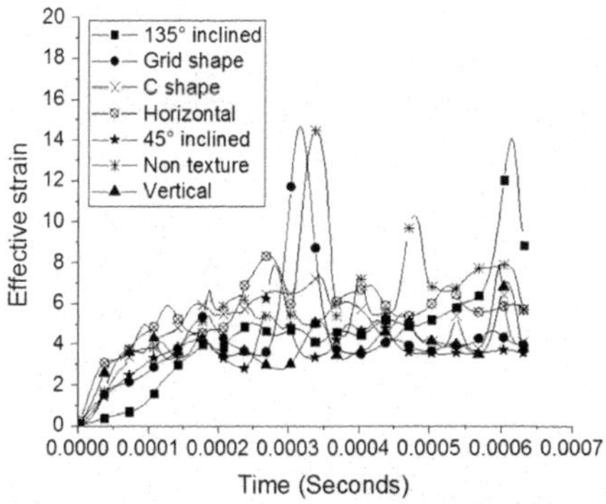

(a) Cutting speed 65 mm/min, feed 0.1 mm/rev, depth of cut 0.5 mm

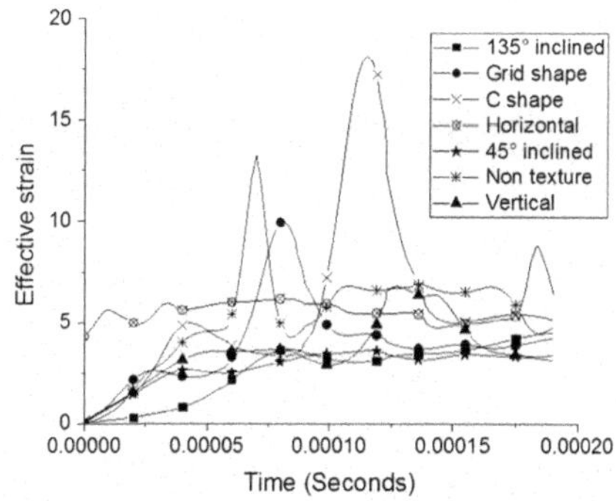

(b) Cutting speed 112mm/min, feed 0.1 mm/rev, depth of cut 0.5 mm

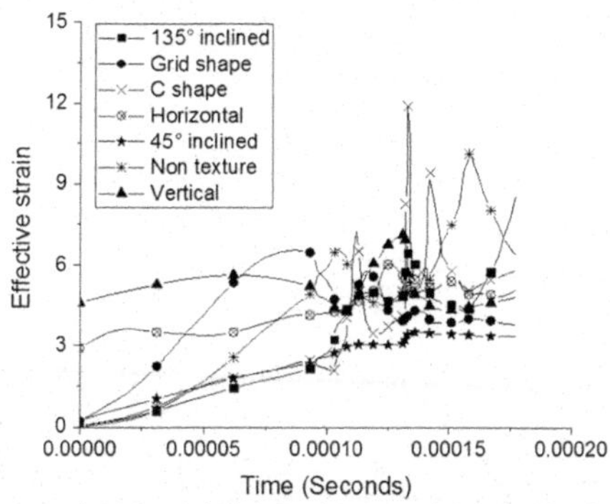

(c) Cutting speed 124 mm/min, feed 0.1 mm/rev, depth of cut 0.5 mm

Fig. 13 Effects of process parameters of different type of microgroove patterns on effective strain total

0.0564 to 0.493 s with an increase in the cutting speed. The rate of wear on the geometry of the cutting tool in terms of time was known as tool wear-worn geometry. The wear on the geometry of the cutting tool with respect to time was observed from the simulation. Moreover, with reference to Fig. 18, it has been found that with the increase in cutting speed, the tool wear decreased.

4 Conclusions

In this study, 3D FEM modeling and simulation of titanium alloy with non-textured and textured microgroove cutting tool were investigated to study the effects of cutting speed on the cutting forces, cutting temperature, and effective stress and strain. The

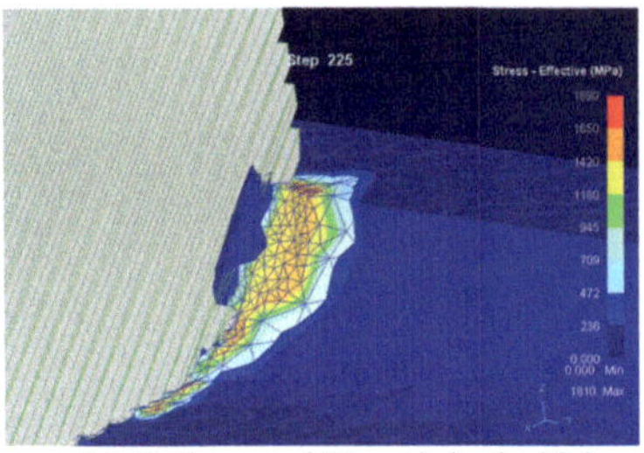

(a) Cutting speed 65 mm/min, feed 0.1 mm/rev, depth of cut 0.5 mm

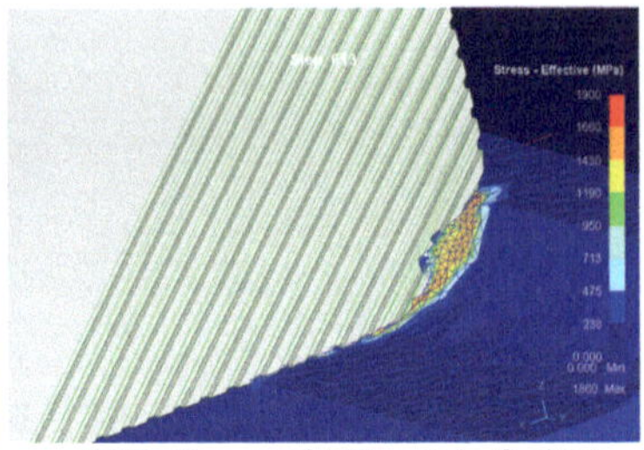

(b) Cutting speed 112mm/min, feed 0.1 mm/rev, depth of cut 0.5 mm

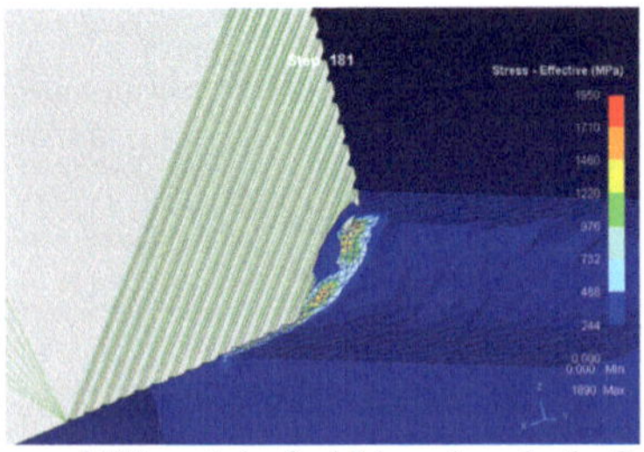

(c) Cutting speed 124 mm/min, feed 0.1 mm/rev, depth of cut 0.5 mm

Fig. 14 Effects of cutting speed on effective stress

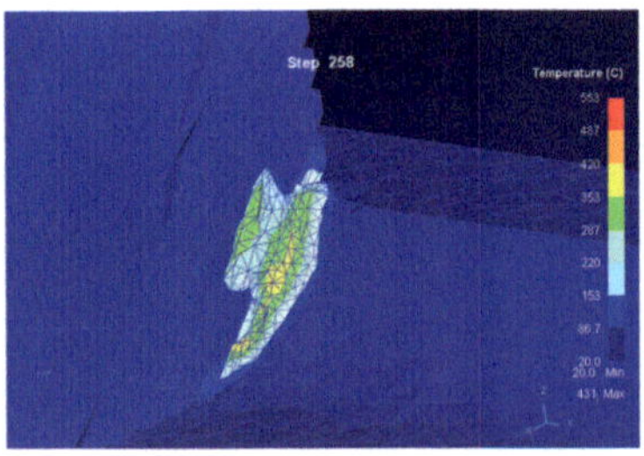

(a) Cutting speed 65 mm/min, feed 0.1 mm/rev, depth of cut 0.5 mm

(b) Cutting speed 112mm/min, feed 0.1 mm/rev, depth of cut 0.5 mm

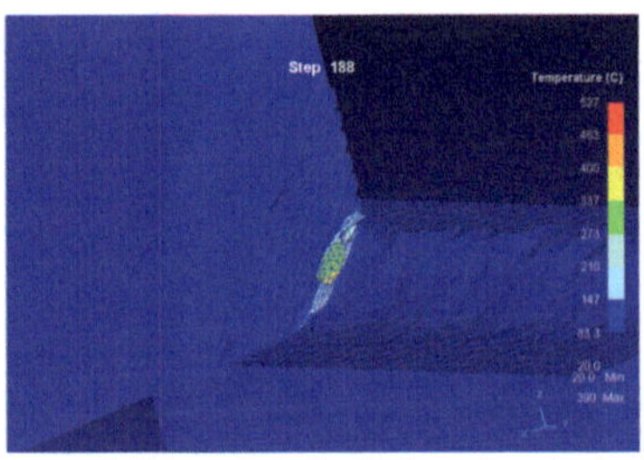

(c) Cutting speed 124 mm/min, feed 0.1 mm/rev, depth of cut 0.5 mm

Fig. 15 Effects of cutting speed on cutting temperature

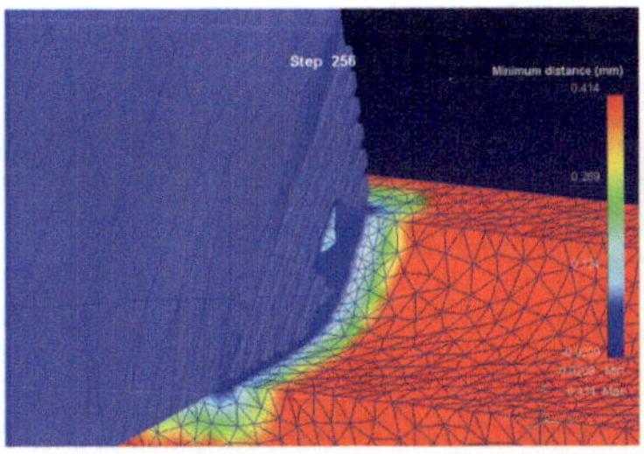

(a) Cutting speed 65 mm/min, feed 0.1 mm/rev, depth of cut 0.5 mm

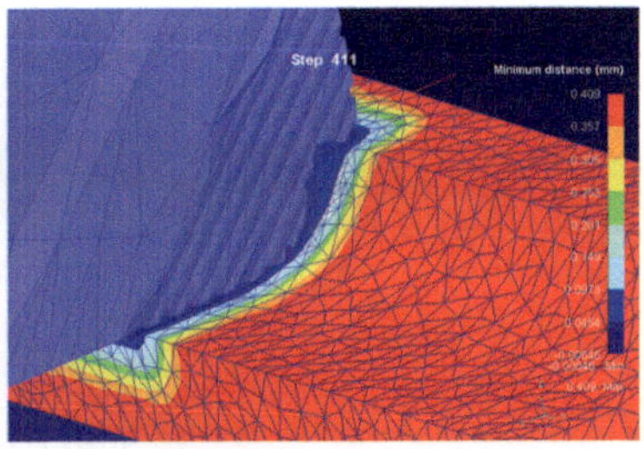

(b) Cutting speed 112mm/min, feed 0.1 mm/rev, depth of cut 0.5 mm

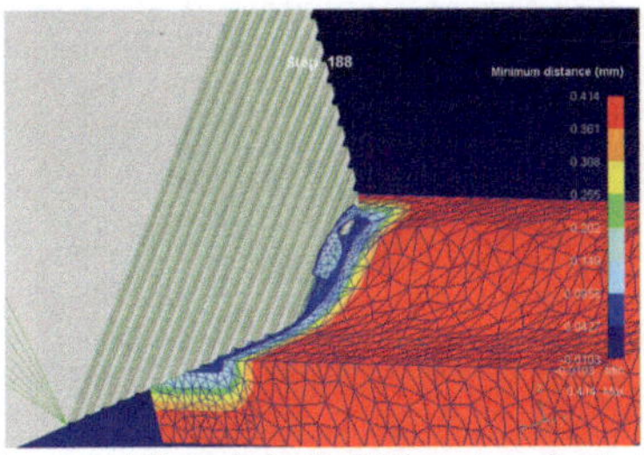

(c) Cutting speed 124 mm/min, feed 0.1 mm/rev, depth of cut 0.5 mm

Fig. 16 Effects of cutting speed on minimum distance

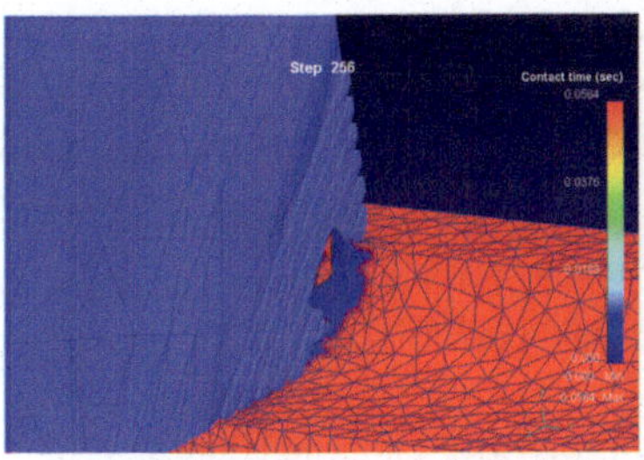

(a) Cutting speed 65 mm/min, feed 0.1 mm/rev, depth of cut 0.5 mm

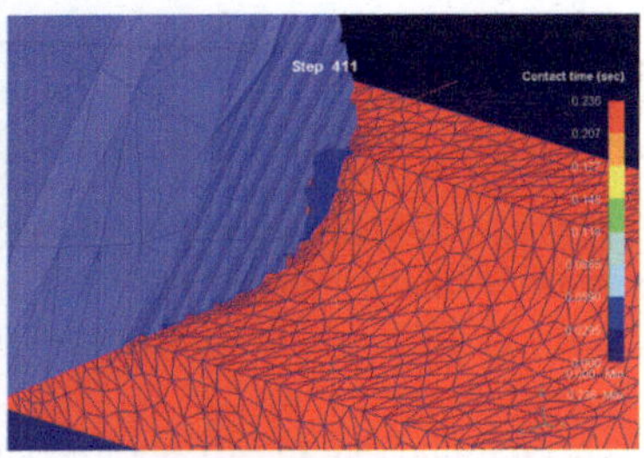

(b) Cutting speed 112mm/min, feed 0.1 mm/rev, depth of cut 0.5 mm

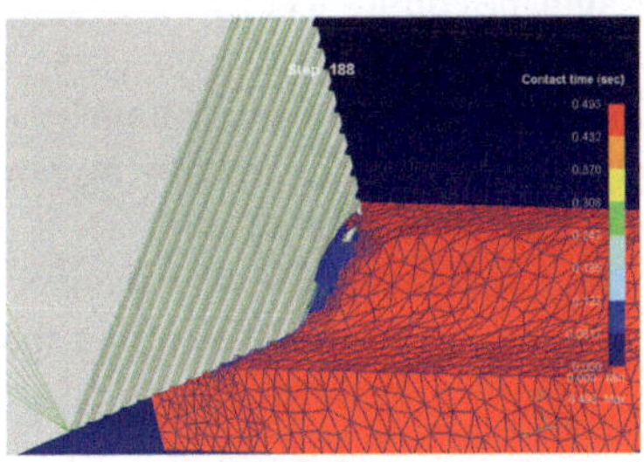

(c) Cutting speed 124 mm/min, feed 0.1 mm/rev, depth of cut 0.5 mm

Fig. 17 Effects of cutting speed on contact time

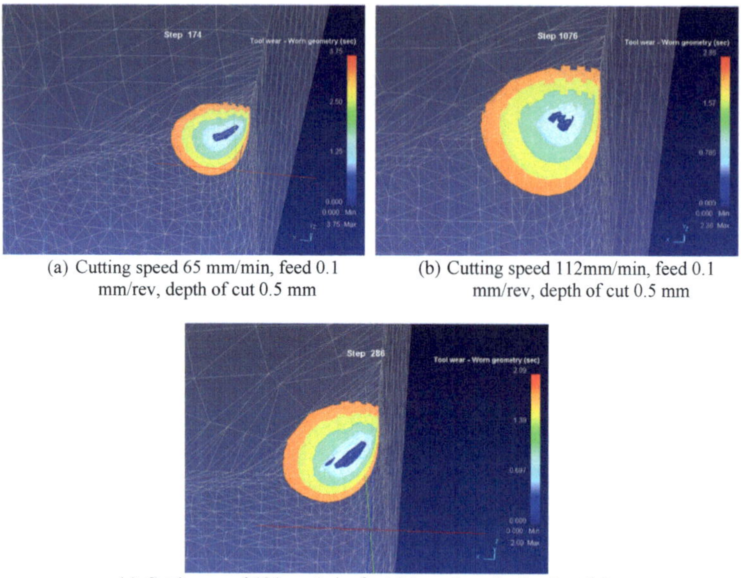

(a) Cutting speed 65 mm/min, feed 0.1 mm/rev, depth of cut 0.5 mm

(b) Cutting speed 112mm/min, feed 0.1 mm/rev, depth of cut 0.5 mm

(c) Cutting speed 124 mm/min, feed 0.1 mm/rev, depth of cut 0.5 mm

Fig. 18 Effects of cutting speed on tool wear-worn geometry

effect of different microgroove patterns of the cutting tool on machinability criteria has been investigated. The key findings of this research are summarized below:

- There was a decrease in cutting force, effective stress, temperature, and effective strain by using cutting tool with the microgroove pattern. The microgroove-textured cutting tool showed better cutting performance as compared to non-textured to cutting tool.
- It has been found that the 45° inclined microgroove patterned cutting tool resulted in an overall better cutting performance as compared to other microgrooved cutting tools. Further, by increasing the cutting speed from 65 to 124 mm/min, the effective stress of the 45° inclined microgroove patterned cutting tool increased.
- Moreover, it has been observed that the contact time between the cutting tool and workpiece during machining of 45° inclined microgroove patterned cutting tool increased with an increase in the cutting speed.
- Also, the tool wear-worn geometry of the 45° inclined microgroove patterned cutting tool decreases with the increase in the cutting speed.

References

1. Prakash C, Kansal HK, Pabla BS, Puri S, Aggarwal A (2016) Electric discharge machining–A potential choice for surface modification of metallic implants for orthopedic applications: a review. Proc Inst Mech Eng B J Eng Manuf 230(2):331–353
2. Prakash C, Kansal HK, Pabla BS, Puri S (2015) Processing and characterization of novel biomimetic nanoporous bioceramic surface on β-Ti implant by powder mixed electric discharge machining. J Mater Eng Perform 24(9):3622–3633
3. Aliyu AA, Abdul-Rani AM, Ginta TL, Prakash C, Axinte E, Razak MA, Ali S (2017) A review of additive mixed-electric discharge machining: current status and future perspectives for surface modification of biomedical implants. Adv Mater Sci Eng 1–23
4. Ezugwu EO, Wang ZM (1997) Titanium alloys and their machinability—a review. J Mater Process Technol 68(3):262–274
5. Arrazola PJ, Garay A, Iriarte LM, Armendia M, Marya S, Le Maitre F (2009) Machinability of titanium alloys (Ti6Al4 V and Ti555. 3). J Mater Process Technol 209(5):2223–2230
6. Cui C, Hu B, Zhao L, Liu S (2011) Titanium alloy production technology, market prospects and industry development. Mater Des 32(3):1684–1691
7. Bruschi S, Bertolini R, Ghiotti A (2017) Coupling machining and heat treatment to enhance the wear behaviour of an additive manufactured Ti6Al4 V titanium alloy. Tribol Int 116:58–68
8. Soro N, Brassart L, Chen Y, Veidt M, Attar H, Dargusch MS (2018) Finite element analysis of porous commercially pure titanium for biomedical implant application. Mater Sci Eng A 725:43–50
9. Barthes J, Ciftci S, Ponzio F, Knopf-Marques H, Pelyhe L, Gudima A, Kientzl I, Bognár E, Weszl M, Kzhyshkowska J (2018) The potential impact of surface crystalline states of titanium for biomedical applications. Crit Rev Biotechnol 38(3):423–437
10. Besinis A, Hadi SD, Le H, Tredwin C, Handy R (2017) Antibacterial activity and biofilm inhibition by surface modified titanium alloy medical implants following application of silver, titanium dioxide and hydroxyapatite nanocoatings. Nanotoxicology 11(3):327–338
11. Campanelli LC, Bortolan CC, da Silva PSCP, Bolfarini C, Oliveira NTC (2017) Effect of an amorphous titania nanotubes coating on the fatigue and corrosion behaviors of the biomedical Ti-6Al-4 V and Ti-6Al-7Nb alloys. J Mech Behav Biomed Mater 65:542–551
12. Cheng A, Goodwin WB, Deglee BM, Gittens RA, Vernon JP, Hyzy SL, Schwartz Z, Sandhage KH, Boyan BD (2018) Surface modification of bulk titanium substrates for biomedical applications via low-temperature microwave hydrothermal oxidation. J Biomed Mater Res Part A 106(3):782–796
13. Liu H, Niinomi M, Nakai M, Obara S, Fujii H (2017) Improved fatigue properties with maintaining low Young's modulus achieved in biomedical beta-type titanium alloy by oxygen addition. Mater Sci Eng, A 704:10–17
14. Mohammed MT, Diwan AA, Ali OI (2018) Study the formation of porous surface layer for a new biomedical titanium alloy. Mater Res Express 5(3):036526
15. Prakash C, Kansal H, Pabla B, Puri S (2017) Experimental investigations in powder mixed electric discharge machining of Ti–35Nb–7Ta–5Zrβ-titanium alloy. Mater Manuf Processes 32(3):274–285
16. Richard C (2017) Innovative surface treatments of titanium alloys for biomedical applications. In: Materials science forum. Trans Tech Publ, pp 1570–1575
17. Rifai A, Tran N, Lau DW, Elbourne A, Zhan H, Stacey AD, Mayes EL, Sarker A, Ivanova EP, Crawford RJ (2018) Polycrystalline diamond coating of additively manufactured titanium for biomedical applications. ACS Appl Mater Interfaces 10(10):8474–8484
18. Trevisan F, Calignano F, Aversa A, Marchese G, Lombardi M, Biamino S, Ugues D, Manfredi D (2018) Additive manufacturing of titanium alloys in the biomedical field: processes, properties and applications. J Appl Biomater Funct Mater 16(2):57–67
19. Valiev R, Sabirov I, Zemtsova E, Parfenov E, Dluhoš L, Lowe T (2018) Nanostructured commercially pure titanium for development of miniaturized biomedical implants. In: Titanium in medical and dental applications. Elsevier, pp 393–417

20. Obikawa T, Kamio A, Takaoka H, Osada A (2011) Micro-texture at the coated tool face for high performance cutting. Int J Mach Tools Manuf 51(12):966–972
21. Enomoto T, Sugihara T (2011) Improvement of anti-adhesive properties of cutting tool by nano/micro textures and its mechanism. Procedia Eng 19:100–105
22. Lei S, Devarajan S, Chang Z (2009) A study of micropool lubricated cutting tool in machining of mild steel. J Mater Process Technol 209(3):1612–1620
23. Yang Y, Su Y, Li L, He N, Zhao W (2015) Performance of cemented carbide tools with micro-grooves in Ti-6Al-4 V titanium alloy cutting. Int J Adv Manuf Technol 76(9–12):1731–1738
24. Ma J, Duong NH, Lei S (2015) 3D numerical investigation of the performance of micro-groove textured cutting tool in dry machining of Ti-6Al-4 V. Int J Adv Manuf Technol 79(5–8):1313–1323
25. Sharma V, Pandey PM (2016) Geometrical design optimization of hybrid textured self-lubricating cutting inserts for turning 4340 hardened steel. Int J Adv Manuf Technol 1–15
26. Xie J, Luo M-J, He J-L, Liu X-R, Tan T-W (2012) Micro-grinding of micro-groove array on tool rake surface for dry cutting of titanium alloy. Int J Precis Eng Manuf 13(10):1845–1852
27. Maity K, Pradhan S (2018) Investigation of FEM simulation of machining of titanium alloy using micro-groove cutting insert. Silicon 1–11
28. Fluhrer J (2006) Deform 3D User's manual version 6.0. Scientific Forming Technologies Corporation, Columbus, OH
29. Rahman M, San WONGY, Zareena AR (2003) Machinability of titanium alloys. JSME Int J Ser C 46(1):107–115
30. Rahman M, Wang ZG, Wong Y-S (2006) A review on high-speed machining of titanium alloys. JSME Int J Series C 49(1):11–20

Dr. Swastik Pradhan is acting as Assistant professor in the department of Mechanical Engineering, Lovely Professional University, Punjab, India. He has 5 years of research experience. His research area broadly includes metal cutting with special interest over high speed machining of materials. He also specializes on modeling and simulation with multi-objective optimization. He holds a Master Certificate in CAD/CAM. He has completed his Ph.D. Degree in Mechanical Engineering at National Institute of Technology, Rourkela India. He holds a Master's Degree in Mechanical engineering with specialization in Manufacturing Processes and systems from Kalinga Institute of Industrial Technology, Bhubaneswar, India in the year 2012. Prior to this, he received his Bachelor's Degree from Ghanshyam Hemalata Institute of Technology and Management, Puri, India in 2009.

Dr. Kalipada Maity is acting as professor in the Department of Mechanical Engineering, National Institute of Technology (NIT), Rourkela, India. He has 30 years of teaching and research experience. His areas of research include micromachining, micro-forming, machining, extrusion, hot-machining and study of machinability using coated and cryo treated tools. He is acting as coordinator for different R & D projects related to micro-machining. He is guiding a number of Ph.D., M-Tech (Res) and M-tech on different emerging areas related to production engineering. He is a member of different professional bodies such as FIE, CSI, IM2 (Institution of micro-machining), and MISTE. He earlier had acted as Head of Department of Mechanical Engineering, NIT, Rourkela for three years. He has participated in a number of national and international conferences. He has visited different countries (UK, USA, Sweden, Italy, Germany, Australia, and France) for training, conferences and visiting professor.

Dr. Sunpreet Singh is Assistant Professor in the School of Mechanical Engineering, Lovely Professional University, Jalandhar, India. He has received Ph.D. in Mechanical Engineering from Guru Nanak Dev Engineering College, Ludhiana, India. His area of research is Additive Manufacturing and application of 3D printing for development of new biomaterials for clinical applications. He has contributed extensively in Additive Manufacturing literature with publications appearing in Journal of Manufacturing Processes, Composite Part: B, Rapid Prototyping Journal, Journal of Mechanical Science and Technology, Measurement, International Journal of Advance Manufacturing Technology, and Journal of Cleaner Production. He authored 10 book chapters and monographs. He is working with joint collaboration with Prof. Seeram Ramakrishna, NUS Nanoscience & Nanotechnology Initiative and Prof. Rupinder Singh, Manufacturing Research Lab, GNDEC, Ludhiana.

Dr. Chander Prakash is Associate Professor in the School of Mechanical Engineering, Lovely Professional University, Jalandhar, India. He has received Ph.D. in Mechanical Engineering from Panjab University, Chandigarh, India. His area of research is biomaterials, rapid prototyping & 3-D printing, advanced manufacturing, modeling, simulation, and optimization. He has more than 11 years of teaching experience and 6 years research experience. He has contributed extensively to the world in the Titanium and Magnesium based implant literature with publications appearing in Surface and Coating Technology, Materials and Manufacturing Processes, Journal of Materials Engineering and Performance, Journal of Mechanical Science and Technology, Nanoscience and Nanotechnology Letters, Proceedings of the Institution of Mechanical Engineers, Part B: Journal of Engineering Manufacture. He authored 50 research papers and 10 book chapters. He is also editor of 3 Book: "Current Trends in Bio-manufacturing", Springer Series in Advanced Manufacturing, Springer International Publishing AG, Gewerbestrasse 11, 6330 Cham, Switzerland., Dec. 2018; "3D Printing in Biomedical Engineering", Book series Materials Horizons: From Nature to Nanomaterials, Springer International Publishing AG, Gewerbestrasse 11, 6330 Cham, Switzerland., August 2019, and "Biomaterials in Orthopaedics and Bone Regeneration—Design and Synthesis", Book series Materials Horizons: From Nature to Nanomaterials, Springer International Publishing AG, Gewerbestrasse 11, 6330 Cham, Switzerland., March 2019. He is also Guest Editor of 3 Journals: Guest Editor of Special Issue of "Functional Materials and Advanced Manufacturing", Facta Universitatis, Series: Mechanical Engineering (Scopus Index), Materials Science Forum (Scopus Index), and Special Issue on "Metrology in Materials and Advanced Manufacturing", Measurement and Control (SCI indexed).

Laser-Assisted Jet Electrochemical Machining of Titanium-Based Biomedical Alloy

Anup Malik, Alakesh Manna, Chander Prakash and Sunpreet Singh

Abstract Titanium alloy (Ti-6Al-4V) is best among various metallic materials for making medical implants due to their longer life and surface topology to promote osseointegration. Ti-6Al-4V has excellent mechanical properties as well as superior biological properties along with higher corrosion resistance. For an implant material, its machining plays an important role to make it fit for the medical applications. This chapter mainly focuses on machining and surface characteristics of Ti-6Al-4V when machined with laser-assisted jet electrochemical machining process. The Taguchi methodology-based design of experiments L_{25} (5^5) employed to study the effect of various parameters of the developed LAJECM setup on various response characteristics is investigated and explained with the help of S/N ratio, analysis of variance, and scanning electron micrograph.

Keywords ANOVA · Hybrid machining · Laser · Taguchi methodology · Ti-6Al-4V

Abbreviations

ANOVA	Analysis of variance
DC	Duty cycle
EC	Electrolyte concentration
ECM	Electrochemical machining

A. Malik (✉)
Department of Mechanical Engineering, Malaviya National Institute of Technology, Jaipur 302017, Rajasthan, India
e-mail: anupmalik321@gmail.com; anup.mech@mnit.ac.in

A. Manna
Department of Mechanical Engineering, Punjab Engineering College, Chandigarh 160012, India

C. Prakash · S. Singh
School of Mechanical Engineering, Lovely Professional University, Phagwara 144411, Punjab, India

C. Prakash et al. (eds.), *Biomanufacturing*,
https://doi.org/10.1007/978-3-030-13951-3_9

EFR	Electrolyte flow rate
HUAT	Hot ultrasonically assisted turning
IEG	Inter-electrode gap
JECM	Jet electrochemical machining
LAJECM	Laser-assisted jet electrochemical machining
LAM	Laser-assisted machining
LAMM	Laser-assisted micro-milling
MRR	Material removal rate
SV	Supply voltage
UAT	Ultrasonic-assisted turning
UVAD	Ultrasonic vibration-assisted drilling
UVAG	Ultrasonic vibration-assisted grinding
UVAT	Ultrasonic vibration-assisted tapping

1 Introduction

Variety of metallic biomaterials such as stainless steel, titanium, cobalt-chromium, and magnesium has been used for biomedical application [1–4]. Among all, titanium metal is widely used for orthopedic implants and accessories [5–7]. Titanium materials are also called as "wonder metal" because its very less density and strength can be increased significantly by alloying certain elements like aluminum, vanadium, chromium, manganese, tin, and molybdenum and also applying heat treatment process [8]. The presence of certain alloying elements in titanium improves its strength-to-weight ratio, high tensile strength, higher toughness, higher corrosion resistance, higher fatigue strength, higher erosion resistance, and biocompatibility which is the most important one. The common Ti alloy used for biomedical applications is Ti-6Al-4V also called as Grade 5. Titanium has very vast application areas in pure as well as in alloyed form. Some applications of the commercially known grades of titanium are as follows: Grade 1, 2, 2H, and 3 are generally used for corrosion resistance applications. Grade 4 and 5 are mostly used for biomedical and aerospace applications. Grade 6 is used in jet engines and airframe applications. Grade 7 and 7H are used for corrosion resistance at high pH and low-temperature applications. Grade 9 is used in athletic equipment and aircraft tubing for hydraulics applications. Grade 11, 16, and 18 are used in highly corrosive environment applications. Ti and its alloys are extensively used in aerospace, chemical and petrochemical, power, automotive, pharmaceutical and food processing, medical implants, and surgical devices. Ti6Al4V (Grade 5) is widely used Ti alloy because of its superior properties among other Ti alloys [9].

2 Machining of Ti Alloy by Hybrid Machining

Significant machining is required for application of titanium and its alloys. But their machining is not easy because of low thermal conductivity, elastic modulus, and high chemical reactivity even at high temperature so they are also referred to as difficult-to-machine materials. Low thermal conductivity does not allow dissipation of heat from the machining zone which results in more tool wear (such as abrasive, adhesive, and diffusive) and leads to catastrophic tool failure. Low elastic modulus allows workpiece deflection during machining which results in the poor surface finish during machining. High chemical reactivity even at higher temperature results in adhesion between tool and workpiece which causes wear and tool failure [8, 10–12]. So, Table 1 shows the list of different hybrid machining methods used previously for the efficient machining of Ti alloy.

3 Electrochemical Machining and LAJECM

Electrochemical machining is non-conventional machining method [30]. ECM extensively is used to produce products as well as parts for various industries such as aerospace, automobile, defense, nuclear, and medical (e.g., implants) [31–33]. Mechanical and thermal stress-free parts with high surface finish can be easily produced with the help of ECM. The mechanism of metal removal is ionic displacement in ECM, in which low DC voltage with high current density is applied between cathode (i.e., tool) and anode (i.e., workpiece), separated by small gap in which electrolyte is pumped. The various functions performed by the electrolyte used in the ECM process are as follows: (i) Complete the electric circuit between cathode and anode; (ii) allow required chemical reaction to occur; (iii) carried away heat and reaction products are generated during chemical reactions from the machining zone. The dissolution process in ECM is very complex in nature, which does not allow predicting the machining behavior for getting the maximum ECM benefits therefore restricting its use universally. The precision and dimensional accuracy of ECM process greatly affect due to unwanted stray machining [31]. Various investigations are carried by different researchers to minimize stray machining effect to control ECM dissolution process, which improves ECM efficiency as follows: (i) use of an insulated electrode; (ii) use of smaller inter-electrode gaps; (iii) use of low concentration electrolytes; (iv) use of pulsed current; (v) use of laser assistance, etc. [31, 32, 34–43]. LAJECM is a hybrid machining method; a low power laser beam is combined with an electrolytic jet to form a non-contact tool electrode. Laser beam is aligned coaxially with the electrolytic jet to enhance the rate of electrochemical dissolution from anode (i.e., workpiece), which in turn improves the machining precision along with its efficiency. The LAJECM experimental setup has been designed and developed for machining of Ti-6Al-4V. The LAJECM setup consists of four subsystems: (i) laser assistance system; (ii) electrolyte supply system; (iii) power supply system; and (iv)

Table 1 Summary of Ti alloy machined using hybrid machining methods

Authors	Method	Ti alloy type	Result findings
Sun et al. [13]	LAM	Pure Ti	The author compared LAM with conventional machining for pure Ti alloy and claimed a drastic reduction of cutting forces, smoother machined surface with a change in chip formation mechanism
Dandekar et al. [14]	LAM	Ti-6Al-4V	Machinability of Ti alloy improved from low cutting speed to medium–high cutting speed that is 60–107 m/min during LAM over conventional machining. Overall cost savings of nearly 30% in machining cost are also claimed by the authors
Hedberg et al. [15]	LAM	Ti-6Al-4V	Authors claimed a decrease in residual stress by 10% for the parts developed by laser-assisted milling over conventional machining. There is also 30% decrease in feed force required during cutting
Sun et al. [16]	LAM	Ti-6Al-4V	Feed force requirement decreases very sharply when temperature increased from 200 to 450 °C in front of the cutting zone after it will become constant. So, optimum temperature range suggested by authors during LAM of Ti alloy is 230–350 °C
Sun et al. [17]	LAM	Ti-6Al-4V	Authors claimed that two types of chips are formed during LAM of Ti-6Al-4V depending upon the cutting speed. While machining at low cutting speed segmented chips with geometry ratio >1 with sharpened teeth formed as compared to geometry ratio <1 formed when machining operation carried out at high cutting speed
Ding et al. [18]	LAMM	Ti-6Al-4V	Authors showed a reduction or elimination of buildup edge formed during LAMM of hard materials through finite element model analysis when workpiece material is preheated prior to cutting tool
Shen and Ding [19]	LAMM	Ti-6Al-4V	Authors claimed improvement in the surface integrity in LAMM process with a reduction in the edge burr formation

(continued)

Table 1 (continued)

Authors	Method	Ti alloy type	Result findings
Hong and Ding [12]	Cryogenic machining	Ti-6Al-4V	Authors claimed diffusivity of titanium to tool material is highly reduced when machining is done in cryogenic cooled conditions and also double working speed of machining can be achieved over conventional machining
Strano et al. [20]	Cryogenic machining	Ti-6Al-4V	There is a significant reduction in requirement of cutting forces, improvement in tool life and reduction in the coefficient of friction at the machining interface when Ti-6Al-4V alloy is cryogenically machined over wet machining
Rashid et al. [21]	LAM	Ti-6Cr-5Mo-5V-4Al β-titanium alloy	Authors machined β-titanium alloy and claimed optimum results achieved between cutting speed (25–100 m/min) and feed rates (0.15–0.25 mm/rev) when laser power of 1200 W is used during LAM
Rashid et al. [22]	LAM	Ti-6Cr-5Mo-5V4Al β-titanium alloy	Authors claimed a significant reduction in cutting force requirement in LAM of β-titanium alloy, when laser power 1200–1600 W is used, and laser power beyond 1600 W is not recommended by the authors for cutting speeds below 25 m/min because of more tool wear occurred due to chip pileup
Rashid et al. [23]	LAM	Ti-10V-2Fe-3Al β-titanium alloy	Authors claimed that laser assistance during machining of β-titanium alloy reduces cutting pressures when cutting speed increased above 50 m/min resulting in thermal softening of material due to the laser heat and also reduces cutting force along with flow stresses
Rashid et al. [24]	LAM	Ti10V2Fe3Al alloy	The laser power of 800–1200 W and cutting speed between 55 and 100 m/min are best suited for the LAM of Ti10V2Fe3Al alloy. Authors also claimed 10% reduction in cutting force requirement when Ti10V2Fe3Al alloy machined with 1600 W laser power and 55–140 m/min cutting speeds. High power of laser beam is recommended by the authors when working at higher cutting speeds to evade chatter and chip pileup

(continued)

Table 1 (continued)

Authors	Method	Ti alloy type	Result findings
Pujana et al. [25]	UVAD	Ti-6Al-4V	The feed force decreased nearly 10–20% in ultrasonic-assisted drilling over conventional drilling claimed by the authors. There is also increase in temperature and reduction in force requirement along with increasing vibration amplitude as reported by the authors
Nik et al. [26]	UVAG	Ti-6Al-4V	Authors claimed reduction in grinding force when ultrasonic vibration is applied to the workpiece and there is also decrease in surface roughness heights when compared with the conventional grinding
Pawar et al. [27]	UVAT	Ti-6Al-4V	Authors claimed that the torque and cutting temperature reductions are 19.1 and 20.3%, respectively, when tapping of Ti-6Al-4V alloy is done with ultrasonic vibration-assisted tapping over conventional tapping. There is also improvement in the surface finish and machinability reported during UVAT over conventional tapping
Maurotto et al. [28]	UAT	β-titanium alloy (Ti-15-3-3-3)	Authors developed an ultrasonic-assisted turning setup and utilized for the machining of β-titanium alloy. Reductions in cutting force during turning along with improved surface quality noticed by the authors for machining of β-titanium alloy during UAT over conventional turning
Muhammad at el. [29]	HUAT	β-Ti alloy	A novel hybrid machining process HUAT is presented by the authors for the machining of β-Ti alloy and claimed 80-85% reductions in the cutting force requirement when machining operations are carried out with HUAT compared to conventional machining

workpiece feed system. The detailed description of the LAJECM setup can be found in the reference [43]. Figure 1 shows the actual diagram of laser-assisted jet electrochemical machining setup developed and used for experimentation [40–43]. The assistance of laser beam in JECM increases localization effect during machining; due to this localization effect, machining rate and precision are increased in the LAJECM process compared to JECM. The machining areas exposed to the laser beam have higher current density and lesser stray machining effects, which leads to more pré-

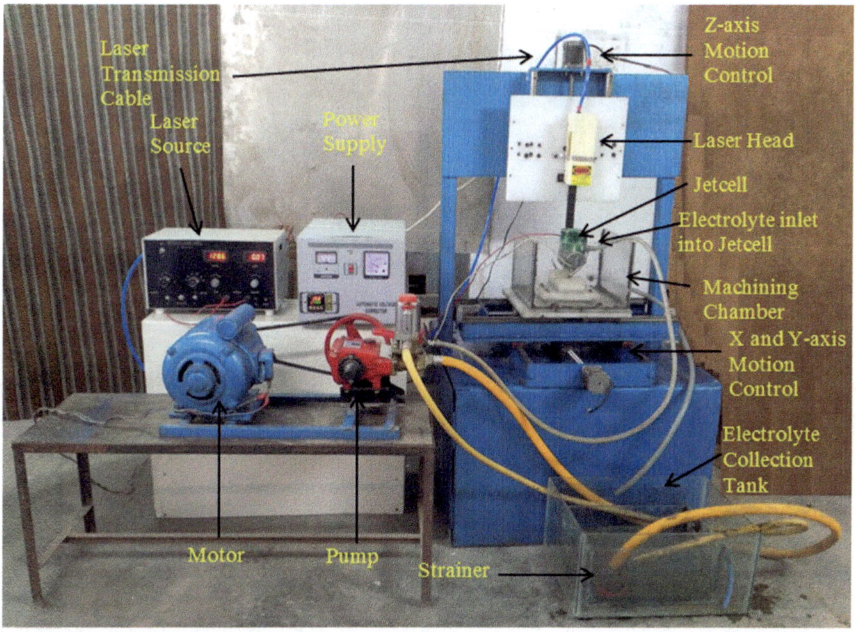

Fig. 1 Actual diagram of laser-assisted jet electrochemical machining setup

cised and faster machining. Therefore, material removal is more in axial direction as compared to the lateral direction. LAJECM process has various advantages along with some limitations. The limitations of LAJECM are as follows: (i) aligning the axis of laser beam with electrolytic jet and (ii) scattering of laser beam by reaction by-products. Laser beam which is visible during machining avoids above-mentioned problems.

4 Results and Discussion

4.1 Experimental Planning, Results, and ANOVA

Taguchi method-based design of experiments L_{25} (5^5) orthogonal array based on a series of experiments has been carried out, and acquired results were used for investigation. The S/N ratio (dB) is the ratio of signal to noise, where signal represents the desirable value and noise represents undesirable value, i.e., ratio of mean to square deviation for the output characteristics [44–51]. Table 2 represents the different parameters and their levels considered for experiments.

Figure 2 shows S/N ratio graph for different factor levels for MRR (mg/min). From Fig. 2, it is clear that for maximum MRR, the optimal parametric combination

Table 2 L_{25} (5^6) orthogonal array, actual parametric values, and test results acquired during machining of Ti-6Al-4V workpiece sample

Exp. no.	Parametric levels and their actual values					Mean value of MRR (mg/min)
	SV, X_1 (V)	*EC*, X_2 (g/l)	*IEG*, X_3 (mm)	*EFR*, X_4 (l/min)	*DC*, X_5 (%)	
1.	80	20	2	0.2	50	1.224
2.	80	50	3	0.4	60	2.228
3.	80	80	4	0.6	70	2.568
4.	80	110	5	0.8	80	3.598
5.	80	140	6	1.0	90	4.312
6.	120	20	3	0.6	80	3.219
7.	120	50	4	0.8	90	3.723
8.	120	80	5	1.0	50	3.014
9.	120	110	6	0.2	60	1.569
10.	120	140	2	0.4	70	4.536
11.	160	20	4	1.0	60	3.274
12.	160	50	5	0.2	70	2.124
13.	160	80	6	0.4	80	2.836
14.	160	110	2	0.6	90	5.621
15.	160	140	3	0.8	50	4.220
16.	200	20	5	0.4	90	2.654
17.	200	50	6	0.6	50	2.212
18.	200	80	2	0.8	60	4.639
19.	200	110	3	1.0	70	5.559
20.	200	140	4	0.2	80	4.332
21.	240	20	6	0.8	70	3.121
22.	240	50	2	1.0	80	5.773
23.	240	80	3	0.2	90	4.426
24.	240	110	4	0.4	50	3.612
25.	240	140	5	0.6	60	4.567

is $A_5B_5C_1D_5E_5$, i.e., at 240 V *SV*, 140 g/l *EC*, 2 mm *IEG*, 1.0 l/min *EFR*, and 90% *DC*.

Table 3 represents the ANOVA for MRR. This ANOVA table is prepared based on the experimental results shown in Table 2 during LAJECM of Ti-6Al-4V workpiece sample. According to the ANOVA, the parameter *EC* is the most significant parameter with "F test" value 31.63 and 23.79% contribution. The other parameters such as *EFR*, *IEG*, *SV*, and *DC* are the significant parameters with 29.11, 26.72, 24.83, and 19.66 "F" test values and 21.90, 20.01, 18.67, and 14.78% contribution, respectively.

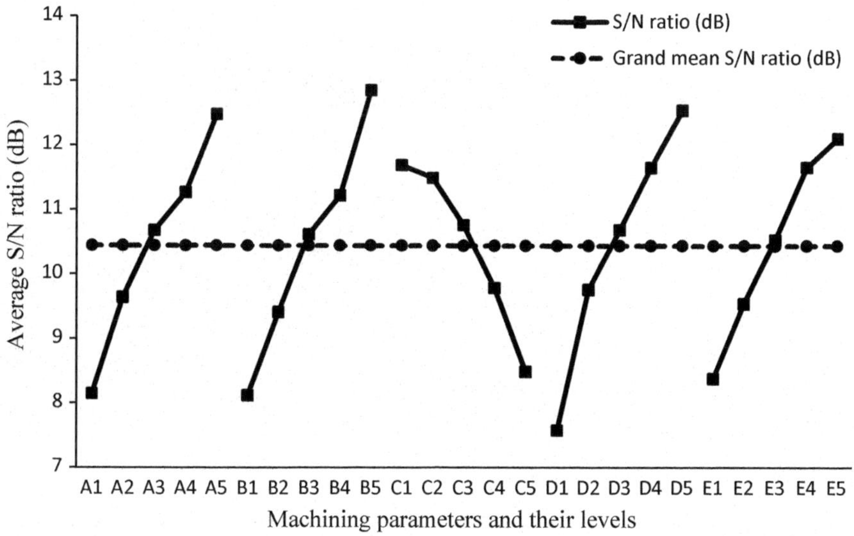

Fig. 2 Average S/N ratio (dB) by factor level for MRR of Ti-6Al-4V workpiece sample

Table 3 ANOVA and "F" test for MRR of Ti-6Al-4V workpiece sample

Sr. no.	Control parameters	Sum of squares	Degree of freedom	Variance	"F" test value	% of contribution
1	Supply voltage (*SV*)	6.861	4	1.715	24.83	18.67
2	Electrolyte concentration (*EC*)	8.742	4	2.185	31.63	23.79
3	Inter-electrode gap (*IEG*)	7.383	4	1.846	26.72	20.01
4	Electrolyte flow rate (*EFR*)	8.046	4	2.011	29.11	21.90
5	Duty cycle (*DC*)	5.432	4	1.358	19.66	14.78
6	Error	0.2764	54	0.005		0.85
7	Total	36.7420	74			100.00

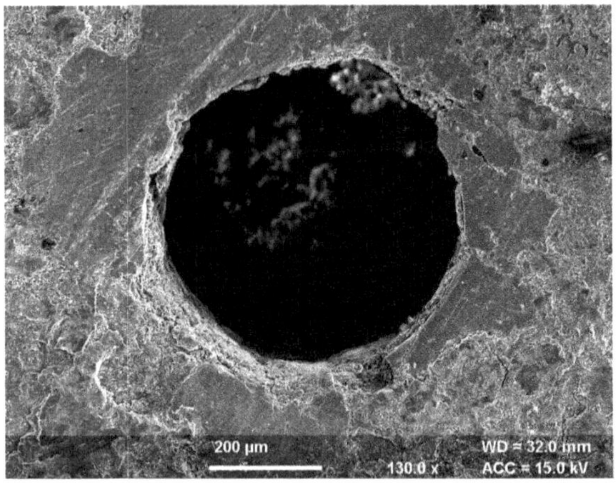

Fig. 3 SEM photograph of a drilled hole on Ti-6Al-4V workpiece sample

4.2 Scanning Electron Micrographs of Ti-6Al-4V Workpiece Sample

In this section, SEM photograph of the drilled hole generated during LAJECM of Ti-6Al-4V workpiece sample is explained in detail. Brief discussions on the experimental results are very important to highlight the research outcome and reported in this section. Figure 3 shows SEM photograph of a drilled hole produced during LAJECM of Ti-6Al-4V workpiece sample. This micro-hole is generated with parametric settings at 160 V *SV*, 140 g/l *EC*, 3 mm *IE*G, 0.8 l/min *EF*R, and 50% *DC*. The shape of the machined hole is not circular and having disturbed edge due to higher *SV* and *EC*. There are visible marks of surface damage due to the higher *EC* resulting in higher current density in the machining zone.

Figure 4 shows SEM photograph and EDX analysis of a drilled hole produced during LAJECM of Ti-6Al-4V workpiece sample. This micro-hole is generated with parametric settings at 120 V *SV*, 110 g/l *EC*, 6 mm *IE*G, 0.2 l/min *EF*R, and 60% *DC*. Figure 4a shows the SEM photograph of the top view of the machined hole. The hole produced is quite circular in nature, and there is no presence of overcut on the periphery of the machined hole. There are also no evident marks of spark-affected zone; it may be due to low *SV* and *DC*. There is some presence of surrounding surface affected due to higher *EC* used. Figure 4b shows the EDX analysis of the section of the machined hole as shown in Fig. 4a. Table 4 shows the element composition of the hole as shown in Fig. 4a.

Figure 5 shows SEM photograph of profile generated on LAJECM of Ti-6Al-4V workpiece sample. This micro-hole is generated with parametric settings at 240 V *SV*, 80 g/l *EC*, 2 mm *IE*G, 20 m/s electrolytic jet velocity, 800 µm electrolytic jet diameter, 400 µm laser beam spot diameter with 0.1 mm/min working table feed rate in X-direction with continuous current supply. Figure 5a shows the profile cut at 100 × magnification, and there are evident marks of small holes or cavity formation

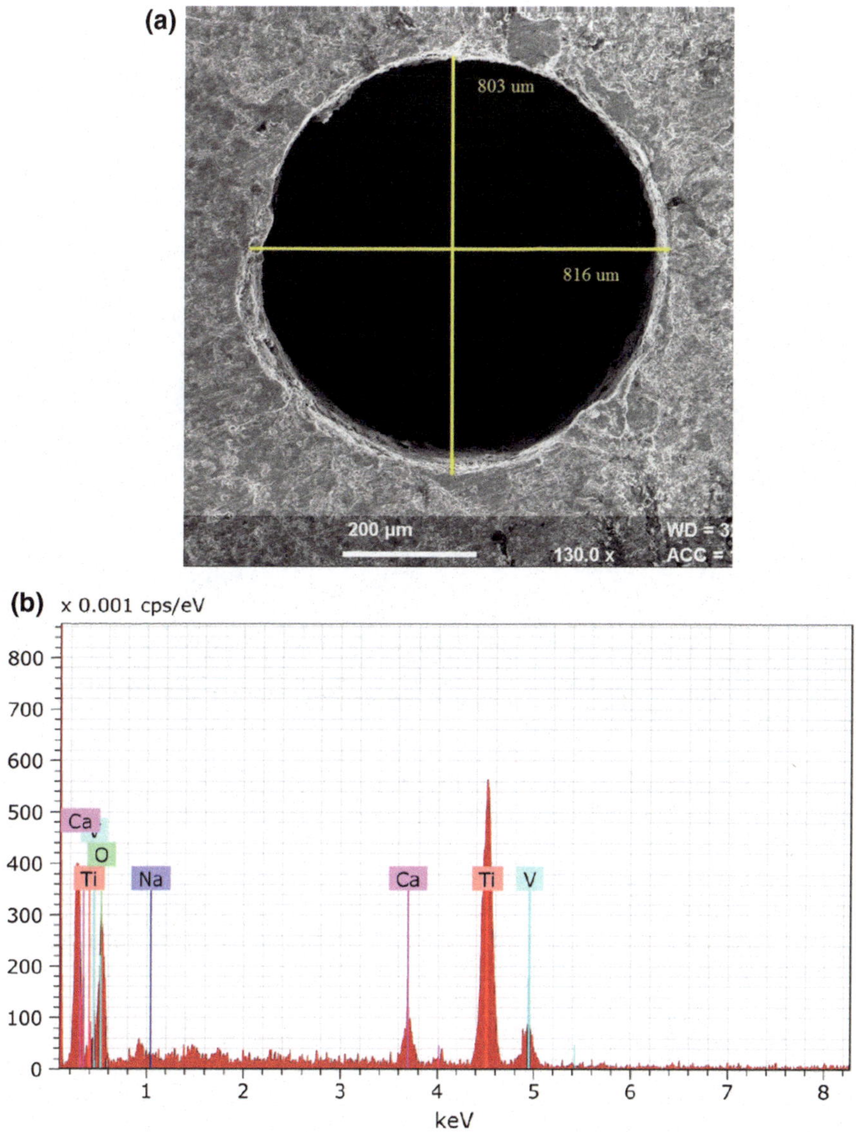

Fig. 4 SEM photograph and EDX analysis of a drilled hole on Ti-6Al-4V workpiece sample

Table 4 Elemental composition shown in EDX analysis

Spectrum: test 787					
Element	Series	Unn. C (wt%)	Norm. C (wt%)	Atom. C (at. %)	Error (3 sigma) (wt%)
Titanium	K-series	51.38	50.16	26.22	5.26
Oxygen	K-series	46.24	45.14	70.62	30.45
Sodium	K-series	0.52	0.51	0.55	0.35
Vanadium	K-series	0.02	0.02	0.01	0.12
Calcium	K-series	4.28	4.17	2.61	0.76
Total		102.42	100.00	100.00	

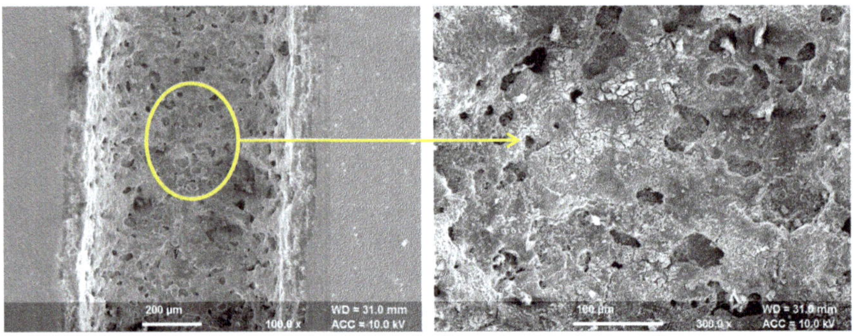

(a) 100 X magnification (b) 300 X magnification

Fig. 5 SEM photograph of profile cut on Ti-6Al-4V workpiece sample

on the machined surface of the profile; it may be due to high *SV* and *EC*. Figure 5b shows the enlarged view (300 × magnification) of the machined profile as shown in Fig. 5a; it is clear that the generated profile surface is not quite good. So, use of low *SV* and *EC* along with low electrolytic jet velocity is recommended for the generation of profiles (micro-contours).

This 3-D graph as shown in Fig. 6 shows the variation of MRR (mg/min) with variation of *SV* (V) and *EC* (g/l). This interaction graph is drawn based on the experimentally obtained results of LAJECM of Ti-6Al-4V workpiece sample at 4 mm *IEG*, 0.6 l/min *EFR*, and 70% *DC*. From Fig. 6, it is clear that increase in the *SV* increases MRR. It is also clear that at high *SV* (240 V) and high *EC* (140 g/l), MRR is higher. Therefore, material removal rate is increased with increase in *SV* and *EC*.

This 3-D graph as shown in Fig. 7 shows the variation of MRR (mg/min) with variation of *SV* (V) and *IEG* (mm). This interaction graph is drawn based on the experimentally obtained results of LAJECM of Ti-6Al-4V workpiece sample at 80 g/l *EC*, 0.6 l/min *EFR*, and 70% *DC*. From Fig. 7, it is clear that increase in the *SV* increases MRR. It is also clear that at high *SV* (240 V) and low *IEG* (2 mm), MRR is higher. Therefore, MRR is increased with increase in *SV* and decrease in *IEG*.

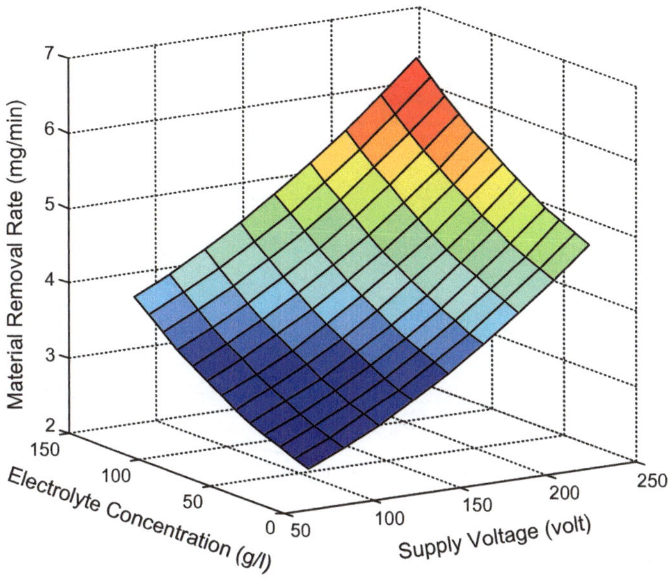

Fig. 6 Interaction effect of *SV* (V) and *EC* (g/l) on MRR

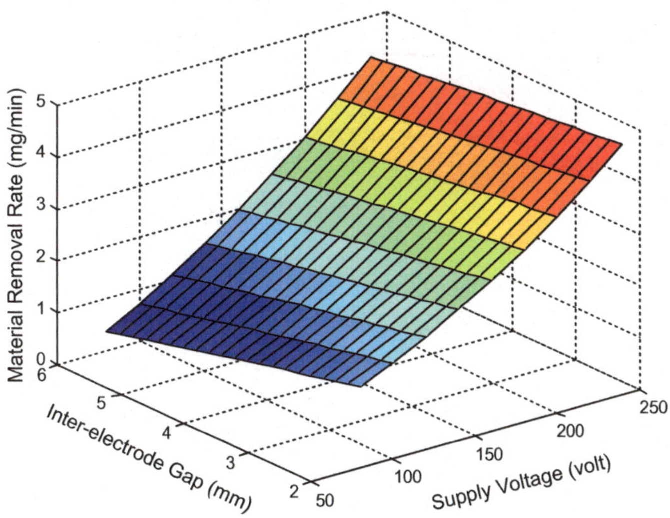

Fig. 7 Interaction effect of *SV* (V) and *IEG* (mm) on MRR

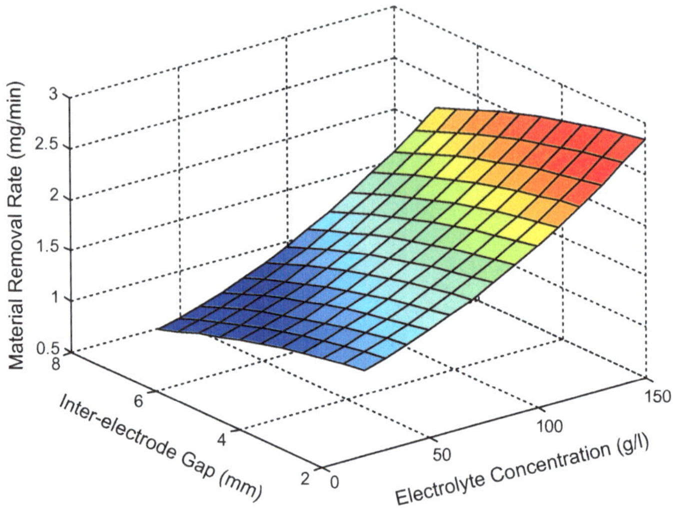

Fig. 8 Interaction effect of *EC* (g/l) and *IEG* (mm) on MRR

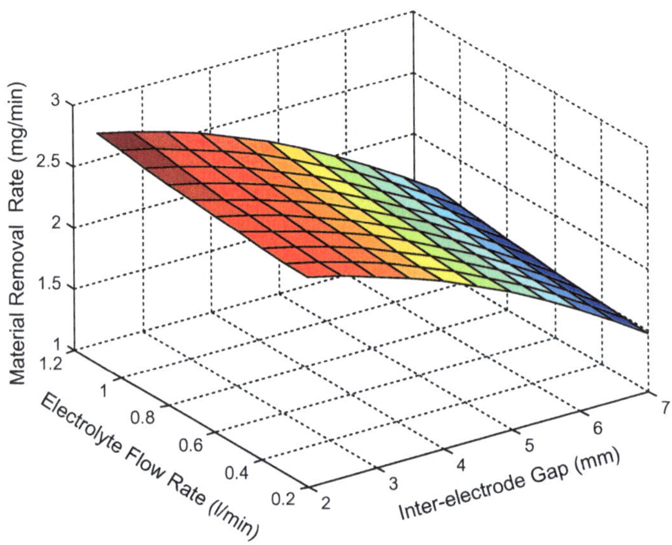

Fig. 9 Interaction effect of *IEG* (mm) and *EFR* (l/min) on MRR

The 3-D graph as shown in Fig. 8 shows the variation of MRR (mg/min) with variation of *EC* (g/l) and *IEG* (mm). This interaction graph is drawn based on the experimentally obtained results of LAJECM of Ti-6Al-4V workpiece sample at 160 V *SV*, 0.6 l/min *EFR*, and 70% *DC*. From Fig. 8, it is clear that increase in the *EC* increases MRR. It is also clear that at high *EC* (140 g/l) and low *IEG* (2 mm), MRR is higher. Therefore, MRR is increasing with increasing *EC* and decreasing *IEG*.

This 3-D graph as shown in Fig. 9 shows the variation of MRR (mg/min) with variation of *IEG* (mm) and *EFR* (l/min). This interaction graph is drawn based on the experimentally obtained results of LAJECM of Ti-6Al-4V workpiece sample at 160 V *SV*, 80 g/l *EC*, and 70% *DC*. From Fig. 9, it is clear when there is increase in *IEG* then there is decrease in *MRR*, but when there is increase in *EFR* then *MRR* is increased. It is also clear that at low *IEG* (2 mm) and high *EFR* (1.0 l/min), MRR is higher. Therefore, MRR is increased with decrease in *IEG* and increase in *EFR*.

5 Conclusion

An LAJECM setup has been indigenously developed and utilized for machining of Ti-6Al-4V workpiece sample. The assistance of laser with JECM increases the MRR. Further studies may be conducted to check some other response characteristics, i.e., taper, overcut, surface roughness, and heat-affected zone. In the future, detailed studies should be carried out to investigate the effect of different parameters during LAJECM of some other materials used for biomedical applications, i.e., Ti–35Nb–7Ta–5Zr β-titanium alloy, Mg-based alloys.

References

1. Geetha M, Singh AK, Asokamani R, Gogia AK (2009) Ti based biomaterials, the ultimate choice for orthopaedic implants—a review. Prog Mater Sci 54:397–425
2. Niinomi M, Nakai M, Hieda J (2012) Development of new metallic alloys for biomedical applications. Acta Biomater 8:3888–3903
3. Prakash C, Singh S, Sidhu SS, Singh S, Verma K (2018) Synthesis and characterization of Mg-Zn-Mn-HA composite by spark plasma sintering process for orthopedic applications. Vacuum 155:578–584. https://doi.org/10.1016/j.vacuum.2018.06.063
4. Prakash C, Kansal HK, Pabla BS, Puri S, Aggarwal A (2016) Electric discharge machining—a potential choice for surface modification of metallic implants for orthopedic applications: a review. Proc IMechE Part B: J Eng Manuf 230:331–353
5. Prakash C, Kansal HK, Pabla BS, Puri S (2015) Potential of powder mixed electric discharge machining to enhance the wear and tribological performance of β-Ti implant for orthopedic applications. J Nanoeng Nanomanuf 5(4):261–269
6. Prakash C, Kansal HK, Pabla BS, Puri S (2015) Processing and characterization of novel biomimetic nanoporous bioceramic surface on β-Ti implant by powder mixed electric discharge machining. J Mater Eng Perform 24:3622–3633

7. Prakash C, Kansal HK, Pabla BS, Puri S (2017) Experimental investigations in powder mixed electrical discharge machining of Ti-35Nb-7Ta-5Zr β-Ti alloy. Mater Manuf Process 32(3):274–285. https://doi.org/10.1080/10426914.2016.1198018
8. Donachie MJ (2000) Titanium: a technical guide. ASM International, Geauga County, OH
9. Gupta K, Laubscher RF (2017) Sustainable machining of titanium alloys: a critical review. Proc IMechE Part B: J Eng Manuf 231(14):2543–2560
10. Davim PJ (2014) Machining of titanium alloys. Springer, Heidelberg
11. Pramanik A (2014) Problems and solutions in machining of titanium alloys. Int J Adv Manuf Tech 70:919–928
12. Hong SY, Ding Y (2001) Cooling approaches and cutting temperatures in cryogenic machining of Ti-6Al-4V. Int J Mach Tools Manuf 41(10):1417–1437
13. Sun S, Harris J, Brandt M (2008) Parametric investigation of laser-assisted machining of commercially pure titanium. Adv Eng Mater 10(6):565–572
14. Dandekar CR, Shina YC, Barnes J (2010) Machinability improvement of titanium alloy (Ti–6Al–4V) via LAM and hybrid machining. Int J Mach Tool Manuf 50(2):174–182
15. Hedberg GK, Shin YC, Xu L (2015) Laser-assisted milling of Ti-4Al-6V with the consideration of surface integrity. Int J Adv Manuf Tech 79(9):1645–1658
16. Sun S, Brandt M, Barnes JE et al (2011) Experimental investigation of cutting forces and tool wear during laser-assisted milling of Ti-6Al-4V alloy. ProcIMechE Part B: J Eng Manuf 225(9):1512–1527
17. Sun S, Brandt M, Dargusch MS (2010) The effect of a laser beam on chip formation during machining of Ti6Al4V alloy. Metall Mater Trans A 41:1573–1581
18. Ding H, Shen N, Shin YC (2012) Thermal and mechanical modeling analysis of laser-assisted micromilling of difficult-to-machine alloys. J Mater Process Technol 212:601–613
19. Shen N, Ding H (2013) Thermo-mechanical coupled analysis of laser-assisted mechanical micromilling of difficult-to-machine metal alloys used for bio-implant. Int J Precis Eng Manuf 14(10):1677–1685
20. Strano M, Chiappini E, Tirelli S et al (2013) Comparison of Ti6Al4V machining forces and tool life for cryogenic versus conventional cooling. ProcIMechE Part B: J Eng Manuf 227(9):1403–1408
21. Rashid RAR, Sun S, Wanga G et al (2012) An investigation of cutting forces and cutting temperatures during laser-assisted machining of the Ti–6Cr–5Mo–5V–4Al beta titanium alloy. Int J Mach Tool Manuf 63:58–69
22. Rashid RAR, Sun S, Wanga G et al (2012) The effect of laser power on the machinability of the Ti-6Cr-5Mo-5V-4Al beta titanium alloy during laser assisted machining. Int J Mach Tool Manuf 63:41–43
23. Rashid RAR, Bermingham MJ, Sun S et al (2013) The response of the high strength Ti–10V–2Fe–3Al beta titanium alloy to laser assisted cutting. Precis Eng 37(2):461–472
24. Rashid RAR, Sun S, Palanisamy S et al (2014) A study on laser assisted machining of Ti10V2Fe3Al alloy with varying laser power. Int J Adv Manuf Tech 74:219–224
25. Pujana J, Rivero A, Celaya A et al (2009) Analysis of ultrasonic-assisted drilling of Ti-6Al-4V. Int J Mach Tool Manuf 49(6):500–508
26. Nik MG, Movahhedy MR, Akbari J (2012) Ultrasonic-assisted grinding of Ti-6Al-4V alloy. Proc CIRP 1:353–358
27. Pawar S, Patil S, Joshi S et al (2014) An experimental analysis of ultrasonic vibration assisted tapping of Ti-6Al-4V. SAE technical paper, 28-0024
28. Maurotto A, Muhammad R, Roy A et al (2013) Enhanced ultrasonically assisted turning of a b-titanium alloy. Ultrasonics 53:1242–1250
29. Muhammad R, Maurotto A, Demiral M et al (2014) Thermally enhanced ultrasonically assisted machining of Ti alloy. CIRP J Manuf Sci Technol 7(2):159–167
30. McGeough JA (1974) Principles of electrochemical machining, 1st edn. Springer, United States, pp 1–255
31. Rajurkar KP, Zhu D, McGeough JA, Kozak J, De Silva AKM (1999) New developments in electro-chemical machining, CIRP Ann Manuf Technol 48(2):567–580

32. McGeough JA, Pajak PT, De Silva AKM, Harrison DK (2003) Recent research and developments in electrochemical machining. Int J Electric Mach 8:1–14
33. Pajak PT, De Silva AKM, McGeough JA, Harrison DK (2004) Modelling the aspects of precision and efficiency in laser-assisted jet electrochemical machining (LAJECM). J Mater Process Technol 149:512–518
34. De Silva AKM, Altena HSJ, McGeough JA (2003) Influence of electrolyte concentration on copying accuracy of precision-ECM. CIRP Ann Manuf Technol 52(1):165–168
35. De Silva AKM, Pajak PT, Harrison DK, McGeough JA (2004) Modelling and experimental investigation of laser assisted jet electrochemical machining. CIRP Ann Manuf Technol 53(1):179–182
36. Ruszaj A, Czekaj J, Chuchro M, Skrabalak G (2001) Electrochemical machining with short interelectrode voltage pulses. In: Proceedings of the international symposium for electromachining, ISEM XIII, vol 1, pp 1–538
37. De Silva AKM, Pajak PT, McGeough JA, Harrison DK (2011) Thermal effects in Laser-assisted jet electrochemical machining. CIRP Ann Manuf Technol 60:243–246
38. Pajak PT, De Silva AKM, Harrison DK, McGeough JA (2006) Process energy analysis for aluminium alloy and stainless steel in laser-assisted jet electrochemical machining. In: Proceedings of the Institution of Mechanical Engineers, Part B: Journal of Engineering Manufacture, vol 220 (Special Issue Paper), pp 405–412
39. Pajak PT, De Silva AKM, Harrison DK, McGeough JA (2006) Precision and efficiency of laser-assisted jet electrochemical machining. Precis Eng 30:288–298
40. Malik A, Manna A (2017) Study on precision microholes using pulsed laser with jet electrochemical machining. Arab J Sci Eng 43(9):4593–4608. https://doi.org/10.1007/s13369-017-2998-9
41. Malik A, Manna A (2018) Multi-response optimization of laser assisted jet electrochemical machining parameters based on grey relational analysis. J Braz Soc Mech Sci Eng 40:148. https://doi.org/10.1007/s40430-018-1069-9
42. Malik A, Manna A (2019) An insight into laser-assisted jet electrochemical machining process. In: Dixit U, Joshi S, Davim J (eds) Application of lasers in manufacturing. Lecture Notes on Multidisciplinary Industrial Engineering. Springer, Singapore. https://doi.org/10.1007/978-981-13-0556-6_7
43. Malik A, Manna A (2018) Investigation on the laser-assisted jet electrochemical machining process for improvement in machining performance. Int J Adv Manuf Technol 96:3917–3932. https://doi.org/10.1007/s00170-018-1846-8
44. Antil P, Singh S, Manna A (2017) Electrochemical discharge drilling of SiC reinforced polymer matrix composite using Taguchi's grey relational analysis. Arab J Sci Eng 43(3):1257–1266 (Springer)
45. Manna A, Malik A (2016) Micro-drilling of Al/Al_2O_3-MMC on developed ECMM. In: Proceedings of the World Congress on Engineering 2016 (WCE 2016), vol 2. London, U.K., pp. 652–657
46. Manna A, Malik A (2014) An experimental investigation on designed and fabricated WECSM setup during micro slicing of e-glass fibre epoxy composite. In: All India Manufacturing Technology, Design and Research Conference 2014 (AIMTDR-2014), IIT Guwahati, vol 26, issue 1, pp 625-1–625-6
47. Manna A, Malik A (2016) Investigation of electrochemical behavior during drilling of EN-31 Steel on developed laser assisted jet electrochemical machine. In: All India Manufacturing Technology, Design and Research Conference 2016 (AIMTDR-2016), College of Engineering Pune, vol 27, pp 583–587
48. Antil P, Singh S, Manna A (2018) Genetic algorithm based optimization of ECDM process for polymer matrix composite. Mater Sci Forum 928:144–149
49. Malik A, Manna A (2016) An experimental investigation on developed WECSM during micro slicing of e-glass fibre epoxy composite. Int J Adv Manuf Technol 85:2097–2106
50. Antil P, Singh S, Singh P (2018) Taguchi's methodology based parametric analysis of material removal rate during ECDM of PMCs. Proc Manuf 26:469–473 (Elsevier)

51. Malik A, Manna A (2017) Travelling wire electrochemical spark machining: an overview. In: Kibria G, Bhattacharyya B, Davim J (eds) Non-traditional micromachining processes. Materials forming, machining and tribology. Springer, Cham. https://doi.org/10.1007/978-3-319-52009-4_11

Anup Malik is working as Assistant Professor in the Department of Mechanical Engineering, Malaviya National Institute of Technology, Jaipur, Rajasthan, India. He received his bachelor of technology from Kurukshetra University. He received his master of engineering and doctorate from Punjab Engineering College. He has authored more than five chapters and more than ten research papers in various international publishers.

Alakesh Manna is working as Professor and Former Head of Department, Punjab Engineering College, Chandigarh, India. He has 18 years of industrial experience and 14 years of teaching experience. He has published 220 research papers in international and national journals of repute and in the proceedings of the international and national conferences. His research areas include advanced and micro-manufacturing, fabrication of MMCs, hybrid and ceramics materials, traditional, non-traditional, advanced, hybrid and micro-machining. He is also Reviewer of many international and national journals of repute. He has guided 17 Ph.D. and 26 postgraduate theses. He is guiding six Ph.D. research scholars and two ME scholars at present. He has delivered about 150 expert lectures on different topics. He also chaired many technical sessions in the international and national conferences. He has organized many faculty development programs and national conferences. He is a recipient of gold medals for best and excellent teaching thrice and various medals, memento, certificates for his excellent past service during Indian Railway. He has received best research paper awards two times from Institution of Engineers, India. He is a recipient of Prof. D. N. Trikha Excellence in Research Publication Award-2010.

Chander Prakash is Associate Professor in the School of Mechanical Engineering, Lovely Professional University, Jalandhar, India. He has received Ph.D. in mechanical engineering from Panjab University, Chandigarh, India. His areas of research are biomaterials, rapid prototyping and 3-D printing, advanced manufacturing, modeling, simulation, and optimization. He has more than 11 years of teaching experience and 6 years of research experience. He has contributed extensively to the world in the titanium- and magnesium-based implant literature with publications appearing in Surface and Coating Technology, Materials and Manufacturing Processes, Journal of Materials Engineering and Performance, Journal of Mechanical Science and Technology, Nanoscience and Nanotechnology Letters, Proceedings of the Institution of Mechanical Engineers, Part B: Journal of Engineering Manufacture. He authored 50 research papers and 10 chapters. He is also Editor of three books: "Current Trends in Bio-manufacturing," Springer Series in Advanced Manufacturing, Springer International Publishing AG, Gewerbestrasse 11, 6330 Cham, Switzerland., December 2018; "3D Printing in Biomedical Engineering," Book series Materials Horizons: From Nature to Nanomaterials, Springer International Publishing AG, Gewerbestrasse 11, 6330 Cham, Switzerland., August 2019; and "Biomaterials in Orthopaedics and Bone Regeneration—Design and Synthesis," Book series Materials Horizons: From Nature to Nanomaterials, Springer International Publishing AG, Gewerbestrasse 11, 6330 Cham, Switzerland., March 2019. He is also Guest Editor of three journals: Special Issue of "Functional Materials and Advanced Manufacturing," Facta Universitatis, Series: Mechanical Engineering (Scopus Index), Materials Science Forum (Scopus Index), and Special Issue on "Metrology in Materials and Advanced Manufacturing," Measurement and Control (SCI Indexed).

Sunpreet Singh is Assistant Professor in the School of Mechanical Engineering, Lovely Professional University, Jalandhar, India. He has received Ph.D. in mechanical engineering from

Guru Nanak Dev Engineering College, Ludhiana, India. His areas of research are additive manufacturing and application of 3D printing for the development of new biomaterials for clinical applications. He has contributed extensively to the additive manufacturing literature with publications appearing in Journal of Manufacturing Processes, Composite Part: B, Rapid Prototyping Journal, Journal of Mechanical Science and Technology, Measurement, International Journal of Advanced Manufacturing Technology, and Journal of Cleaner Production. He authored ten chapters and monographs. He is working in joint collaboration with Prof. Seeram Ramakrishna, NUS Nanoscience & Nanotechnology Initiative, and Prof. Rupinder Singh, Manufacturing Research Lab, GNDEC, Ludhiana.

Effective Heat Treatment for Improvement in Diamond-like Carbon Coatings for Biomedical Applications

Abdul Wasy Zia

Abstract The chapter briefly introduces the diamond-like carbon (DLC) coatings, their manufacturing techniques, revenue and applications. The DLC coatings are actively involved in the biomedical industry. Therefore, a dedicated section describes the biocompatibility of these coatings concisely. Different ways for improvement in DLC coatings performance are being researched nowadays, and heat treatment is attractive among them. The chapter introduces a fresh approach of two-step heat treatment method to improve DLC properties. A conventional heat treatment usually reduces the coating properties like sp^3 fraction, hardness and increased the friction coefficient, whereas the two-step heat treatment has represented about 20% improvement in hardness and about 10% improvement in toughness simultaneously when deposited on the selected bias voltage.

Keyword Diamond-like carbon · Coating · Heat treatment · Biomaterials · Hardness · Toughness · Tribology

1 Introduction

Diamond-like carbon (DLC) coatings are industrially acknowledged due to unique multifunctional characteristics. These coatings have excellent mechanical properties such as high hardness, high Young's modulus and superior tribological performance such as near-zero friction coefficient and higher wear resistance. At the same time, DLC coatings are chemically inert, thermally stable up to 300 °C, oxidation resistant, anti-reflective and biocompatible. Due to such unique combinations, DLC coatings are widely being used for mechanical [1] and automotive components [2], cutting tools [3, 4], computer [5] and other electronic parts [6] and biomedical devices [7]. The DLC coatings are 40 years old and discovered by Schmellenmeier [8] in 1953 in the form of the hard, scratch resistance carbon film. The 1960s and 1970s were remained dedicated to graphite-like carbon coatings and their manufacturing

A. W. Zia (✉)
City University of Hong Kong, Kowloon, Hong Kong SAR
e-mail: abdul.wasy@ymail.com

C. Prakash et al. (eds.), *Biomanufacturing*,
https://doi.org/10.1007/978-3-030-13951-3_10

techniques. These coatings got a reputation in early 1980 due to prominent diamond-like features. Since 1980, the DLC is intensively being studied to improve their performance, to reduce their limitations and to explore new applications. Currently, new markets are emerging such as audio devices and biomedical products. Non-stick surgical tools, non-stick and inert coating for artery stunts, protective coatings for hip joints are typical examples of DLC coatings in the biomedical sector. The DLC revenue has increased by 150% since 2009 [9]. Particularly in the biomedical industry, the overall DLC revenue was $14 M in 2009 which rose to $78 M in 2015 with a forecast of 33% compound annual growth rate which is a true inspiration for intensive research in DLC area.

DLC coatings mainly contain sp^2- and sp^3-bonded [10] carbon atoms which decide their diamond-like or graphite-like characteristics. Generally, DLC coatings are amorphous, and their higher quality corresponds to the higher sp^3 faction of carbon atoms in the coating. The sp^3 fraction depends on the manufacturing techniques, such as physical vapour deposition (PVD), chemical vapour deposition (CVD) and detonation nano-diamond (DND) [11]. For DLC, the PVD method is common and includes sputtering, filtered cathode vacuum arc (FCVA), pulsed laser deposition (PLD), laser irradiation, etc. Each technique has similar fundamentals of carbon evaporation from the graphite target and then transported to the substrate to deposit a carbon film. However, each technology has different features and operational conditions. The ion energies are higher in FCVA techniques than sputtering which influences the plasma characteristic and coating structure. Similarly, the unbalanced magnetron sputtering has a closed magnetic field. Therefore, it produces a higher degree of ionization and gives better results than a balanced magnetron sputtering system. Besides the manufacturing methods, the manufacturing conditions such as bias voltage, chamber pressure, target current and substrate-to-target distance influence the coating structure as well.

2 DLC Coatings for Biomedical Sector

DLC coatings have proved biocompatibility [12, 13]. The carbon coatings are known since 1967 to be used for medical prosthesis [14]. These coatings are neither cytotoxic nor anti-thrombogenic, therefore, worked for both in vivo and in vitro application. These coatings have intensively explored and well served for orthopaedic and dental implants. Since, DLC coatings have high hardness, high residual stresses. Therefore, they exhibit poor toughness and do not fit for knee and hip joints in pure form as they may fail on lower impact and fatigue loads. Thus, pure DLC coatings are additionally doped with foreign elements (metallic, ceramic or gaseous elements) to boost their functionality. For example, the DLC coatings are generally doped with metal elements to enhance ductility, to reduce residual stress and to improve the toughness for better mechanical performance. In the same way, the DLC coatings are doped with silicon for orthopaedic, silver for antibacterial, hydrogen for cardiovascular, gold for cancer therapy, zirconium for osteogenesis, camphor for candidia-

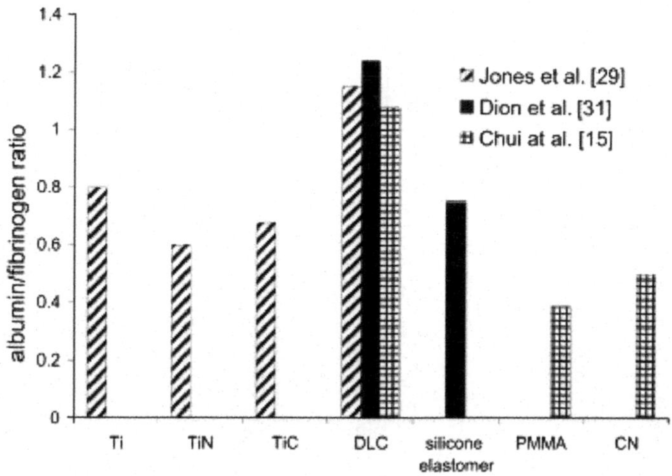

Fig. 1 Biocompatibility of DLC coatings among different materials indicated on the bases of albumin/fibrinogen ratio [16]. Reproduced with permissions

sis proliferation combat, fluorene with hydrogen for anti-thrombogenic and nitrogen hydride for neural interfaces [15]. Figure 1 demonstrates the biocompatibility of DLC coatings in between different types of coatings on the bases of albumin/fibrinogen ratio. It can be observed that DLC coatings have maximum albumin/fibrinogen ratio among titanium, titanium nitride, titanium carbide, silicon elastomers, poly-methyl-methacrylate and carbon nitride. Since the DLC coatings have further subcatagories as a-C, ta-C which depends on the atomic structure of carbon atoms present in the coatings and the ratio of sp^2 and sp^3 fractions as discussed in the earlier sections. Thus, Fig. 1 indicates higher biocompatibility of DLC coatings among different materials while Figs. 2 and 3 present the recipes to improve the DLC biocompatibility with different methods further. Figure 2 shows the interfacial tension in between foreign biomaterial and blood for different types of DLC coatings.

Usually, blood cells maintain the interfacial adhesion of 2–3 dyne/cm with blood plasma. Therefore, a foreign biomaterial shall have interfacial adhesion in the similar range [17] for best performance. It can be observed that pure ta-C DLC (enrich carbon chain structure and sp^3 carbon bonds) have interfacial adhesion of ~8 dyne/cm which usually increases for low-temperature isotropic pyrolytic carbon and a-C:H DLC (enrich carbon aromatic rings and sp^2 bonds and hydrogen fraction) above 10 dyne/cm. However, it can be observed that if a-C:H DLC are additionally doped with phosphorus, the interfacial adhesion in between coating and blood plasma was ~5 dyne/cm which is close to the natural biomedical design.

Similarly, plasma etching [19] also changes the coating properties. Figure 3 presents the effects of nitrogen and oxygen plasma treatment to boost the haemocompatibility of pure and silicon-doped DLC coatings. It can be observed that platelets adhere to ~90% area of pure a-C:H coating. However, the adherence percentage

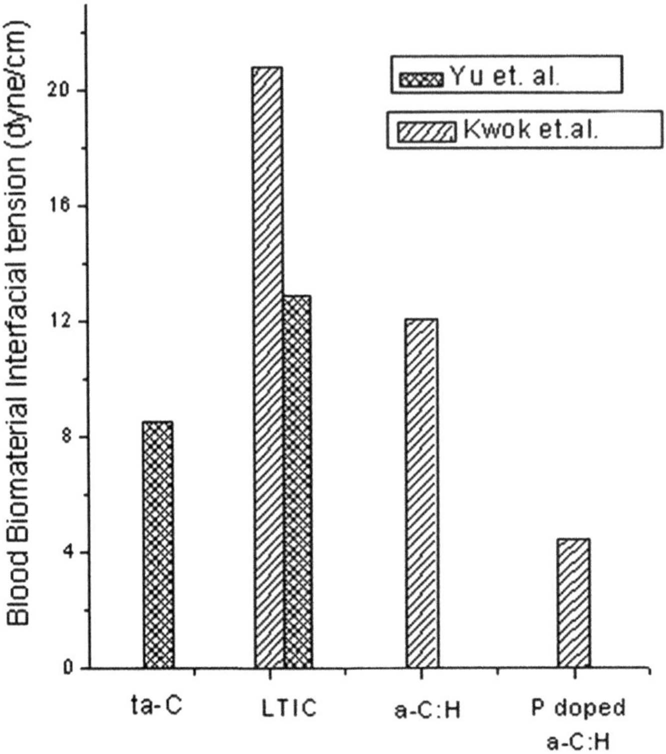

Fig. 2 Biocompatibility of different types of DLC coatings indicated as a function of interfacial adhesion in between blood plasma and coating surface [18]. Reproduced with permissions

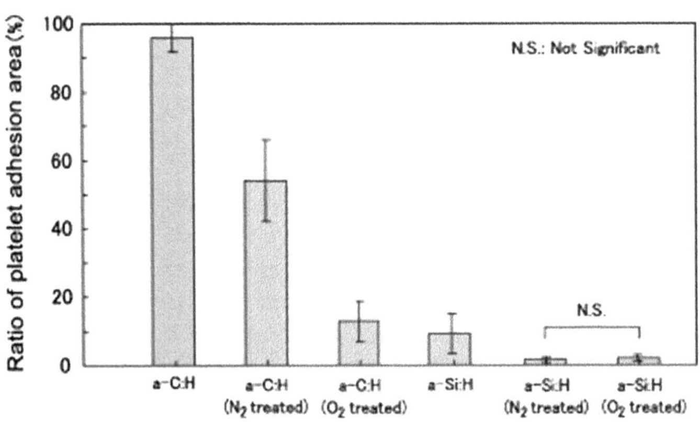

Fig. 3 Haemocompatibility of pure and Si-doped DLC coatings without plasma treatment and with nitrogen and oxygen plasma treatment indicated as a function of platelet adhesion percentage on coating area [20]. Reproduced with permissions

can be reduced to the range of 45–65% if the same coating is treated with nitrogen plasma and further reduced from 20–30% if treated with oxygen plasma. Similar to phosphorus-doped DLC, if the DLC coatings are doped with silicon and additionally treated with nitrogen and oxygen plasmas, the platelet adherence could be reduced lower than 10% of coating area. Hence, it can be summarized that the DLC coatings have better biocompatibility and haemocompatibility than the above-described materials, and it can be further improved by doping with additional elements like phosphorus and silicon and also with nitrogen and oxygen plasma treatments.

3 Current Trends for Improvement in DLC Properties

Since the pure DLC coatings have high hardness, low friction and higher wear resistance, but foreign element doping usually reduces the mechanical strength. For example, the hardness may reduce from 13 to 9 GPa, friction coefficient may increase from ~0.10 to ~0.20 and the wear rate usually increases by 5 times after silver doping up to 13% in DLC [21]. In this case, the silver has improved the biocompatibility of the DLC coating, but it has also reduced the mechanical life of the coating at the same time. Similar trends can be found in other studies [22–27]. Therefore, the researchers are equally working to improve DLC performance other than doping methods, such as DLC surface texturing [28], multi-layer architectures [29], dual-phase carbon materials [30] and heat treatment.

The heat treatment has proved benefits for some of the coating materials, where a change in material phases during heat treatment leads to the improvement in coating properties. The researchers working on carbon coatings also borrowed the same idea and tried to improve the DLC performance with heat treatment methods. It could be summarized that the heat treatment has given a slight improvement to DLC, such as an increase in hydrophobicity [31], as shown in Fig. 4 where the contact angle may increase from ~85° to ~110° which happened due to change in surface roughness. Otherwise, most of the properties like hardness, anti-reflection, coating life, etc. are reduced with heat treatment. Therefore, the researchers are still investigating an effective heat treatment method of carbon coatings which should improve the overall coating performance instead of minimizing the DLC features. The following sections will highlight the demerits of conventional one-step heat treatment on DLC performance followed by the benefits of two-step heat treatment which is a fresh area in DLC coatings.

4 Heat Treatment of DLC Coatings

Figure 5 demonstrates the effect of heat treatment on microstructure in transmission electron microscopy (TEM) images. It can be observed that the DLC coatings retained their amorphous structure up to 500 °C. The heating at elevated temperatures transforms the amorphous structure into graphite [32] clustering. Once the coating loses its DLC characteristics, it will lose its performance as well. Since the graphite is soft, therefore, the coating hardness will reduce, and the friction coefficient will increase which usually depends on the size of the graphite clusters [33]. Hence, the heat treatment of DLC is valid as long as it retains its diamond-like features rather than graphite-like characteristics. Since the transmission electron microscopy (TEM) is a detailed, systematic and time-consuming test, therefore, Raman spectroscopy could be an attractive alternative to estimate the change in the microstructure of DLC coating after heat treatment. The Raman spectroscopy is a quick tool, where a laser of specific wavelength interacts with material and provides the structure information in a few minutes. Figure 6 presents the Raman spectra of DLC coatings tested with and without heat treatment. The DLC spectra usually have two peaks, IG around 1560 cm^{-1} and ID around 1350 cm^{-1}. The peak positions classify the carbon materials such as a-C DLC have IG peak position around 1520 cm^{-1} which have dominant aliphatic rings of carbon atoms, whereas IG peak position ~1575 cm^{-1} indicates the ta-C which generally have a large amount of carbon atoms connected in chain structures.

The IG peak position at 1580 cm^{-1} presents the nano-graphitic and at 1600 cm^{-1} presents the graphite structure. The ID peak position has various interpretation based on the material and experimental conditions. It is generally used to interpret the defects, amount of graphite [34] and cluster size present in the DLC coatings. Referring to Fig. 6, it can be observed that the DLC coating without heating does not have ID shoulder peak which starts appearing after heat treatment and become prominent for DLC coating heat treated at 400 °C which clearly indicates the increase in a graphitic fraction in DLC coatings. Meanwhile, the right shift of IG peaks also shows the increase in a graphitic fraction. The more details on Raman spectroscopy of carbons materials could be found in Ferrari and Robertson three-stage model and related work [35, 36]. The Raman spectra usually present the qualitative change in

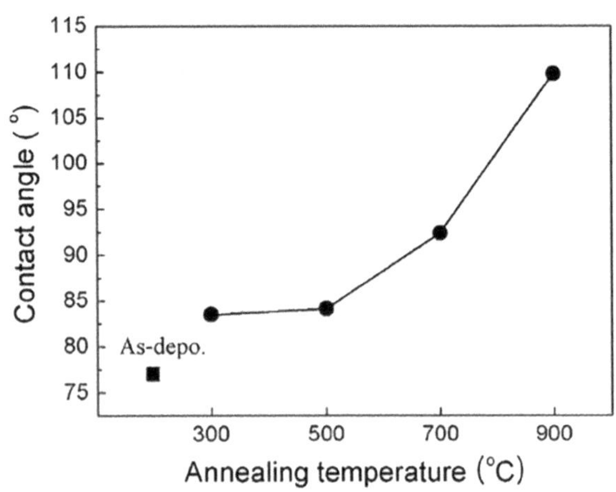

Fig. 4 Effect of heat treatment on the contact angle of DLC surface as a function of annealing temperature [31]. Reproduced with permissions

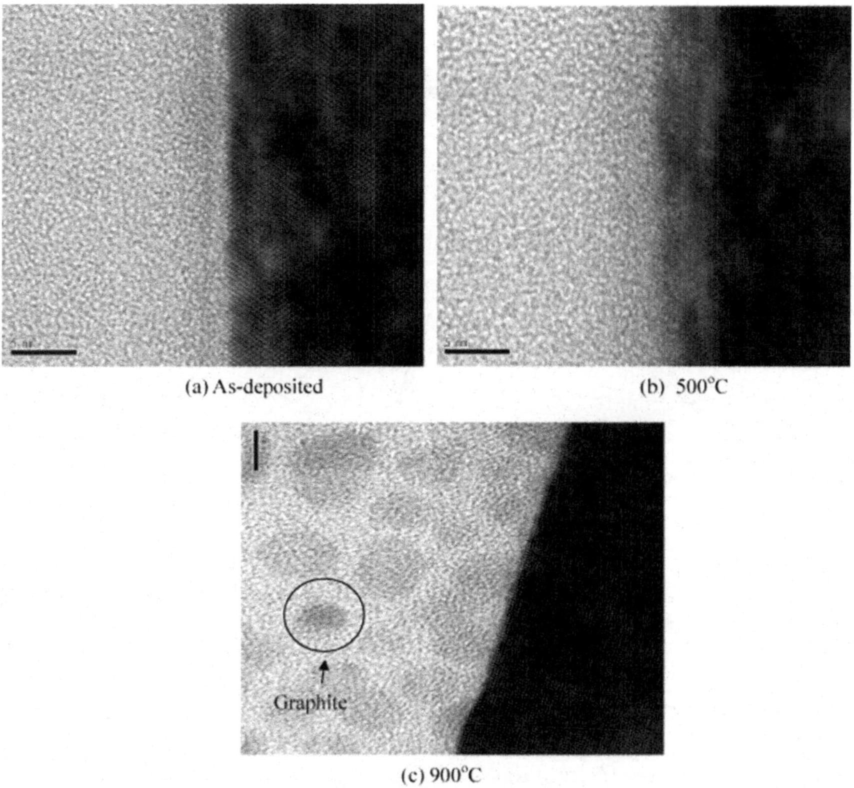

Fig. 5 Transmission electron microscopy (TEM) images of the DLC films, without heat treatment, i.e. as deposited (**a**) and with heat treatment, annealed, at 500 °C (**b**), and 900 °C (**c**) [31]. Reproduced with permissions

coating structure. However, the spectra could be processed through XPSPEAK to obtain IG and ID peak positions that are managed to get ID/IG ratios and full-width of half-maximum (FWHM) values. The ID/IG ratios interpret the average crystallite size of graphite structure, coating hardness and approximate ratio of sp^3 and sp^2 carbon atoms available in the coatings. Similarly, higher FWHM values correspond to an amorphous structure, and lower FWHM values indicate the crystalline structure which helps to differentiate the amount of graphite in the DLC coating qualitatively. The deconvolution of Raman spectra is usually performed with Gaussian multi-peak fitting which calculates ID/IG by iterating the area under curves, a Lorentzian fitting which considers the ID and IG peak intensities or the hybrid Lorentzian-Gaussian fittings.

The change in coating structure mainly occurs due to the transformation in atomic bonding. The DLC coatings are characterized by higher sp^3 fractions. The heat treatment provides additional energy to carbon atoms which enhance their kinetic energy, thermal migration and weakens the atomic bonding. Therefore, a transformation of

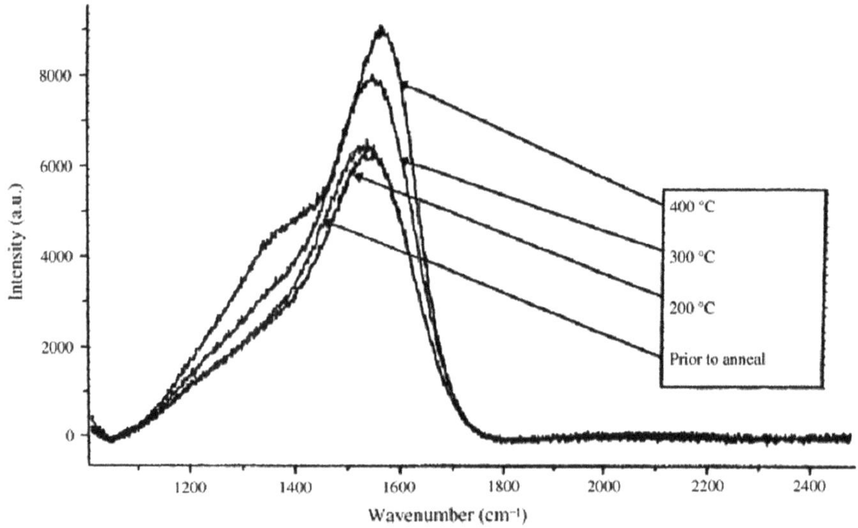

Fig. 6 Effect of heat treatment on change in the atomic structure of a-C:H DLC coatings indicated by Raman spectroscopy. The image presents the Raman spectra of pristine, 200, 300 and 400 °C heat-treated DLC coatings [37]. Reproduced with permissions

sp^3 carbon bonds to *sp*^{2} carbon bonds occurs with an increase in heating temperature. The increase in *sp*^{2} bonds develops the graphite clustering, and overall coatings structure starts changing from diamond-like to graphite-like coating. The *sp*^{3} and *sp*^{2} fractions could be found precisely with X-ray photoelectron spectroscopy (XPS) and electron energy loss spectroscopy (EELS). Referring to Fig. 7, it presents the variation in *sp*^{3} fraction and *sp*^{2}-to-*sp*^{3} ratio of carbon atoms in the DLC coating as a function of annealing temperatures. It can be observed that the *sp*^{3} fraction remains in the similar range ~70% up to 400 °C and then started to decrease with an increase in annealing temperature. At the same time, the *sp*^{2} bonds increased, and *sp*^{2}/*sp*^{3} ratio increases after 400 °C with an increase in temperature. Generally, the DLC lose its features around 400 °C, but the transformation temperature could be slightly more or less, depending on the deposition technique and experimental conditions as observed 500 °C for the coating as shown in Fig. 5. The heat treatment dramatically influences the mechanical properties such as hardness and Young's modulus. Since the diamond-like structure transforms into the graphite-like structure, therefore, the coating hardness reduces with the heating duration and heating temperature.

Figure 8a presents the hardness variation of DLC coatings with annealing temperature. It can be observed that the impact of heating temperature is dominant than DLC composition or the heating environment. The DLC hardness always reduces with an increase in annealing temperature either for pure DLC or DLC doped with MoS_2 and also whether the annealing was either performed in air or the low-pressure environment. The diamond is most hard while the graphite is a soft material. The diamond and graphite have a hardness of 10 and 1–2 at Mohs scale, respectively. The

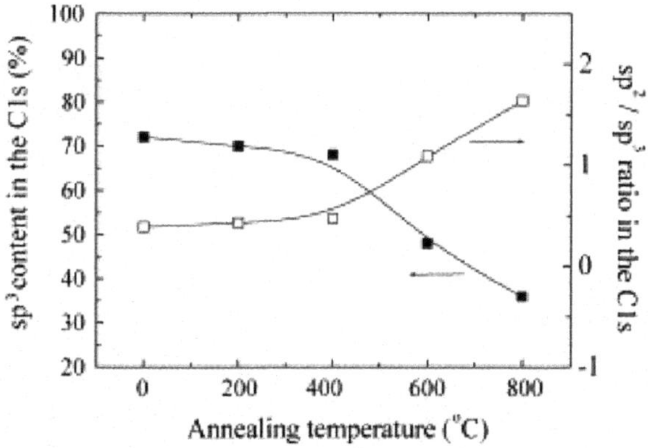

Fig. 7 Change in sp^3 carbon fraction and sp^2-to-sp^3 ratio as a function of annealing temperature. The change in sp^2-to-sp^3 ratio alters the DLC coating features [38]. Reproduced with permissions

increase in temperature transforms the diamond-like sp^3 atomic structure into sp^2 clustering and promotes graphitization which overall reduces the coating hardness. Generally, the pre-DLC coatings lose their hardness after 200 °C and drastically reduced after 300 °C when heated in air. The pure DLC coatings have a hardness of ~13.5 GPa which remain similar up to 200 °C and then reduced to ~11.5 GPa at 300 °C and further reduced to ~7 GPa at 400 °C. The Young's modulus also has a similar trend which reduced from ~150 to ~75 GPa by heating at 400 °C. The increase in temperature has rearranged the nanostructure by increasing the thermal migration of vacancies and carbon atoms and localized plastic flow. Consequently, the ductility increases, and residual stress and Young's modulus reduce as a function of temperature. The MoS_2 doping in DLC has enhanced the thermal resistance. It was observed that the hardness of MoS_2 doped DLC retained up to 400 °C (double than pure DLC) and then started to reduce down. The pristine MoS_2 doped DLC has a hardness of ~8.5 GPa which remain similar up to 400 °C and then reduced to ~7.5 GPa by heating at 600 °C. Referring to Fig. 8b, A similar trend was observed for pure and MoS_2 doped DLC coatings when heated at low pressure. However, their thermal resistance was higher in vacuum conditions than the airy environment. The pure DLC coatings remain stable up to 300 °C (while in the air it was 200 °C), and then, hardness abruptly reduces by further increasing the annealing temperature. A ~13.5 GPa hardness of pure DLC coatings remain similar up to 300 °C and then reduced to ~9 GPa at 500 °C and further reduced to ~7 GPa at 500 °C. MoS_2 doped DLC coatings were found less affected by the heating environment as they have similar variation in hardness when annealed either in the air or vacuumed conditions. Similar trends for the decrease in DLC coating hardness upon heat treatment can be found in other studies [39].

The transformation of the DLC structure with heat treatment also influences the friction coefficient. The microstructural change also changes the surface morphology, especially the surface roughness which increases the run-in period. Therefore, the coefficient of friction also increased. The friction coefficient also depends on the size of the graphite clusters [33]. Since the graphite clustering increases with annealing

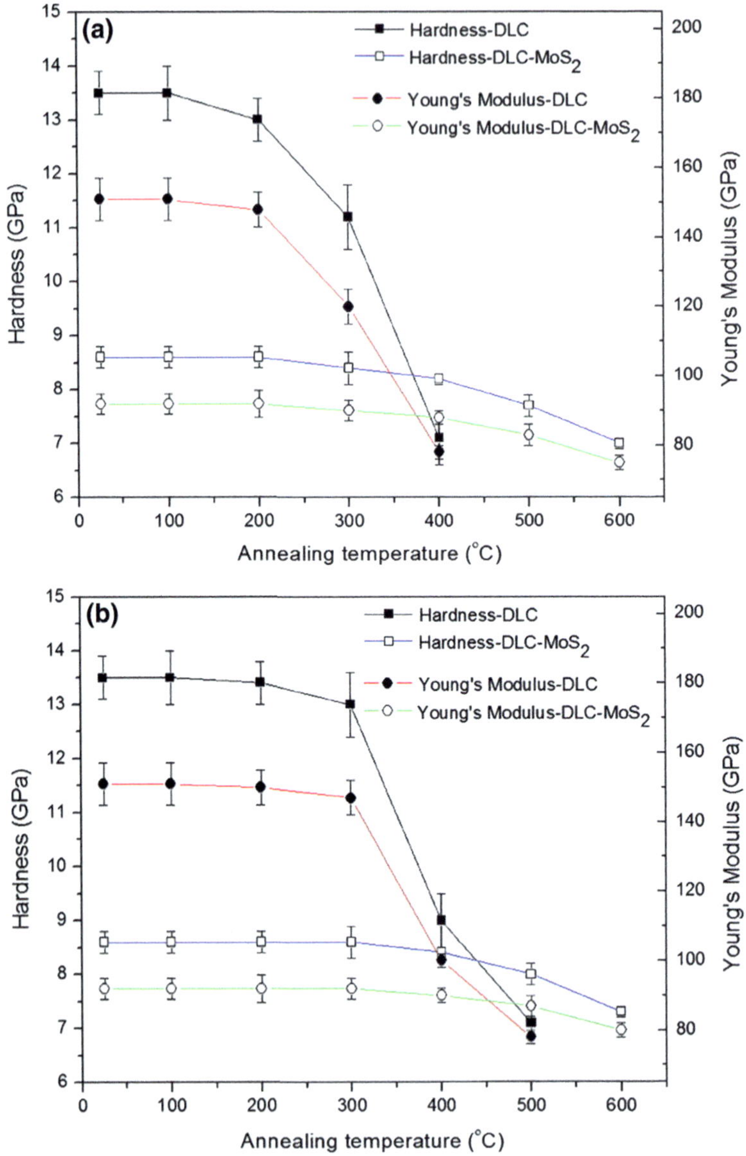

Fig. 8 Change in hardness and Young's modulus of pure DLC and MoS_2 doped DLC coatings as a function of annealing temperature performed in air (**a**) and low-pressure environment (**b**) [40]. Reproduced with permissions

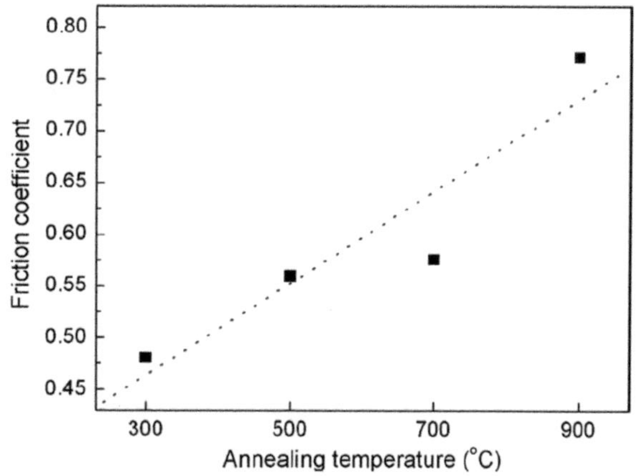

Fig. 9 Change in the friction coefficient of DLC coatings in relation to the annealing temperature [31]. Reproduced with permissions

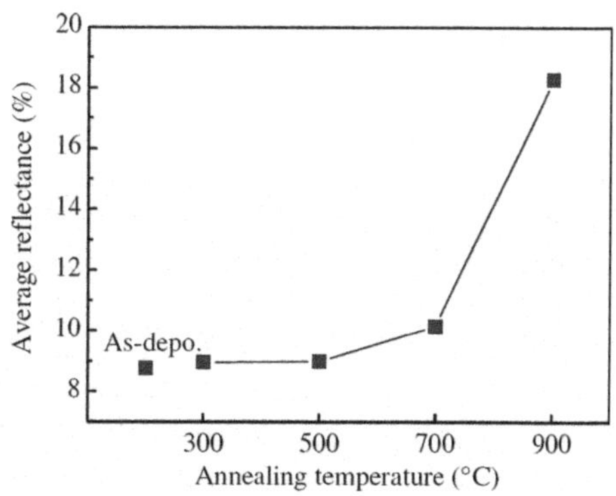

Fig. 10 Change in reflectance of DLC coatings as a function of annealing temperatures [32]. Reproduced with permissions

temperature, therefore, it could be the reason for a directly proportional relationship in between the friction coefficient and an annealing temperature as shown in Fig. 9. Generally, the DLC coatings have a friction coefficient between 0.08 and 0.2. It could be observed that friction coefficient of DLC coating increased to ~0.47 at 300 °C, ~0.55 in the temperature range of 500–700 °C and further increased to ~0.78 at the annealing temperature of 900 °C. For comparison, it can be noted that a bare silicon wafer generally has a friction coefficient of ~0.8, and mirror polished stainless steel usually have a friction coefficient of ~0.5 when tested with a steel ball.

Similarly, the heat treatment also influences the optical properties. The DLC coatings are recognized with excellent anti-reflection properties. However, the anti-reflection also reduces with increase in heating temperature. Referring to Fig. 10, it can be seen that the DLC coatings have ~9% reflection at room temperature which remains similar up to 500 °C. The reflection increases up to ~10% at 700 °C and abruptly increased to ~18% at 900 °C.

Conclusively, if the conventional heat treatment of DLC coatings gives a slight benefit for the self-cleaning application, then it destroys the other DLC features such as sp^3 fraction, hardness, friction coefficient and anti-reflection properties.

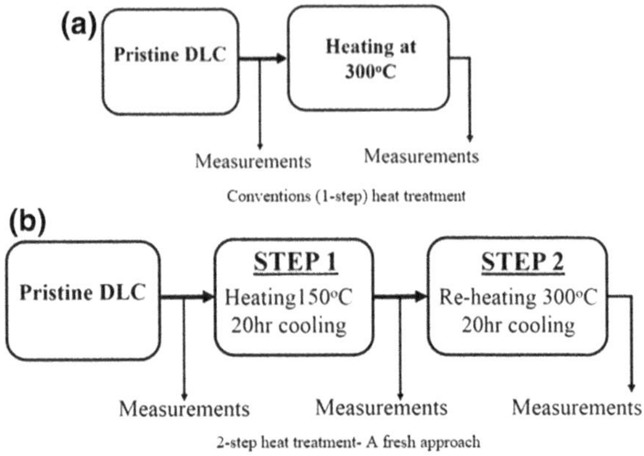

Fig. 11 Flow chart of a conventional one-step method (**a**) and recently practised two-step method (**b**) of heat treatment of DLC coatings

5 Effective (Two-Step) Heat Treatment for DLC Coatings

A fresh two-step heat treatment method has recently practised [41] to improve the structural, mechanical and tribological characteristics of DLC coatings. Generally, the graphitic shift in DLC properties is prominent around 300 °C such as the decline in hardness from ~14 to ~8 GPa [42–44] or fractures energy [45] from ~10 to ~2.0 J/m^2. Therefore, the effect of conventional (one-step) and two-step heat treatment has investigated at 300 °C temperature. Figure 11 summarize the flow chart of traditional one-step and newly practised two-step heat treatment of DLC coatings. In one-step heat treatment process, the coatings are directly heated at 300 °C, while in two-step heat treatment process, the DLC coatings are first heat-treated at 150 °C followed by 20 h furnace annealing and then again the same specimens reheated at 300 °C. The atomic structure, graphitization rate and thermal migration of carbon atoms/material flow inside the DLC coating are greatly influenced by heating temperature. It is presumed that the change in heating pattern will affect the coating performance as well.

5.1 *Atomic Structure Analysis*

Figure 12 presents the Raman spectroscopy results of one-step and two-step heat-treated DLC coatings deposited in the range of bias voltage from −40 to −140 V. ID, and IG peak positions indicate the atomic structure of carbon atoms available in the coatings. For example, diamond have standard ID peak at 1332 cm^{-1}, and graphite has IG peak at 1600 cm^{-1}. Similarly, DLC coatings having more carbon aliphatic rings present IG peak position at 1520 cm^{-1} while the dominant chain structure present IG peak position at 1575 cm^{-1}. These standard indicators are widely reported in the literature. Each bias voltage delivers a unique combination of sp^2 and sp^3 fractions in carbon coatings. The heat treatment alters the atomic structure, i.e. sp^2 and sp^3

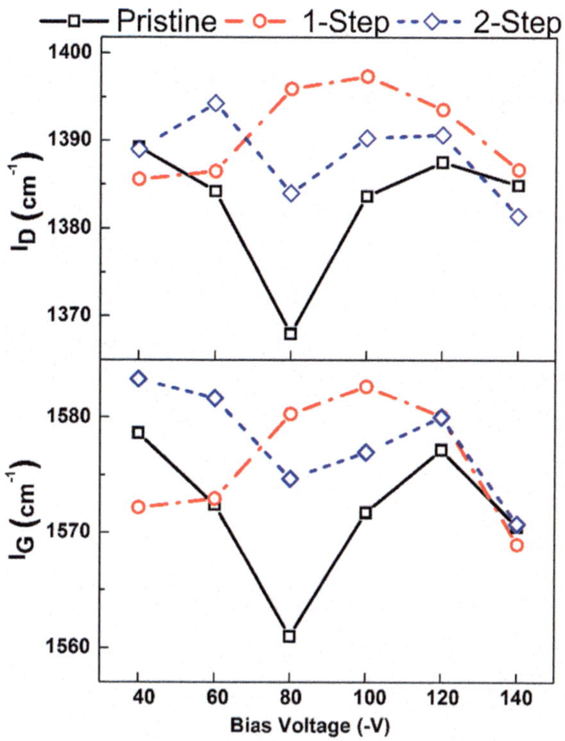

Fig. 12 ID and IG peak position obtained after one-step and two-step heat treatment process [41]. Reproduced with permissions

ratio, which leads to the change in coating properties. There could be a formation of new sp^2 or sp^3 carbon bonds or the transformation in between sp^2 and sp^3 bonds. Similarly, the carbon clustering or the change in the ring structure and chain structure all depends on the thermal conditions and composition of sp^2 and sp^3 fractions in pristine coatings. The said changes are estimated with ID and IG peak positions and help to understand the effect of systematic parameters like bias voltage or temperature on coating structure.

It can be observed that the I_D and I_G peaks positions shifted down from 1390 to 1370 cm^{-1} and 1578 to 1560 cm^{-1} by increasing the bias from −40 to −80 V. Hereafter, both ID and IG peak positions again increased with an increase in bias and reached to 1387 and 1577 cm^{-1} at −120 V bias. After one-step heat treatment, a new trend was developed which presents the formation of new bonds for −40 and −140 V bias only, while excessive graphitization has observed for the coatings deposited in between −80 and −120 V. Meanwhile, the trend was almost consistent for −60 V coatings. The two-step heat treatment process gave a mixed trend. The specimens which had higher sp^2 fractions got more graphitization like −40 and −60 V than those who had a high sp^3 fraction like −80 and −100 V. The results of both techniques were similar for −120 V. However, −140 V has an exceptional behaviour where there was no graphitization occurred, and possibly new carbon bonds were

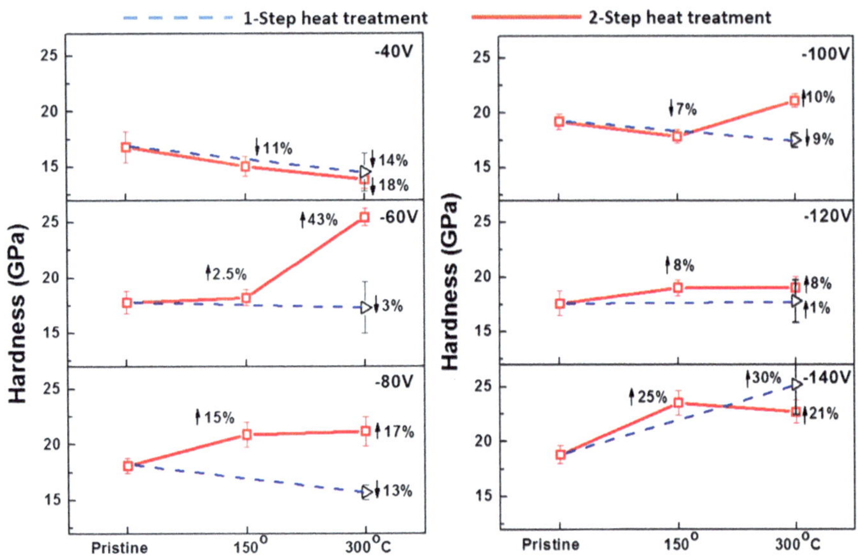

Fig. 13 Change in hardness of DLC coatings as a function of one-step and two-step heat treatment at 300 °C [41]. Reproduced with permissions

formed. In general, the results reflect that the two-step heat treatment has produced less graphitization than one-step heat treatment.

5.2 *Hardness and Fracture Toughness Measurement*

Figures 13 and 14 present the change in hardness and fracture toughness after one-step and two-step heat treatment process. Figure 13 demonstrates the change in hardness as a function of one-step 300 °C and two-step 300 °C heat treatment for a bias voltage of −40, −60, −80, −100, −120 and −140 V coatings. For pristine coatings, the minimum hardness was ~ 17 GPa at −40 V while the maximum hardness was ~ 19.2 GPa appeared at −100 V. It has been observed that the hardness of pristine DLC coating slightly increases from ~ 16 to ~ 19 GPa by increasing the bias from −40 to −80 V, and hereafter, overall it decreased again. For one-step 300 °C heat treatment, almost 2–3 GPa hardness was reduced for all samples except −140 V. The −140 V has unusual behaviour, and hardness was increased from ~18 to ~ 25 GPa after one-step 300 °C heat treatment. The two-step heat treatment has a reverse trend where the hardness of all specimens increased at least ~2 GPa for each sample except −40 V. A decrease in ~3 GPa of hardness for −40 V specimen is likely due to excessive graphitization as observed in Raman spectroscopy result.

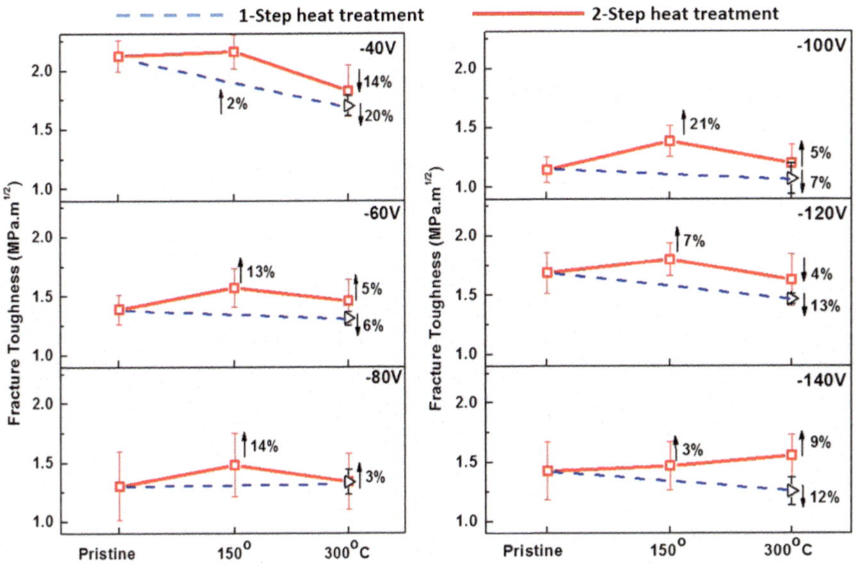

Fig. 14 Change in fracture toughness variation of DLC coatings tested with one-step and two-step heat treatment methods at 300 °C [41]. Reproduced with permissions

In this series, the best results were presented by −60 V where hardness increased from ~17.8 to ~ 25.5 GPa. The trends give a quick overview of each sample to compare the outputs of one-step and two-step heat treatment process. The DLC coating hardness decreased for all the specimens deposited at a bias of −40 to −120 V except for −140 V in 1–step 300 °C heat treatment. In this series, the maximum decline was 18% for −40 V, and the maximum increase was 30% for −140 V. For the two-step heat treatment process, only −40 V showed a decrease of 18% hardness due to an excessive amount of sp^2 fraction and graphitization. Otherwise, it retained or increased for all specimens. The maximum increase of 43% in hardness was observed for −60 V. In general, the hard materials are brittle and possess poor toughness. It was observed that the pristine −40 V sample have a maximum fracture toughness of 2.1 MPa.m$^{1/2}$ which reduces to a minimum of 1.1 MPa.m$^{1/2}$ by increasing the bias voltage up to −100 V, and then it started to increase again. For one-step 300 °C, the fracture toughness has decreased for all specimens regardless of substrate bias. The toughness decreased to 1.7 MPa.m$^{1/2}$ at −40 V and 1.0 MPa.m$^{1/2}$ for −100 V. Whereas, two-step 300 °C has demonstrated a mixed trend, where the toughness either retained or increased for some of the specimens, like −60 and −140 V. The maximum improvement was observed for the sample deposited at −140 V where the toughness was increased from 1.4 to 1.6 MPa.m$^{1/2}$. Figure 14 gives the overview for change in fracture toughness of same coatings tested with one-step and two-step 300 °C heat treatments. It can be observed that toughness reduced for all specimens after one-step heat treatment. The maximum decline of 20% was recorded for −40 V. Two-step heat treatment had a mixed trend where −40 V had a huge decline of 14%.

Otherwise, all specimens presented a slight increase in toughness or at least kept the same toughness as pristine specimens. For this method, a maximum improvement of 9% was observed for −140 V specimens.

5.3 *Tribological Studies*

Figure 15 presents the wear track profiles of DLC coatings tested before and after one-step and two-step heat treatments. The reciprocating wear test was performed for 5400 cycles at 6 mm/s reciprocating velocity and 3 mm stroke length with AISI steel ball of 4 mm diameter. It can be observed that the track depth of pristine DLC

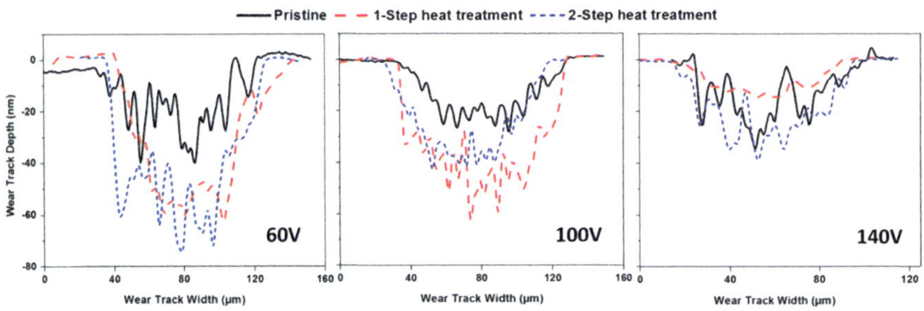

Fig. 15 Wear profile of pristine, one-step and two-step heat-treated DLC coatings [41]. Reproduced with permissions

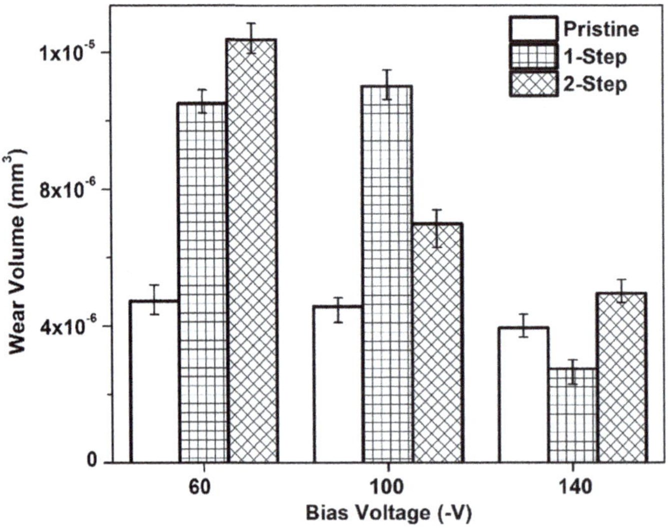

Fig. 16 Wear volume of pristine and heat-treated DLC coatings [41]. Reproduced with permissions

was maximum for −60 V samples which gradually reduced with an increase in bias voltage to −140 V. At atmospheric conditions, the DLC wear resistance is known to increase with hardness. Therefore, it can be observed that the increase in bias voltage increases the hardness, and the wear track depth also reduces simultaneously. For the thermal environment, the effect of bias voltage on hardness and wear resistance solely depends on the change in the composition of sp^2 and sp^3 fractions as a function of temperature. The track depths sufficiently increased after one-step and two-step heat treatment for −60 V. However, it is interesting to see that the incremental increase in wear track depth was lower for the specimens deposited at 100 V in comparison with −60 V. The difference further reduced from −140 V, and track depth for the pristine, one-step and two-step heat-treated specimens was in the same range. Figure 16 highlights the wear loss of pristine, one-step 300 °C heat-treated and two-step 300 °C heat-treated specimens which were deposited at −60, −100 and −140 V. It can be observed that the wear volume gradually reduced with an increase in bias voltage. After heat treatment, the wear volume increased in general despite one-step or two-step heat treatment methods. However, it can also be observed that the increases in wear volume were relatively higher for low bias coatings like −60 V than the coatings which were deposited at high bias like −140 V. The wear volume of −60 V pristine coating increased from ~4×10^{-6} mm^3 to 9×10^{-6} mm^3 after one-step 300 °C heat treatment and up to ~10×10^{-6} mm^3 after two-step 300 °C heat treatment, which was a huge increase of 100 and 130%, respectively. However, in the case of −100 V pristine, this percentage remained the same for one-step 300 °C heat treatment but reduced to 55% for two-step 300 °C heat treatment. −140 V DLC coatings have mixed trends where the wear volume reduced ~20% after one-step 300 °C heat treatment, but 20% increased after two-step 300 °C heat treatment.

6 Concluding Remarks

The chapter briefly introduces DLC coatings, their unique properties, manufacturing methods, global revenue and applications. Recently, the DLC coatings are trending in the biomedical sector. Therefore, a dedicated section gives a brief overview of the biocompatibility of DLC coatings. In addition, the doping of foreign element in DLC matrix and plasma treatment to further improve the biocompatibility of DLC coatings has covered. The DLC coatings are widely being researched to enhance their life and functionalities. It has been observed that conventional heat treatment may improve some of the properties like surface hydrophobicity, but it reduces most of the major properties like sp^3 fraction, hardness, anti-reflection and increase the friction coefficient at the same time. The chapter introduces the alternative two-step heat treatment method which has been observed relatively better than the conventional one-step heat treatment method. It has observed that the DLC heat treatment at 150 °C and then reheating at 300 °C can improve hardness up to 20% and toughness up to 10% simultaneously when deposited at a bias voltage of −140 V. Generally, the hardness and toughness reduce due to graphitization upon heat treatment in a single step at 300 °C during the same experimental conditions. Since the two-step heat treatment of DLC coating is a fresh area wherein the fundamental mechanisms are being explored in more details and expected to enhance the coating performance and coating life concurrently.

References

1. Hainsworth SV, Uhure NJ (2007) Diamond like carbon coatings for tribology: production techniques, characterisation methods and applications. Int Mater Rev 52:153–174
2. Götze A, Makowski S, Kunze T, Hubner M, Zellbeck H, Weihnacht V, Leson S, Beyer E, Joswig J, Seifert G, Abrasonis G, Posselt M, Fassbender J, Möller W, Gemming S, Krause M (2014) Tetrahedral amorphous carbon coatings for friction reduction of the valve train in internal combustion engines. Adv Engg Mater 16:1226–1233
3. Decker TG, Lundie GP, Pappas DL, Welty RP, Parent CR (1994) Amorphous diamond coating of blades. US Patent US5799549A
4. Fukui H, Okida J, Omori N, Moriguchi H, Tsuda K (2004) Cutting performance of DLC coated tools in dry machining aluminum alloys. Surf Coat Technol 187:70–76
5. Fung MK, Lai KH, Lai HL, Chan CY, Wong NB, Bello I, Lee CS, Lee ST (2000) Diamond-like carbon coatings applied to hard disks. Diamond Relat Mater 9:815–818
6. Tibrewala A (2006) Piezoresistive effect in diamond-like carbon films. Cuvillier Verlag, Gottingen
7. Dearnaley G, Arps JH (2005) Biomedical applications of diamond-like carbon (DLC) coatings: a review. Surf Coat Tech 200:2518–2524
8. Schmellenmeier H (1953) Die Beeinflussung on festen obserflachen durch eine ionisierte Gasatmosphare. Exp Tech Phys 1:49–68
9. McWilliams A (2010) Diamond, diamond-like and CBN films & coating products. BCC Research LLC, USA
10. Suski CA, de-Oliveira C (2012) Coating and applications In: Davim JP (ed) Tribology in manufacturing technology, materials forming, machinging and tribology, Springer, Heidelberg
11. Bogdanowicz R (2015) Advancements in diamond-like carbon coatings. In: Makhlouf ASH, Scharnweber D (eds) Handbook of nanoceramic and nanocomposite coatings and materials, Elsevier, pp 183–205
12. Grill A (2003) Diamond-like carbon coatings as biocompatible materials - an overview. Diamond Relat. Mater 12(2):166–170
13. Thomson LA, Law FC, Rushton N, Franks J (1991) Biocompatibility of diamond-like carbon coating. Biommaterials 12:37–40
14. Bokros JC, Ellis WH (1970) Method of preparing an intravascular defect by implanting a pyrolytic carbon coated prosthesis. US Patent No 3,526,005
15. Paul R (2017) Diamond-like-carbon coatings for advanced biomedical applications. Glob J Nano 2(5):555598
16. Hauert R (2003) A review of modified DLC coatings for biological applications. Diamond Relat Mater 12:583–589
17. Ruckensten E, Gourisanker SV (1984) A surface energetic criterion of blood compatibility of foreign surfaces. J Colloid Interface Sci 101:436–451
18. Roy RK, Lee KR (2007) Biomedical applications of diamond-like carbon coatings: a review. J Biomed Mater Res Part B: App Biomater 83:72–84
19. Zia AW, Wang YQ, Lee S (2015) Effect of physical and chemical plasma etching on surface wettability of carbon fiber-reinforced polymer composites for bone plate applications. Adv Poly Technol 36(1):21480

20. Roy RK, Choi HW, Yi JW, Moon MW, Lee KR, Han DK, Shin JH, Kamijo A, Hasebe T (2009) Hemocompatibility of surface-modified, silicon-incorporated, diamond-like carbon films. Acta Biomater 5:249–256
21. Manninen NK, Ribeiro F, Escudeiro A, Polcar T, Carvalho S, Cavaleiro A (2013) In fluence of Ag content on mechanical and tribological behavior of DLC coatings. Surf Coat Technol 232:440–446
22. Chang YY, Wang DY (2006) Structural and electrical properties of Cr doped a-C: H films synthesized by a cathodic-arc activated deposition process. Surf Coat Technol 200:3170–3174
23. Tang XS, Wang HJ, Feng L, Shao LX, Zou CW (2014) Mo doped DLC nanocomposite coatings with improved mechanical and blood compatibility properties. Appl Surf Sci 311:758–762
24. Liu N, Zhu H, Wei Q, Long H, Deng Z, Yu Z, Xie Y, Wang J, Zhou LMK (2017) A niobium and nitrogen co-doped dlc film electrode and its electrochemical properties. J Electrochem Soc 164:H1091–H1098
25. Ma G, Gong S, Lin G, Zhang L, Sun G (2012) A study of structure and properties of Ti-doped DLC film by reactive magnetron sputtering with ion implantation. Appl Surf Sci 258:2045–3050
26. Ray SC, Pong WF, Papakonstantinou P (2016) Iron, nitrogen and silicon doped diamond like carbon (DLC) thin films: a comparative study. Thin Solid Film 640:42–47
27. Safaie P, Eshaghi A, Bakhshi SR (2016) Optical properties of oxygen doped diamond-like carbon thin films. J Alloys compd 672:426–432
28. Arslan A, Masjuki HH, Kalam MA, Varman M, Mufti RA, Mosarof MH, Khuong LS, Quazi MM (2016) Surface texture manufacturing techniques and tribological effect of surface texturing on cutting tool performance: a review. Crit Rev Solid State Mater Sci 41(2016):447–481
29. Lin Y, Zia AW, Zhou ZF, Shum PW, LI KY (2017) Top of form development of diamond-like carbon (DLC) coatings with alternate soft and hard multilayer architecture for enhancing wear performance at high contact stress. Surf Coatings Technol 320:7–12
30. Zia AW, Zhou ZF, LI KY (2017) A preliminary wear studies of isolated carbon particles embedded diamond-like carbon coatings. Tribol Int 114:42–44
31. Choi WS, Joung YH, Heo J, Hong B (2012) Friction force microscopy study of annealed diamond-like carbon film. Mater Res Bull 47:2780–2783
32. Choi WS, Hong B (2008) The effect of annealing on the properties of diamond-like carbon protective antireflection coatings. Renew Energy 33:226–231
33. Katiyar JK, Sinha SK, Kumar A (2016) Friction and wear durability study of epoxy-based polymer (SU-8) composite coatings with talc and graphite as fillers. Wear 362:199–208
34. Robertson J (2001) Amorphous and non-crystalline carbons. In: Delhaes P (ed) Graphite and precursors. Gordon and breach science publishers, The Netherlands
35. Ferrari AC, Robertson J (2000) Interpretation of Raman spectra of disordered and amorphous carbon. Phys Rev B 61:14095–14107
36. Robertson J (2002) Diamond-like amorphous carbon. Mater Sci Eng 37:129–281
37. Ogwu AA, Lamberton RW, Morley S, Maguire P, McLaughlin J (1999) Characterisation of thermally annealed diamond like carbon (DLC) and silicon modified DLC films by Raman spectroscopy. Phys B 269:335–344
38. Li JJ, Gu CZ, Peng HY, Wu HH, Zheng WT, Jin ZS (2005) Field emission properties of diamond-like carbon films annealed at different temperatures. App Surf Sci 251:236–241
39. Nobili L, Guglielmini A (2013) Thermal stability and mechanical properties of fluorinated diamond-like carbon coatings. Surf Coat Technol 219:144–150
40. Niakan H, Zhang C, Hu Y, Szpunar JA, Yang Q (2014) Thermal stability of diamond-like carbon–MoS2 thin films in different environments. Thin Solid Films 562:244–249
41. Zia AW, Zhou ZF, Shum PW, LI KY (2017) The effect of two-step heat treatment on hardness, fracture toughness, and wear of different biased diamond-like carbon coatings. Surf Coat Technol 320:118–125
42. Meškinis Š, Čiegis A, Vasiliauskas A, Šlapikas K, Gudaitis R, Yaremchuk I, Fitio V, Bobitski Y, Tamulevičius S (2016) Annealing effects on structure and optical properties of diamond-like carbon films containing silver. Nanoscale Res Lett 11:146

43. Buršíková V, Navrátil V, Zajíčková L, Janča J (2002) Temperature dependence of mechanical properties of DLC/Si protective coatings prepared by PECVD. Mater Sci Eng A 324:251–254
44. Li H, Xu T, Wang C, Chen J, Liu H (2006) Annealing effect on the structure, mechanical and tribological properties of hydrogenated diamond-like carbon films. Thin Solid Films 515:2153–2160
45. Choi HW, Gage DM, Dauskardt RH, Lee KR, Oh KH (2009) Effects of thermal annealing and Si incorporation on bonding structure and fracture properties of diamond-like carbon films. Diamond Relat Mater 18:615–619

Dr. Abdul Wasy Zia is experienced in surface coatings and technology. He has earned his master's degree in mechanical engineering from Changwon National University, South Korea and his doctoral degree from City University of Hong Kong. Recently, he is employed at City University of Hong Kong, and previously he has been working at Technical University Ilmenau, Germany. His research focus is hard carbon coatings their design, manufacturing strategies, and performance evaluation especially solid mechanics including tribology.

Innovative Surface Engineering Technique for Surface Modification of Mg Alloy for Orthopedic Application

Muhammad Al'Hapis Abdul Razak, Ahmad Majdi Abdul-Rani and Abdul' Azeez Abdu Aliyu

Abstract Biodegradable materials are required for the development of bone fixation accessories. Mg was recently identified to be biodegradable material and a better alternative for temporary implants application. However, the Mg alloy corrosion rate must be suitable for the application as biodegradable orthopedic implants. Controlling the Mg alloy corrosion rate is not easy. It is hypothesized that the Mg alloy-machined surface corrosion rate is possible to be controlled by manipulating the powder mixed electro-discharge machining (PMEDM) setting parameters. Therefore, this chapter aims at generating a method on how to determine the correct combination of PMEDM setting parameters to obtain a specific corrosion rate of Mg alloy. An opened-loop PMEDM dielectric circulation system is employed in the experiments. The setting parameters involved in the experiments include Zn particles concentration (P_{con}), peak current (P_{c}), pulse-on time (T_{on}) and pulse-off time (T_{off}). The experiment results reveal that the Zn particles suspended in the dielectric fluid providing a bridge for uniform discharge energy on the Mg alloy. It results in a lower surface roughness which then leads to a lower corrosion rate due to smaller exposed surface area to the solution. Since the response output is influenced by the interaction among the setting parameters, an equation to determine the corrosion rate of Mg alloy-machined surface is generated by manipulating the value combination of those four setting parameters. With all the evidence presented in this chapter, it is proven that the Mg alloy-machined surface corrosion rate can be controlled by manipulating the PMEDM setting parameters.

Keywords Biodegradable orthopedic implants · Corrosion rate · Mg alloy · PMEDM · Zn particles

---

M. A. A. Razak (✉) · A. M. Abdul-Rani · A. A. A. Aliyu
Mechanical Engineering Department, Universiti Teknologi PETRONAS, 32610 Bandar Seri Iskandar, Perak, Malaysia
e-mail: alhapis@unikl.edu.my

M. A. A. Razak
Manufacturing Section, Universiti Kuala Lumpur Malaysian Spanish Institute, Kulim Hi-Tech Park, 09000 Kulim, Kedah, Malaysia

C. Prakash et al. (eds.), *Biomanufacturing*,
https://doi.org/10.1007/978-3-030-13951-3_11

1 Introduction and Background

Biodegradable materials are required for the development of bone fixation accessories [1]. This is due to the fact that non-degradable material-based accessories were removed after the bone healing, which is a typical process. In order to deal with the same, Mg-based biodegradable materials has gained interest in the development of screws and plates. Mg as was recently identified to be biodegradable material and a better alternative for temporary implants application due to its biocompatibility, potential for in vivo degradation, and non-toxic [2–4]. Mg alloys are highly potential to be applied in fractured bone support devices such as staples, rods, screws, and pins [5, 6]. On average, the human body contains 35 g Mg per 70 kg body weight. Our body needs 375 mg of Mg every day [7]. The density of Mg alloys is between 1.75 and 1.85 g/cm^3 and pure Mg is 1.738 g/cm^3. These are quite the same to the density of human cortical bone, which is 1.75 g/cm^3 [8]. Mg alloys also possessing fracture toughness between 17.6 and 20.7 MPa m$^{1/2}$ and mechanical tensile strength between 185 and 230 MPa are suitable for bone fixation device [9, 10]. The development of Mg-based bio-absorbable micro-clip and Mg wires was reported recently [11, 12]. Other than possessing sufficient strength, the corrosion over appropriate time was also important in designing biodegradable orthopedic implants [13–15].

Despite the fact that Mg alloys have the ability to dissolve in biological environments [16–19], the corrosion rate of the fabricated Mg alloys need a proper investigation before it can be applied as biodegradable orthopedic implants because impermanent implants should permit an ample time for newly generated tissue to replace the fractured bone [20–23]. The drawback of Mg alloys is possessing high corrosion rate. In vitro study demonstrates that the uncoated Mg alloy corrodes 12 mg per two months [24]. This disadvantage causes the implant to crack before the fractured bone recuperates [25, 26]. Furthermore, the different healing rate of fractured bone between the children and adults demands dissimilar corrosion rate of the implants. An impermanent implantation is commonly employed for more than six months to the children aged less than six years and more than ten months to the adults [27–29]. However, controlling the Mg alloy surface corrosion rate is not easy without changing its mechanical properties. The corrosion rate of Mg alloy can be reduced by alloying and surface modification/coating [1, 30, 31]. Alloying is a very helpful method for microstructure tailoring to improve the corrosion resistance of Mg alloy. The alloying elements can be applied for biomedical device-manufacturing purposes, including Zn, Ca, Mn, Zr, Ag, and Sr [32, 33]. Other than the potential to improve the mechanical properties of Mg alloy, these elements have less or no toxicity. However, surface modification/coating is the easiest way to improve the corrosion resistance of Mg alloy [34, 35]. Surface treatment will improve the mechanical properties of the implants by tailoring the coating, composition, and structure. The application of protective thin films by surface modification/coating/deposition exhibited the most promising successful practice to regulate the degradation rate of Mg alloys [36]. Therefore, this initiates a scope and stipulates logical enthusiasm to regulate the degradation behavior of Mg-based alloys with the point of expanding the implant's performance by means of various surface alteration/treatment methods.

2 Powder Mixed EDM

EDM technology has been through some stages of improvement. Recently, many researchers in the field of advanced manufacturing technology take an initiative to study new techniques to improve the EDM efficiency. One of the techniques is called PMEDM [37–39]. In PMEDM process, the dielectric fluid is mixed with conductive particles as shown in Fig. 1 to create a conductive bridge across the spark gap.

PMEDM was reported as an alternative to conventional coating/surface modification methods or techniques for the bio-implants. Recently, Prakash et.al deposited biomimetic nano-hydroxyapatite (*n-HA*) coating on biodegradable Mg–Zn–Mn alloy by PMEDM process to enhance its mechano-corrosion integrity [40]. PMEDM will facilitate in the homogeneous discharge energy distribution which creates uniform material corrosion from the workpiece and ensues in shallow craters which benefits to the improvement of machined surface quality [41–44]. PMEDM method is also highly potential for modifying the machined surface properties, topography, and element composition; thus the machined surface corrosion rate will be reduced [41, 45–48]. Hence, the formation of microcracks on the machined surface [49] produced by the conventional EDM process as presented in Fig. 2 which may cause disruption of the surface oxide layer and speeds up the progress of corrosion [50, 51] can be reduced.

The advantages of PMEDM method may be applied to reduce the surface roughness and corrosion rate of Mg alloy-machined surface by applying Zn particles as

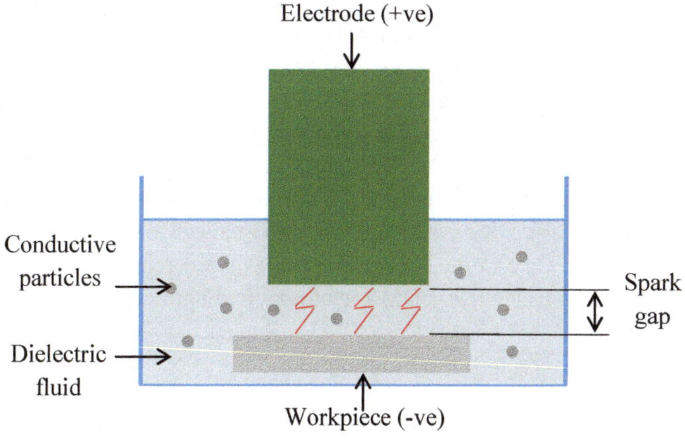

Fig. 1 Particles-mixed dielectric fluid in the PMEDM process

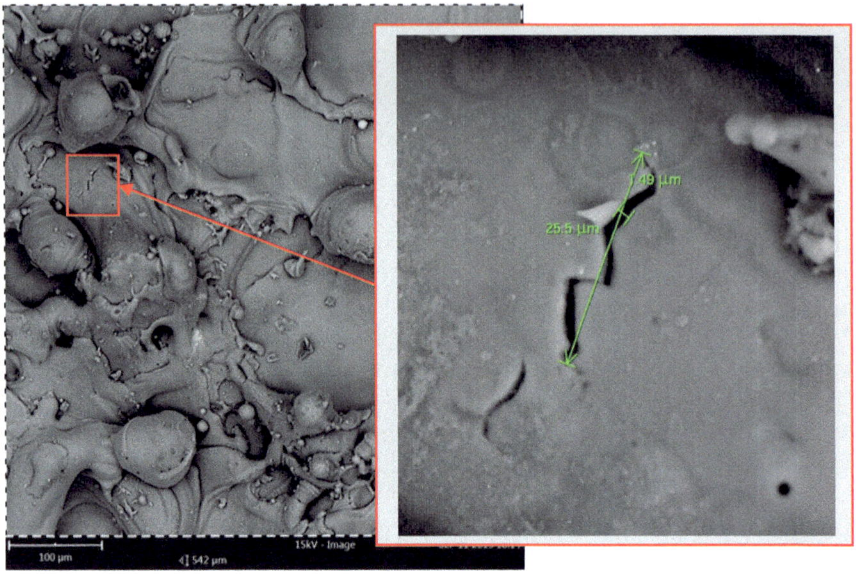

Fig. 2 Microcrack on the machined surface processed by conventional EDM

the mixture in the dielectric fluid. Zn is one of the common alloying elements for Mg alloys [52]. Mg alloys with high content of Zn is found hardly producing hydrogen gas [15]. The free hydrogen gas will reduce because the Zn ions will bind with free OH^- anions forming $Zn(OH)_2$. It is known that Zn is a trace element in the human body. The increases in Zn composition of Mg alloy will increase its yield strength. The extremely high localized temperature up to 10,000 °C [41] during the spark erosion process may allow the Zn particles present at the cutting zone to melt and lodged on the Mg alloy-machined surface. The electrode potential of Zn is −0.76 V and Mg is −2.37 V. Since the electrode potential of Zn is higher than Mg, the corrosion rate of Mg alloy-machined surface may be decreased. Based on the advantages of PMEDM over the conventional EDM method as reported in the literature, it is hypothesized that the Mg alloy-machined surface corrosion rate is possible to be controlled by manipulating the PMEDM setting parameters. Therefore, this chapter aims at generating a method to determine the correct combination of PMEDM setting parameters to obtain a specific corrosion rate of Mg alloy. To the best knowledge of the authors, till date, very limited research studies were reported on machining of Mg alloy by EDM/W-EDM/PM-EDM for biomedical applications.

3 Experimental Procedures

The specimen used in the experiment was an AZ31 Mg alloy in the shape of a cube with a constant size of 11 × 9 × 5 mm. The tool electrode used was copper. The Zn particles with 99.9% purity as determined using X-ray fluorescence (XRF) spectrometer were added in the hydrocarbon dielectric fluid in different concentrations. The total cutting depth was 2 mm. The Mitsubishi EA8 die sinker EDM was used in the experiment, and an opened-loop PMEDM dielectric circulation system was developed for this project as shown in Fig. 3 with three stirrers equipped in the reservoir.

This research involves four setting parameters which were P_{con}, P_c, T_{on}, and T_{off} with three different values as shown in Table 1. Response surface methodology (RSM) was applied in the design of experiments. Earlier, preliminary experiments were conducted to determine the range of the value since there was no reference on EDM setting parameters to machine Mg alloy has been published to the authors' knowledge.

The in vitro corrosion tests were carried out complying the ASTM G59-97 [53] with the application of the simulated body fluid [54]. The CH1608E electrochemical analyzer was used in analyzing the corrosion rate of Mg alloy. The outcomes of corrosion tests were analyzed using Design–Expert software. The machined surface was then further analyzed using Mitutoyo SurfTest SV-3000, Phenom ProX desktop scanning electron microscope with energy-dispersive X-ray spectroscope

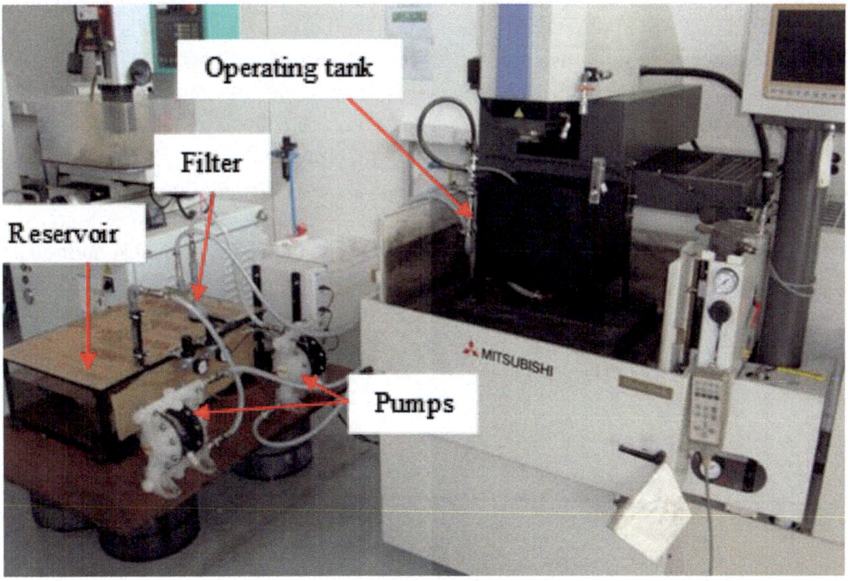

Fig. 3 Die sinker EDM with fabricated opened-loop PMEDM dielectric circulation system

Table 1 PMEDM setting parameters

Parameter	Unit	Value
P_{con}	g/l	1, 2 and 3
P_c	A	38, 47 and 55
T_{on}	μs	16, 32 and 64
T_{off}	μs	128, 256 and 512

(SEM/EDS) and Bruker S8 Tiger XRF spectrometer. Responses from the analysis of variance (ANOVA) were fitted into the quadratic mathematical model to establish the final equation to calculate the potential corrosion rate (Table 2).

4 Results and Discussion

4.1 Surface Morphology

An intensive experiment has been conducted by applying setting parameters combination of 80 V gap voltage, 47 A P_c, 16 μs T_{on}, and 512 μs T_{off} using conventional EDM and PMEDM methods. The value combination was defined in the authors' previous research work as reported in [38] as the optimum combination to obtain the lowest surface roughness using conventional EDM method. With 1 g/l P_{con} applied in the PMEDM experiment, the different physical characteristics of the Mg alloy-machined surface obtained from both methods were analyzed. The obtained surface roughness of the specimen machined using conventional EDM method was 5.805 μm. On the other hand, the obtained surface roughness of the specimen machined using PMEDM method was 5.279 μm. It brought into a reduction of 9.06% of surface roughness. The smoother surface found on the specimen machined using the PMEDM method was influenced by shorter peak and ravine, as well as the smaller size of globules on the machined surface. The SEM images as shown in Fig. 4a, b show clearly that the globules created on the Mg alloy-machined using conventional EDM method are larger than PMEDM method. In PMEDM method, the existence of Zn particles at the cutting zone causes an expansion of plasma channel which contributes into homogeneous distribution of electric sparks and ensues into shallow craters and lower surface roughness.

4.2 Analysis of Elemental Composition

Figure 5 presents the EDX spectrum of Mg alloy-machined using conventional EDM method. It indicates 1.8% of Zn composed on the specimen surface. The Zn content is found higher on the specimen surface machined using the PMEDM method as

Table 2 PMEDM setting parameters combination for specific corrosion rate

Powder concentration (g/l)	Peak current (A)	Pulse-on time (μs)	Pulse-off time (μs)	Corrosion rate (mm/year)
2	38	16	512	0.000183
3	38	16	512	0.000271
2	38	32	512	0.000291
2	55	16	512	0.000310
2	38	64	512	0.000340
2	47	16	512	0.000355
2	38	16	256	0.000359
3	38	32	512	0.000379
1	38	16	512	0.000395
3	55	16	512	0.000398
2	55	32	512	0.000417
3	38	64	512	0.000428
3	47	16	512	0.000443
3	38	16	256	0.000447
2	47	32	512	0.000462
2	55	64	512	0.000466
2	38	32	256	0.000467
2	55	16	256	0.000486
1	38	32	512	0.000503
3	55	32	512	0.000505
2	47	64	512	0.000511
2	38	64	256	0.000516
1	55	16	512	0.000522
2	47	16	256	0.000531
3	47	32	512	0.000550
1	38	64	512	0.000552
3	55	64	512	0.000554
3	38	32	256	0.000555
1	47	16	512	0.000567
1	38	16	256	0.000571
3	55	16	256	0.000574
2	55	32	256	0.000593
3	47	64	512	0.000599
3	38	64	256	0.000604
3	47	16	256	0.000619

(continued)

Table 2 (continued)

Powder concentration (g/l)	Peak current (A)	Pulse-on time (μs)	Pulse-off time (μs)	Corrosion rate (mm/year)
1	55	32	512	0.000630
2	47	32	256	0.000638
2	55	64	256	0.000642
1	47	32	512	0.000675
1	55	64	512	0.000679
1	38	32	256	0.000679
3	55	32	256	0.000681
2	47	64	256	0.000687
1	55	16	256	0.000698

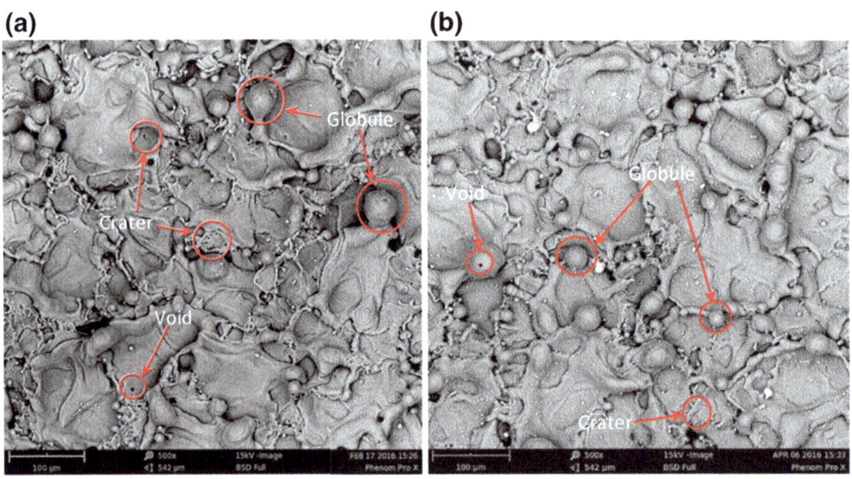

Fig. 4 SEM image of Mg alloy processed using **a** EDM and **b** PM-EDM

presented in Fig. 6 with 5.5%. Since Zn particles are suspended in the dielectric fluid during PMEDM process, the high localized temperature at the cutting zone causes some particles to melt and deposit on the machined surface.

The elemental distribution map of Mg alloy-machined using conventional EDM and PMEDM methods are presented in Fig. 7a, b. The map reveals that even though there are Zn particles melt and deposit on the Mg alloy-machined surface, the distribution of Zn is uneven. The role of Zn particles suspended in the dielectric fluid is not to coat the whole machined surface, but more on assisting in obtaining a smoother surface with better topological properties by providing a highly conductive bridge for homogeneous distribution of spark energy. Generally, Zn is known as a common alloying element for Mg and it is also an essential trace element in the human body.

The presence of Zn on the Mg alloy surface helps to reduce the releases of hydrogen gas when Mg reacts with the body fluid.

4.3 *Effect of Process Parameters on Corrosion Resistance*

All the data collected from the in vitro corrosion test have been thoroughly analyzed statistically. The experiment results reveal that the presence of Zn particles in the cutting zone provides the bridge for uniform discharge energy on the Mg alloy. It results in a lower surface roughness which then leads to a lower corrosion rate due to smaller exposed surface area to the solution. Figure 8 indicates that there are interactions among all four setting parameters during the PMEDM process on the corrosion rate of Mg alloy-machined surface. An interaction occurred when the effect of one setting parameter changes depending on the level of other setting parameters. In this case, all four setting parameters are depending on each other in influencing the response. It is clearly presented in the interaction plot. It occurs due to the combination of setting parameters which will be influencing the size of discharge energy.

Since the response output is influenced by the interaction amongst the setting parameters, an equation to determine the corrosion rate of Mg alloy-machined surface is generated by manipulating the value combination of those four setting parameters. For this purpose, all the collected data were analyzed using ANOVA, which are generated using Design–Expert software. Table 3 presents the ANOVA for PMEDM setting parameters on the corrosion rate of Mg alloy-machined surface. The sig-

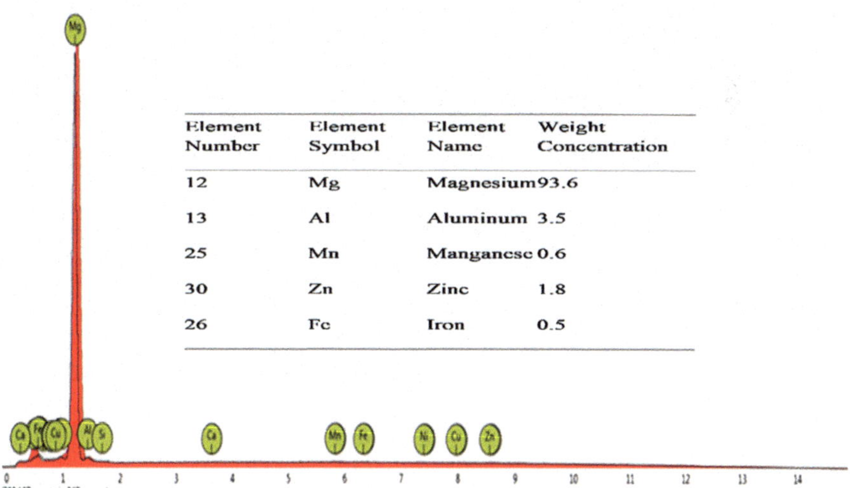

Element Number	Element Symbol	Element Name	Weight Concentration
12	Mg	Magnesium	93.6
13	Al	Aluminum	3.5
25	Mn	Manganese	0.6
30	Zn	Zinc	1.8
26	Fe	Iron	0.5

Fig. 5 EDX spectrum of Mg alloy machined using conventional EDM method

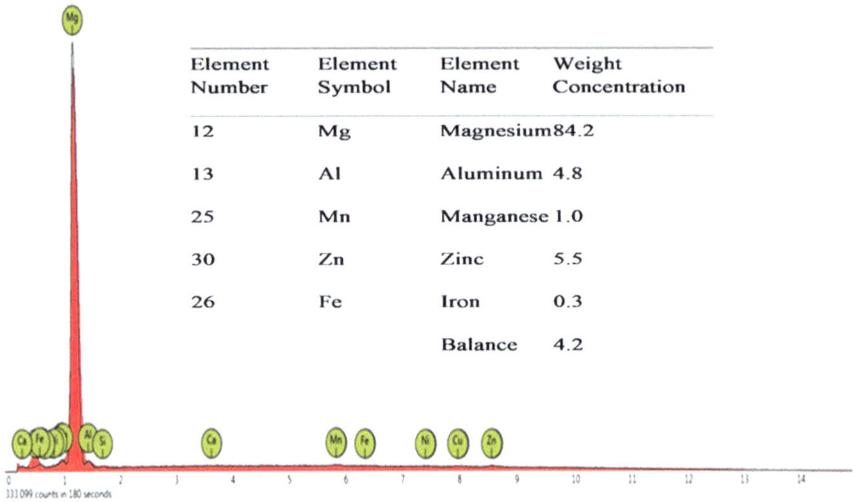

Element Number	Element Symbol	Element Name	Weight Concentration
12	Mg	Magnesium	84.2
13	Al	Aluminum	4.8
25	Mn	Manganese	1.0
30	Zn	Zinc	5.5
26	Fe	Iron	0.3
		Balance	4.2

Fig. 6 EDX spectrum of Mg alloy processed using PMEDM method

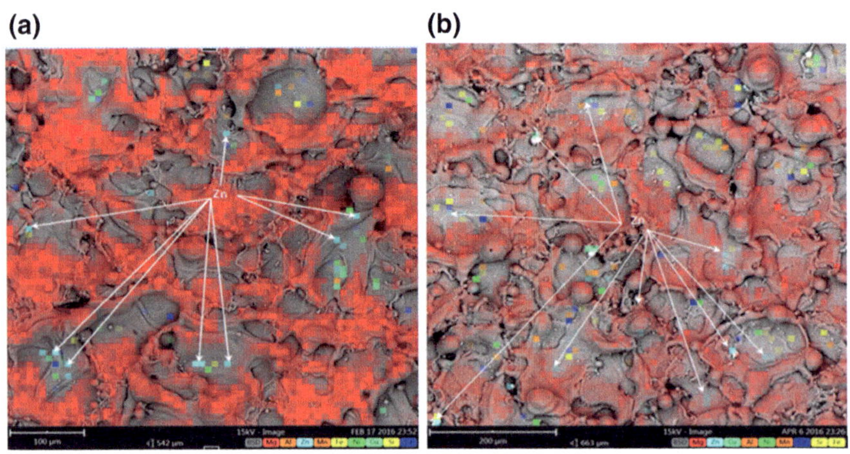

Fig. 7 Elemental distribution map of Mg alloy processed **a** EDM and **b** PMEDM

nificant and insignificant interaction can be identified by referring to the *p*-value. Those sources with *p*-value larger than 0.1 are insignificant. After removing all the insignificant sources, the final ANOVA was generated and the constants for the final equation in terms of actual factors were determined as follows:

$$\text{Corrosion rate}\ (\beta_O) = -7.81276 \times 10^{-4}$$

$$P_{\text{con}}\ (\beta_W) = -6.62901 \times 10^{-4}$$

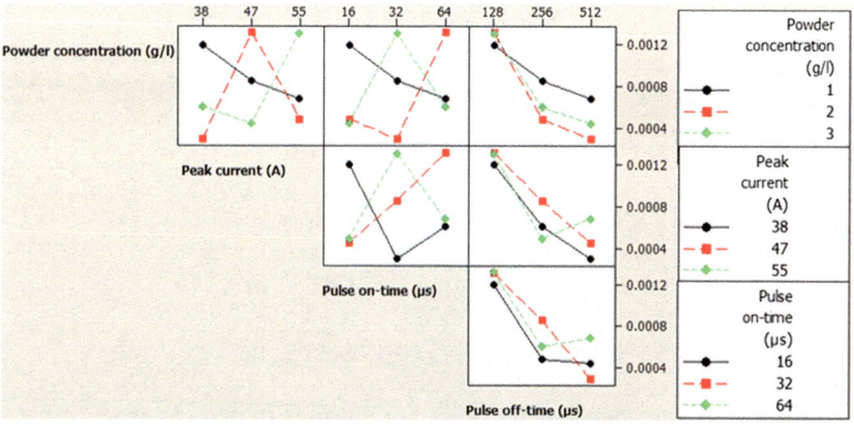

Fig. 8 PMEDM setting parameters interaction plots on the corrosion rate

Table 3 ANOVA for PMEDM setting parameters on the corrosion rate

Source	Sum of squares	df	Mean square	*p*-value
Model	3.50937E−06	14	2.5067E−07	<0.0001
$W - P_{con}$	6.93502E−08	1	6.935E−08	<0.0001
$X - P_c$	7.19852E−08	1	7.1985E−08	<0.0001
$Y - T_{on}$	1.10281E−07	1	1.1028E−07	<0.0001
$Z - T_{off}$	2.90676E−06	1	2.9068E−06	<0.0001
WX	9.15062E−17	1	9.1506E−17	0.9836
WY	2.97307E−15	1	2.9731E−15	0.9069
WZ	2.97307E−15	1	2.9731E−15	0.9069
XY	2.07711E−15	1	2.0771E−15	0.9222
XZ	2.07711E−15	1	2.0771E−15	0.9222
YZ	1.1551E−14	1	1.1551E−14	0.8180
W^2	5.74302E−08	1	5.743E−08	<0.0001
X^2	2.78402E−08	1	2.784E−08	<0.0001
Y^2	7.69767E−09	1	7.6977E−09	<0.0001
Z^2	3.25177E−07	1	3.2518E−07	<0.0001
Residual	2.06857E−12	10	2.0686E−13	
Cor total	3.50938E−06	24		

$$P_c\ (\beta_X) = 1.42548 \times 10^{-4}$$

$$T_{on}\ (\beta_Y) = 1.19244 \times 10^{-5}$$

$$T_{off}\ (\beta_Z) = -9.13054 \times 10^{-6}$$

$$\text{Squared } P_{con}\ (\beta_{WW}) = 1.50184 \times 10^{-4}$$

$$\text{Squared } P_c\ (\beta_{XX}) = -1.45265 \times 10^{-6}$$

$$\text{Squared } T_{on}\ (\beta_{YY}) = -1.08257 \times 10^{-7}$$

$$\text{Squared } T_{off}\ (\beta_{ZZ}) = 1.09936 \times 10^{-8}$$

Equation (1) presents the quadratic mathematical model.

$$\Upsilon = \beta_0 + \beta_W W + \beta_X X + \beta_Y Y + \beta_Z Z + \beta_{WW} W^2 + \beta_{XX} X^2 + \beta_{YY} Y^2 + \beta_{ZZ} Z^2 \quad (1)$$

By inserting the determined constants into the equation, the final equation is shown in Eq. (2), where Υ is the corrosion rate of Mg alloy-machined surface, W is the P_{con}, X is the P_c, Y is the T_{on}, and Z is the T_{off}.

$$\begin{aligned}\Upsilon = {} & \left(-7.81276 \times 10^{-4}\right) + \left(-6.62901 \times 10^{-4}\right)W + \left(1.42548 \times 10^{-4}\right)X \\ & + \left(1.19244 \times 10^{-5}\right)Y + \left(-9.13054 \times 10^{-6}\right)Z + \left(1.50184 \times 10^{-4}\right)W^2 \\ & + \left(-1.45265 \times 10^{-6}\right)X^2 + \left(-1.08257 \times 10^{-7}\right)Y^2 \\ & + \left(1.09936 \times 10^{-8}\right)Z^2 \end{aligned} \quad (2)$$

In the future, this formula can be used for PMEDM pre-machining setup to fabricate implants with a specific required surface corrosion rate. By changing the values of P_{con}, P_c, T_{on}, and T_{off}, the corrosion rate of Mg alloy-machined surface can be calculated. Therefore, a correct combination of PMEDM setting parameters can now be determined to obtain the required corrosion rate. A list of determined PMEDM setting parameters for a specific corrosion rate is provided in the Appendix, at the end of this chapter.

4.3.1 Verification Test

A verification test has been conducted to verify the formula by applying the setting parameters combination of 2 g/l P_{con}, 38 A P_c, 32 μs T_{on}, and 512 μs T_{off}. The forecasted corrosion rate was calculated using the established formula as shown below.

Table 4 Verification test

	P_{con} (g/l)	P_c (A)	T_{on} (μs)	T_{off} (μs)	Corrosion rate (mm/year)
Prediction	2	38	32	512	0.000291
Actual	2	38	32	512	0.000296

$$\begin{aligned} \Upsilon = {} & \left(-7.81276 \times 10^{-4}\right) + \left(-6.62901 \times 10^{-4}\right)2 + \left(1.42548 \times 10^{-4}\right)38 \\ & + \left(1.19244 \times 10^{-5}\right)32 + \left(-9.13054 \times 10^{-6}\right)512 + \left(1.50184 \times 10^{-4}\right)2^2 \\ & + \left(-1.45265 \times 10^{-6}\right)38^2 + \left(-1.08257 \times 10^{-7}\right)32^2 \\ & + \left(1.09936 \times 10^{-8}\right)512^2 \\ \Upsilon = {} & 0.000291 \text{ mm/year} \end{aligned}$$

From the calculation, the predicted corrosion rate of the Mg alloy-machined surface was 0.000291 mm/year. Meanwhile, the actual value obtained from the verification test was 0.000296 mm/year as shown in Table 4. It brought in 98.28% of similarity between the actual corrosion rate from the verification test and the predicted value. Since the similarity index is more than 95%, the credibility of this formula is highly accepted.

5 Conclusion

With all the evidence presented in this chapter, it is proven that the PMEDM method has the potential to be used to control the corrosion rate of Mg alloy-machined surface. In this research, the developed equation to forecast the corrosion rate of Mg alloy-machined surface has been verified by a verification test. It brought in 98.28% of similarity between the corrosion rate obtained from the verification test and the forecasted values. By applying the established equation, the corrosion rate of Mg alloy-machined surface can be forecasted by changing the values of PMEDM setting parameters. In the future, this equation can be used for PMEDM pre-machining setup to fabricate implants with a specific required surface corrosion rate. In terms of the role of Zn particles suspended in the dielectric fluid during PMEDM process, this research reveals that there are Zn particles melt and deposit on the specimen surface, but the distribution is inhomogeneous. The advantage of mixing Zn particles in the PMEDM process is providing a bridge for uniform discharge energy on the machined surface. It ensues in the lower surface roughness of Mg alloy which leads to a smaller exposed surface area to the solution during in vitro corrosion test. It results in a lower corrosion rate, which may benefit the biodegradable orthopedic implants application in the future.

References

1. Uddin M, Hall C, Murphy P (2015) Surface treatments for controlling corrosion rate of biodegradable Mg and Mg-based alloy implants. Sci Technol Adv Mater 16:053501
2. Helsen JA, Jürgen Breme H (1998) Metals as biomaterials. In: Helsen JA, Jürgen Breme H (eds) Metals as biomaterials, vol 1. Wiley-VCH, p 522. ISBN 0-471-96935-4
3. Avedesian MM, Baker H (1999) ASM specialty handbook: magnesium and magnesium alloys. ASM International
4. Witte F (2010) The history of biodegradable magnesium implants: a review. Acta Biomater 6:1680–1692
5. Caborn DN, Urban WP, Johnson DL, Nyland J, Pienkowski D (1997) Biomechanical comparison between BioScrew and titanium alloy interference screws for bone—patellar tendon—bone graft fixation in anterior cruciate ligament reconstruction. Arthrosc: The J Arthrosc Relat Surg 13:229–232
6. Timmer MD, Carter C, Ambrose CG, Mikos AG (2003) Fabrication of poly (propylene fumarate)-based orthopaedic implants by photo-crosslinking through transparent silicone molds. Biomaterials 24:4707–4714
7. Song G, Song S-Z (2007) A possible biodegradable magnesium implant material. Adv Eng Mater 9:298–302
8. Li L, Gao J, Wang Y (2004) Evaluation of cyto-toxicity and corrosion behavior of alkali-heat-treated magnesium in simulated body fluid. Surf Coat Technol 185:92–98
9. NADCA (2009) Alloy data. In: Product specification standard for die castings. ed: North American Die Casting Association
10. Somekawa H, Mukai T (2006) Fracture toughness in a rolled AZ31 magnesium alloy. J Alloys Compound 417:209–213
11. Chng C, Lau D, Choo J, Chui C (2012) A bioabsorbable microclip for laryngeal microsurgery: design and evaluation. Acta Biomater 8:2835–2844
12. Seitz JM, Wulf E, Freytag P, Bormann D, Bach FW (2010) The manufacture of resorbable suture material from magnesium. Adv Eng Mater 12:1099–1105
13. Li Z, Gu X, Lou S, Zheng Y (2008) The development of binary Mg–Ca alloys for use as biodegradable materials within bone. Biomaterials 29:1329–1344
14. Zhang S, Li J, Song Y, Zhao C, Zhang X, Xie C et al (2009) In vitro degradation, hemolysis and MC3T3-E1 cell adhesion of biodegradable Mg–Zn alloy. Mater Sci Eng, C 29:1907–1912
15. Zberg B, Uggowitzer PJ, Löffler JF (2009) MgZnCa glasses without clinically observable hydrogen evolution for biodegradable implants. Nat Mater 8:887–891
16. Ducheyne P (1984) Functional behavior of orthopedic biomaterials: applications. vol 2, CRC
17. Witte F, Fischer J, Nellesen J, Crostack H-A, Kaese V, Pisch A et al (2006) In vitro and in vivo corrosion measurements of magnesium alloys. Biomaterials 27:1013–1018
18. Klocke F, Schwade M, Klink A, Kopp A (2011) EDM machining capabilities of magnesium (Mg) alloy WE43 for medical applications. Proc Eng 19:190–195
19. Andani MT, Moghaddam NS, Haberland C, Dean D, Miller MJ, Elahinia M (2014) Metals for bone implants. Part 1. Powder metallurgy and implant rendering. Acta Biomater 10:4058–4070
20. Park H, Temenoff JS, Mikos AG (2007) Biodegradable orthopedic implants. Eng Funct Skelet Tissues 55–68
21. Nassif L, El Sabban M (2011) Mesenchymal stem cells in combination with scaffolds for bone tissue engineering. Materials 4:1793–1804
22. Olson JL, Atala A, Yoo JJ (2011) Tissue engineering: current strategies and future directions. Chonnam Med J 47:1–13
23. Temenoff JS, Mikos AG (2000) Injectable biodegradable materials for orthopedic tissue engineering. Biomaterials 21:2405–2412
24. Wong HM, Yeung KW, Lam KO, Tam V, Chu PK, Luk KD et al (2010) A biodegradable polymer-based coating to control the performance of magnesium alloy orthopaedic implants. Biomaterials 31:2084–2096

25. Yaszemski MJ, Payne RG, Hayes WC, Langer R, Mikos AG (1996) In vitro degradation of a poly (propylene fumarate)-based composite material. Biomaterials 17:2127–2130
26. Kannan MB, Raman RS (2008) In vitro degradation and mechanical integrity of calcium-containing magnesium alloys in modified-simulated body fluid. Biomaterials 29:2306–2314
27. Islam O, Soboleski D, Symons S, Davidson L, Ashworth M, Babyn P (2000) Development and duration of radiographic signs of bone healing in children. Am J Roentgenol 175:75–78
28. Jackson W, Theologis T, Gibbons C, Mathews S, Kambouroglou G (2007) Early management of pathological fractures in children. Injury 38:194–200
29. Giannoudis PV, Jones E, Einhorn TA (2011) Fracture healing and bone repair. Injury 42:549–550
30. Tiyyagura HR, Mohan T, Pal S, Mohan MK (2018) Surface modification of magnesium and its alloy as orthopedic biomaterials with biopolymers. Fundam Biomater: Metals 197–210
31. Song G, Atrens A (2003) Understanding magnesium corrosion—a framework for improved alloy performance. Adv Eng Mater 5:837–858
32. Amerstorfer F, Fischerauer SF, Fischer L, Eichler J, Draxler J, Zitek A et al (2016) Long-term in vivo degradation behavior and near-implant distribution of resorbed elements for magnesium alloys WZ21 and ZX50. Acta Biomater 42:440–450
33. Song GL, Atrens A (1999) Corrosion mechanisms of magnesium alloys. Adv Eng Mater 1:11–33
34. Uddin MS, Rosman H, Hall C, Murphy P (2017) Enhancing the corrosion resistance of biodegradable Mg-based alloy by machining-induced surface integrity: influence of machining parameters on surface roughness and hardness. The Int J Adv Manuf Technol 90:2095–2108
35. Prakash C, Singh S, Pabla B, Uddin M (2018) Synthesis, characterization, corrosion and bioactivity investigation of nano-HA coating deposited on biodegradable Mg–Zn–Mn alloy. Surfce Coat Technol 346:9–18
36. Singh B, Singh R, Mehta J, Gupta A, Prakash C (2018) Nano-mechanical characterization of Mg–Zn–Mn–Si alloy fabricated by spark plasma sintering for biomedical applications. Mater Today: Proc 5:27742–27748
37. Razak MA, Abdul-Rani AM, Nanimina AM (2015) Improving EDM efficiency with silicon carbide powder-mixed dielectric fluid. Int J Mater, Mech Manuf 3:40–43
38. Razak MA, Abdul-Rani AM, Rao TVVLN, Pedapati SR, Kamal S (2016) Electrical discharge machining on biodegradable AZ31 magnesium alloy using Taguchi method. Proc Eng 148:916–922
39. Prakash C, Kansal H, Pabla B, Puri S (2015) Potential of powder mixed electric discharge machining to enhance the wear and tribological performance of β-Ti implant for orthopedic applications. J Nanoeng Nanomanuf 5:261–269
40. Prakash C, Singh S, Verma K, Sidhu SS, Singh S (2018) Synthesis and characterization of Mg–Zn–Mn–HA composite by spark plasma sintering process for orthopedic applications. Vacuum 155:578–584
41. Ou S-F, Wang C-Y (2017) Effects of bioceramic particles in dielectric of powder-mixed electrical discharge machining on machining and surface characteristics of titanium alloys. J Mater Process Technol 245:70–79
42. Bajaj R, Tiwari AK, Dixit AR (2015) Current trends in electric discharge machining using micro and nano powder materials-a review. Mater Today: Proc 2:3302–3307
43. Kumar N, Ale P, Kumar N, Sharma S (2016) Study of PMEDM efficiency on HCHCr steel using silicon powder in dielectric fluid. Int J Emer Technol Eng Res 4:54–61
44. Prakash C, Uddin M (2017) Surface modification of β-phase Ti implant by hydroaxyapatite mixed electric discharge machining to enhance the corrosion resistance and in-vitro bioactivity. Surf Coat Technol 326:134–145
45. Prakash C, Kansal H, Pabla B, Puri S (2017) Experimental investigations in powder mixed electric discharge machining of Ti–35Nb–7Ta–5Zrβ-titanium alloy. Mater Manuf Process 32:274–285
46. Prakash C, Kansal H, Pabla B, Puri S (2016) Effect of surface nano-porosities fabricated by powder mixed electric discharge machining on bone-implant interface: an experimental and finite element study. Nanosci Nanotechnol Lett 8:815–826

47. Prakash C, Kansal H, Pabla B, Puri S (2016) Multi-objective optimization of powder mixed electric discharge machining parameters for fabrication of biocompatible layer on β-Ti alloy using NSGA-II coupled with Taguchi based response surface methodology. J Mech Sci Technol 30:4195–4204
48. Sales W, Oliveira A, Raslan A (2016) Titanium perovskite (CaTiO3) formation in Ti6Al4V alloy using the electrical discharge machining process for biomedical applications. Surf Coat Technol 307:1011–1015
49. Prakash C, Kansal HK, Pabla B, Puri S (2016) Powder mixed electric discharge machining: an innovative surface modification technique to enhance fatigue performance and bioactivity of β-Ti implant for orthopedics application. J Comput Inf Sci Eng 16:041006
50. Ponappa K, Aravindan S, Rao P, Ramkumar J, Gupta M (2010) The effect of process parameters on machining of magnesium nano alumina composites through EDM. The Int J Adv Manuf Technol 46:1035–1042
51. Walter R, Kannan MB, He Y, Sandham A (2013) Effect of surface roughness on the in vitro degradation behaviour of a biodegradable magnesium-based alloy. Appl Surf Sci 279:343–348
52. ASTM B90/B90M-98 (2001) Standard specification for magnesium alloy sheet and plate. In: Nonferrous metal products, ed. ASTM International, West Conshohocken, PA
53. Standard A (2009) G59–97: standard test method for conducting potentiodynamic polarization resistance measurements. ASTM International, West Conshohocken, PA
54. Bidhendi ARH, Pouranvari M (2011) Corrosion study of metallic biomaterials in simulated body fluid. Metalurgija 17:13–22

Mr. Muhammad Al'Hapis Abdul Razak is working as Lecturer in the Department of Mechanical Engineering, Universiti Kuala Lumpur (UniKL), Malaysia. He is pursuing Ph.D. under the supervision of Dr. Ahmad Majdi bin Abdul Rani, working as Associate Professor in the Department of Mechanical Engineering, Universiti Teknologi Malaysia, Malaysia. He has authored more than 20 research papers in various journal and international conferences.

Dr. Ahmad Majdi Abdul-Rani is working as Associate Professor in the Department of Mechanical Engineering, Universiti Teknologi Malaysia, Malaysia. His areas of research are materials and advanced manufacturing. Moreover, he is working in advanced manufacturing technology-EDM die-sinker; EDM W/cut; rapid prototyping/rapid tooling—MJM/investment cast tool/silicon rubber tooling; reverse engineering—laser digitzer, bio-medical measuring device; CNC; STEP-NC; CAD/CAM; CIM; FMS; deepwater technology—design and prototype fabrication of parts and components for deepwater facilities. He has authored more than 50 research papers. He is a reviewer of various national and international journals.

Mr. Abdul' Azeez Abdu Aliyu is working as Lecturer in Department of Mechanical Engineering, Bayero University Kano, Malaysia. He is pursuing Ph.D. under the supervision of Dr. Ahmad Majdi bin Abdul Rani, working as Associate Professor in the Department of Mechanical Engineering, Universiti Teknologi Malaysia, Malaysia. He has authored more than 10 research papers in various journal and international conferences.

Cortical Bone Adaptation to Mechanical Environment: Strain Energy Density Versus Fluid Motion

Abhishek Kumar Tiwari and Jitendra Prasad

Abstract This chapter presents an in silico study to compare the osteogenic potentials of normal strain-derived strain energy density (SED) and fluid shear. In vivo studies reported that mechanical loading promotes osteogenesis (i.e., new bone formation) at the sites elevated normal strain magnitude. Accordingly, in silico models assumed normal strain-derived SED as an osteogenic stimulus to predict the site-specific new bone formation. Nevertheless, there are in vivo studies where new bone formation is noticed at the sites of minimal normal strain magnitude especially near the neutral axis of bending. It is anticipated that SED as stimulus will have limited success in explaining such new bone distribution. Thus, there is no unifying principle that can relate the new bone formation to mechanical environment. A secondary component of mechanical environment, i.e., canalicular fluid flow derived shear, is reported as a potential stimulus of osteogenesis in the literature; however, their exact role is not well established. Therefore, this chapter presents an in silico model which studies site-specific new bone formation as a function of SED and fluid shear, individually and in their combination. The model simulates experimental new bone formation reported in different in vivo animal loading studies. The chapter also concludes that fluid shear closely fits the new bone formation near the minimal strain sites, and both SED and fluid shear contribute collectively to new bone formation. The findings presented in the chapter may be useful in the design of biomechanical strategies to cure bone loss and also in the improvement of the design of orthopedic implants.

Keywords Bone adaptation · Strain energy density · Fluid shear · New bone formation · Computer model

A. K. Tiwari (✉)
Department of Applied Mechanics, Motilal Nehru National Institute of Technology Allahabad, Allahabad 211004, Uttar Pradesh, India
e-mail: aktiwari@mnnit.ac.in

J. Prasad
Department of Mechanical Engineering, Indian Institute of Technology Ropar, Rupnagar 140001, Punjab, India

C. Prakash et al. (eds.), *Biomanufacturing*,
https://doi.org/10.1007/978-3-030-13951-3_12

1 Introduction

Postmenopausal women, bedridden patients, individuals with spinal cord injury or physical disability, and astronauts in microgravity environment experience bone loss [1]. Bone loss severely damages cortical and trabecular bone microarchitecture. For example, trabecular bone porosity increases and cortical bone area decreases in the case of bone loss which reduces the weight bearing strength of long bones and enhances fracture risks. Pharmaceutical drugs such as bisphosphonates, etc. are typically used to inhibit the bone loss; however, these drugs adversely affect bone-remodeling activities. Bisphosphonate retards the bone resorption which also reduces the self-renewal capability of bone. Therefore, their long-term use is not recommended [2]. Recent efforts in orthopedic research reported that weight-bearing exercises prevent and cure bone loss [3]. Weight-bearing exercise or external mechanical loading introduces mechanical deformations in bone which promotes osteogenesis (i.e., new bone formation). In this direction, several in vivo studies attempted to relate the sites of new bone formation to normal strain distribution [4–7].

2 In Vivo Versus In Silico Models

Invasive and non-invasive animal loading models are widely used to study bone adaptation. Based on loading applied, non-invasive models can be categorized as (i) cantilever bending [8], (ii) four-point bending [9], (iii) three-point bending [10], (iv) axial loading [6], and (v) knee loading [11] (Fig. 1). A variety of species such as rats, mice, roosters, turkey, primates, dogs, horses, sheep, and humans have been involved. These models applied dynamic mechanical loading on forelimbs such as ulna or radius, or hind limbs such as tibia or femur. The new bone formation is observed at mid-diaphyseal cross sections of loaded bone using histology or imaging techniques such as μ-CT [12]. In vivo studies reported that new bone formation often occurs at the sites of maximal normal strain magnitude. For example, Sugiyama et al. [13] applied axial loading on tibiae of C57BL/6J mice. They observed new bone formation at the sites of elevated strain magnitude of mid-diaphyseal cross sections of the tibia. Based on in vivo experiments, several in silico studies established a mathematical relationship between the new bone formation and the strain distribution. Computer models assumed normal strain- or strain-derived quantities such as strain energy density (SED) as the stimulus of osteogenesis to predict site-specific new bone formation [14, 15]. These models, however, may fall short in several instances. For example, several in vivo observed new bone formation at the sites of minimal normal strain magnitude specifically near the neutral axis of bending which are listed in Table 1 and also reported in [16].

3 Strain Energy Density Versus Fluid Motion

Numerical models established SED as a mechanical stimulus to predict the site-specific new bone formation. Moreover, it has also been suggested that the difference between the mechanical stimulus and an osteogenic threshold governs the site-specificity and the amount of new bone formation. This indicates that osteogenesis will occur at those locations where SED surpasses the osteogenic threshold. Therefore, in silico models, which consider SED as the stimulus of osteogenesis, may have limited success in fitting the new bone formation noticed at minimal strain sites reported in the in vivo studies listed in Table 1. This lack of correlation between the sites of new bone formation and the stimulus distribution indicates that secondary components of mechanical environment may have encouraged osteogenesis near the sites of minimal strain magnitude. Canalicular fluid flow and strain gradients are secondary components of mechanical environment. Gross and his co-workers [17] noticed that the sites of new bone formation at turkey ulna cortex poorly correlate with strain energy density distribution; whereas, it strongly correlates with strain gradients. Judex et al. [18] also substantiated that strain gradients regulate new bone formation near the sites of minimal strain magnitude. A computational study of Srinivasan et al. [19] reported that the fluid flow in circumferential direction closely correlates with the new bone formation noticed near the neutral axis of bending at a turkey ulna cross section. Thus, it is evident that secondary components (fluid flow or strain gradients) other than SED could be potential stimuli of new bone formation at such locations; nevertheless, their exact role is required to be investigated.

Qin et al. [20] observed that new bone distribution at avian ulna mid-diaphyseal cross section correlates with intramedullary pressure (ImP) and fluid velocity distributions. Qin et al. [21] also reported that pressure gradients developed within the bone drives interstitial fluid motion. Hu et al. [22] further reported that an increase in the hydraulic stimulation increases local intramedullary pressure and the new bone formation. It is observed that cortical bone experiences habitual bending due to its curvature which develops a pressure gradient. This allows the fluid to flow from compression to tension regions [23–25]. Weinbaum et al. [26] also explained that loading-induced fluid motion is an important mechanical signal for mechanotransduction since it imposes shear stress on osteocyte cells in osteogenic range. Based on these findings, several in silico, in vivo, and in vitro studies investigated the osteogenic potential of bending-induced fluid flow. For example, Carriero et al. [27] recently used a 3D fluorochrome mapping and poroelastic finite element model to investigate the spatial relationship between bone formation and the mechanical stimulus. They observed newly formed bone on periosteal and endocortical envelopes of cortical bone in the regions of high fluid motion. Local fluid velocity distribution closely predicted the regions of new bone formation especially at endocortical surface, whereas strain-dependent stimuli, e.g., strain energy density (SED), failed to explain the same. Tiwari et al. [28] also reported that strain energy, as a stimulus, does not precisely predict the enhancement in new bone formation as noticed in the case of rest-inserted loading, whereas canalicular fluid motion improves the prediction.

Table 1 Literature used for site-specific analysis between new bone distribution and corresponding strain distribution

S. no.	In vivo study	Animal strain, gender, and age	New bone distribution (reference)	Strain distribution (reference)	New bone distribution near neutral axis (present or absent?)
1.	Kuruvilla et al. [29]	C57BL/6, DBA/2, and C3H/He, female mice, 16 weeks	Figure 4A–C in Kuruvilla et al. [29]	Figure 3 in Wagner et al. [30]	Present
2.	LaMothe et al. [31]	C57BL/6, female mice, 16 weeks	Figure 2 in LaMothe et al. [31]	Figure 1 in Srinivasan et al. [32]	Present
3.	Lynch et al. [33]	C57BL/6, female mice, 26 weeks	Figure 3 in Lynch et al. [33]	Figure 6 in Yang et al. [34]	Present
4.	Matsumoto et al. [35] 2008	Sprague-Dawley, Male rats, 8 weeks	Figure 6 in Matsumuto et al. [35]	Figure 5 in Matsumuto et al. [35]	Present
5.	Roberts et al. [36]	Sprague-Dawley, Female rats, 10–11 months	Figure 2 in Roberts et al. [36]	Figure 5 in Akhter et al. [9]	Present
6.	Sakai et al. [37]	C57Bl/6, Male mice, 10 weeks	Figure 1b–e in Sakai et al. [37]	Figure 5 in Matsumuto et al. [35]	Present
7.	Weatherholt et al. [38]	C57Bl/6, female mice, 16 weeks	Figure 4 in Weatherholt et al. [38]	Figure 6 in Patel et al. [39]	Present
8.	Zhang [11]	C57Bl/6, female mice, 14 weeks	Figure 3 in Zhang [11]	Zhang et al. [11]	Present
9.	Mosley et al. [40]	Sprague-Dawley, Female rats, 43 days	Figure 4 in Mosley et al. [40]	Figure 6 in Mosley et al. [40]	Present
10.	Wallace et al. [41]	Juvenile Sheep, 40 days	Figure 1 in Wallace et al. [41]	Figure 2 in Wallace et al. [41]	Present
11.	Judex et al. [18]	Roosters, 1-year old	Figure 4 in Judex et al. [18]	Figure 2 in Judex et al. [18]	Present

(continued)

Table 1 (continued)

S. no.	In vivo study	Animal strain, gender, and age	New bone distribution (reference)	Strain distribution (reference)	New bone distribution near neutral axis (present or absent?)
12.	Gross et al. [17]	Adult male turkeys	Figure 2 in Gross et al. [18]	Figure 4 in Gross et al. [18]	Present
13.	Birkhold et al. [42]	C57Bl/6, female mice, 26 weeks	Figure 2D in Birkhold et al. [42]	Figure 6 in Yang et al. in [43] and Fig. 7 in Razi et al. [44]	Present
14.	Srinivasan et al. [45–47]	C57Bl/6, female mice 10 weeks, 22 months and 22 months in [45–47], respectively	Figures 2c, g, 3 and 5c in Srinivasan et al. [45–47], respectively	Figure 1 in Srinivasan et al. [32]	Present
15.	Srinivasan et al. [45]	Middle-aged turkeys, 2 year	Figure 2 in Srinivasan et al. [45]	Figure 1 in Srinivasan et al. [19]	Present

4 The Lacuna

In silico models preferred SED to fluid motion. Therefore, there are limited studies which attempted to establish the fluid motion as a mechanobiological stimulus. This chapter attempts to investigate the osteogenic potential of a most prevalent stimulus, i.e., SED and a recently reported stimulus, i.e., canalicular fluid flow. This chapter presents a computer model to relate SED and fluid shear distributions to in vivo new bone formation. The model computes the rate of new bone formation as a function of strain energy or fluid shear above their respective osteogenic thresholds. In vivo experimental new bone formation noticed in the studies listed in Table 2 is simulated using the model. It can be observed that cantilever, four-point, three-point bending, and axial loading models are used in these in vivo studies. This chapter also presents an idea for the development of a computationally easy model to simulate site-specific new bone formation.

Table 2 In vivo studies simulated using the model

Case	New bone distribution (reference)	Animal strain, gender, and age	Skeletal site	Mode of loading	Loading protocol (load (N), frequency (Hz), and cycles)
	Figure 3b in Srinivasan et al. [45]	C57BL/6, female mouse, 10 weeks old	Right tibia	Cantilever bending	0.5 N, 1 Hz, 500 cycles
	Figure 2g in Srinivasan et al. [47]	C57BL/6, female mouse, 16 weeks old	Right tibia	Cantilever bending	0.5 N, 1 Hz, 450 cycles
	Figure 2c in Akhter et al. [9]	C57BL/6, female mouse, 16 weeks old	Right tibia	Four point bending	5 N, 2 Hz, 324 cycles
	Figure 2 in Silva and Brodt [10]	SAMR1, male mice, 16–20 weeks old	Right tibia	Three point bending	15.1 N, 2.7 Hz, 300 cycles
	Figure 4 in Weatherholt et al. [38]	C57BL/6, female mouse, 16 weeks old	Right tibia	Axial loading	9 N, 2 Hz, 12 × 360 cycles

5 Mathematical Framework

This section presents the mathematical framework used in model development. The model computes new bone formation as a function of normal strain-derived strain energy density (SED) and fluid flow-induced shear stress. The calculation of SED and fluid shear is explained in the following subsections:

5.1 Calculation of Normal Strain and Strain Energy Density (SED)

Beam theory is used to calculate normal strain distribution at those mid-diaphyseal cross sections of tibia or ulna where new bone formation has been reported in the in vivo studies listed in Table 2. Tibiae or ulnae of C57/BL6 mice or rats are idealized as beams. In cantilever loading, tibia is fixed at the proximal end and the load is applied at the distal end (Fig. 1a), whereas three-point and four-point bending models applied the load on mid-diaphyseal section of the tibia with the supports at the proximal and the distal ends (Fig. 1b, c). Axial loading models supported long bones such as tibia or ulna at the proximal end and applied the load axially at the distal end (Fig. 1d, e). Boundary conditions are defined according to the loading configurations.

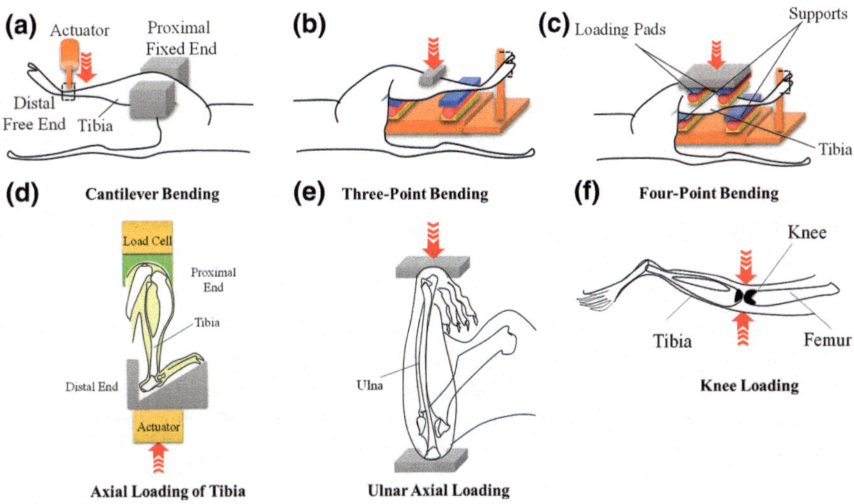

Fig. 1 In vivo animal loading models: **a** cantilever bending of tibia (adapted from Gross et al. [8]), **b** three-point bending (adapted from Silva and Brodt [10]), **c** four-point bending (adapted from Turner et al. [9]), **d** axial loading (adapted from De Souza et al. 2005 [6]), **e** ulnar axial loading (adapted from Sample et al. 2010 [7]), and **f** knee loading (adapted from Zhang et al. 2006 [11])

Histological image data reported in the simulated in vivo studies are used to idealize the cortex. These images are imported in MATLAB (The Mathworks Inc.). The centroid of the cortex is calculated. Equally spaced radial lines neighboring at 3° and passing through the centroid are used to divide the cross section. These lines intersect cortical envelopes (both periosteal and endocortical) at certain points. ImageJ, an open-source software, is used to manually extract these intersection points. These intersection points served as surface coordinates. It is assumed that linear piecewise curves passing through these surface coordinates envelop the cortical surfaces.

Accordingly, all the geometric properties such as cortical area and area moment of inertia are calculated.

Normal strain (ε_i) at cortical bone surfaces (periosteal and endocortical) is calculated as:

$$\varepsilon_i = \frac{M.y_i}{E.I_{\text{N.A.}}} \tag{1}$$

where M is the applied bending moment, y_i is the perpendicular distance from the neutral axis to any surface coordinate "i," $I_{\text{N.A.}}$ is the second area moment of inertia of the cortex about the neutral axis of bending passing through the centroid, and E is Young's modulus of the cortical bone. Cortical bone is considered as a linear-elastic isotropic material based on previous studies in the literature [13]. SED $\left(U_i^{\varepsilon}\right)$ is calculated using normal strain ε_i at each surface coordinate using the following relationships:

$$U_i^{\varepsilon} = \frac{1}{2} E \varepsilon_i^2 \quad (2)$$

5.2 Poroelastic Formulation

Bone can be modeled as a biphasic material due to the presence of various fluids at different levels of porosities. It is believed that applied stress develops pore-pressure gradient which permits fluid transport in bone. This is referred to as *solid-fluid coupling*. In addition, fluid pressure and fluid transport change bone tissue volume. This phenomenon is referred to as *fluid-solid coupling*. Poroelastic formulations are widely used to calculate pore fluid pressure and interstitial fluid motion in the bone under the influence of mechanical loading. Poroelasticity involves both solid and fluid deformations such that it can be expressed as a linear sum of mechanical stresses and pore-fluid pressure developed in bone. Accordingly, constitutive equations between solid matrix deformations [strain tensor (ε_{ij})] and fluid pore pressure (p) can be established as [48, 49]:

$$\varepsilon_{ij} = \frac{1}{2G}\left(\sigma_{ij} - \frac{\nu}{\nu_{\mathrm{u}}}\sigma_{kk}\delta_{ij}\right) + \frac{\alpha}{3K} p\delta_{ij} \quad (3)$$

$$\xi = \frac{\alpha}{K}\frac{\sigma_{kk}}{3} + \frac{\alpha}{KB} p \quad (4)$$

where ε_{ij} and σ_{ij} are strain and stress tensors; σ_{kk} is sum of normal stresses oriented in three Cartesian directions; $G(= E/2(1/\nu))$ is shear modulus and $K(= 2G(1/\nu)/3(1 - 2\nu))$ is bulk modulus of solid bone matrix under drained condition; ν and ν_{u} are Poisson's ratio for bone under drained and undrained conditions, respectively; $\alpha\left(= 1 - \frac{K}{K_{\mathrm{s}}}\right)$ is Biot effective stress coefficient, where K_{s} is bulk modulus of the solid bone matrix; ξ is the variation of fluid content; and B is Skempton coefficient which is defined as:

$$B = \frac{\frac{1}{K} - \frac{1}{K}}{\frac{\phi}{K_{\mathrm{f}}} + \frac{1}{K} - \frac{(1+\phi)}{K_{\mathrm{s}}}} \quad (5)$$

where K_{f} is bulk modulus of the fluid phase, and ϕ is interstitial porosity.

Biot's theory, strain compatibility equation, and diffusion equation are combined to model the poroelastic response. Strain compatibility and diffusion equations considered in the formulation are defined, respectively, as:

$$\nabla^2\left[\sigma_{kk} + \frac{2\alpha(1 - 2\nu)}{(1 - \nu)} P\right] = 0 \quad (6)$$

and

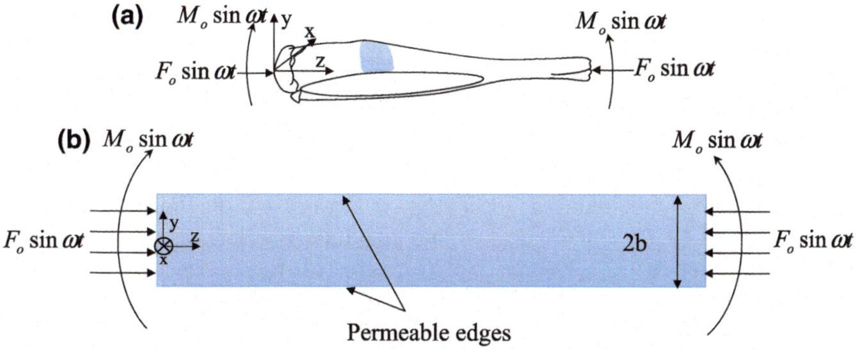

Fig. 2 **a** Mouse tibia subjected to a cyclic axial load and bending moment, and **b** the idealized poroelastic model of a bone tissue separated from mid-diaphysis of the tibia

$$c\nabla^2\left(\sigma_{kk}+\frac{3}{B}p\right)=\frac{\partial}{\partial t}\left(\sigma_{kk}+\frac{3}{B}p\right) \tag{7}$$

where $c=\frac{k}{\mu}\left[\frac{2GB^2(1-\nu)(1+\nu_u)^2}{9(1-\nu_u)(\nu_u-\nu)}\right]$ is diffusion coefficient, μ is the dynamic fluid viscosity, k is intrinsic permeability, and ν_u is undrained Poisson's ratio.

5.3 Boundary Conditions and Governing Equations

In the present work, cortical bone is modelled as poroelastic beam subjected to cyclic mechanical loads, e.g., axial loading $(F(t))$ and bending moment $(M(t))$ (Fig. 2). Cortical bone surfaces are assumed stress-free, drained, and permeable. This indicates fluid motion is possible across the cortical envelopes. Boundary conditions are defined for the bone subjected to both axial loading and bending moment as:

$$p(y=\pm b,t)=0 \tag{8}$$

$$\int_{-b}^{b}\sigma_{zz}\mathrm{d}y=F_o\ \sin\ \omega t \tag{9}$$

$$\int_{-b}^{b}y\sigma_{zz}\mathrm{d}y=-M_o\ \sin\ \omega t \tag{10}$$

where ω is the angular frequency of applied axial load (F_o) and moment (M_o), b is the cortex thickness, y is oriented perpendicular to the neutral axis of bending, and t is the time. Cortical bone has an anisotropic distribution of porosities. Thus, it is intuitive that bone may exhibit different poroelastic behavior in different directions. In the present work, an isotropic porosity distribution is idealized due to non-availability of experimental data on porosity distribution and to reduce computational complexity. In addition, there are in silico studies which have also assumed cortical bone as an isotropic poroelastic material [27].

It is assumed that stress and pore pressure are the function of y-coordinates and time t, respectively. In addition, shear stress is negligible in the bone cortex which implies that $\sigma_{yz}(y,t) = 0$. A plane strain condition is considered along with the x-direction which reduces the constitutive equations (6) and (7) as:

$$\frac{\partial^2}{\partial y^2}\left[\sigma_{zz} + \alpha\frac{(1-2\nu)}{(1-\nu)}p\right] = 0 \tag{11}$$

and

$$c\frac{\partial^2}{\partial y^2}\left[\sigma_{kk} + 3\frac{p}{B}\right] = \frac{\partial}{\partial t}\left[\sigma_{kk} + 3\frac{p}{B}\right] \tag{12}$$

To solve Eqs. (11) and (12), following dimensionless parameters are introduced:

$$y^* = \frac{y}{b};\ t^* = \frac{ct}{b^2};\ \sigma^*_{ij}\frac{2b^2\sigma ij}{M_o};\ p^*\frac{6pb^2}{B(1+\nu_u)M_o};\ H = \frac{(1-\nu)}{(\nu_u-\nu)};\ \Omega = \frac{b^2\omega}{c};\ \Lambda = \frac{3M_o}{bF_o}$$

where H is a dimensionless stress coefficient, Ω is dimensionless frequency, and Λ is axial-bending loading ratio.

Thus, Eqs. (11) and (12) can be defined as:

$$\frac{\partial^2}{\partial y^2}\left[\sigma^*_{zz} + p^*\frac{1}{H}\right] = 0 \tag{13}$$

$$\frac{\partial^2}{\partial y^2}(\sigma^*_{zz} + p^*) = \frac{\partial}{\partial t}(\sigma^*_{zz} + p^*) \tag{14}$$

5.4 Estimation of Pore-Fluid Pressure

Laplace transformation and residue method of inverse transformations are used to solve the differential equations subjected to following boundary conditions:

$$\int_{-1}^{1} \sigma^*_{zz}\mathrm{d}y^* = -2\sin\Omega t^* \tag{15}$$

$$\int_{-1}^{1} y^* \sigma_{zz}^* \mathrm{d}y^* = -\frac{2}{3}\Lambda \sin \Omega t^* \tag{16}$$

The solution for the pore pressure can be expressed as a sum of steady-state $\left(p^*_{\text{steady}}\right)$ and transient $\left(p^*_{\text{tran}}\right)$ responses as follows:

$$p^*(y^*, t^*) = p^*_{\text{steady}}(y^*, t^*) + p^*_{\text{tran}}(y^*, t^*) \tag{17}$$

where

$$p^*_{\text{steady}}(y^*, t^*) = (1 + \Lambda y^*) \sin \Omega t^* - \operatorname{Im}\left\{ \left(\frac{H\sqrt{i\Omega}\cosh\sqrt{i\Omega}y^* - \sinh\sqrt{\Omega i}}{H\sqrt{i\Omega}\cosh\sqrt{i\Omega} - \sinh\sqrt{i\Omega}} + \Lambda \frac{i\Omega\, H \sinh\sqrt{i\Omega}y^* + 3y^*(\sinh\sqrt{i\Omega} - \cosh\sqrt{i\Omega})}{i\Omega\, H \sinh\sqrt{i\Omega} + 3y^*(\sinh\sqrt{i\Omega} - \cosh\sqrt{i\Omega})} \right) \mathrm{e}^{i\Omega t^*} \right\} \tag{18}$$

and

$$p^*_{\text{tran}}(y^*, t^*) = 2\sum_{n=1}^{\infty} \frac{\Omega\,\lambda_n^2 \sin(\lambda_n \cos\lambda_n - \cos(\lambda_n y^*))}{(\lambda_n^4 + \Omega^2)(\lambda_n - \sin\lambda_n \cos\lambda_n)} \mathrm{e}^{-\lambda_n^2 t^*} + 2\Lambda \sum_{n=1}^{\infty} \frac{\Omega\,\mu_n^2(\sin\mu_n - \mu_n\cos\mu_n)(y^*\sin\mu_n - (\sin\mu_n y^*))}{(\mu_n^4 + \Omega^2)(\mu_n^2 + \mu_n \sin\mu_n \cos\mu_n - 2\sin^2\mu_n)} \mathrm{e}^{-\mu_n^2 t^*} \tag{19}$$

in which λ_n and μ_n are defined as:

$$\frac{\tan\lambda_n}{\lambda_n} = H \text{ and } \frac{\tan\mu_n}{\mu_n} = \frac{3}{3 + H\mu_n^2}$$

In vivo and in silico studies have reported that the transient pore-pressure $\left(p^*_{\text{tran}}\right)$ response diminishes within a fraction of a second of loading. Thus, steady-state pore-pressure $\left(p^*_{\text{steady}}\right)$ which typically varies with spatial locations is considered to calculate the fluid motion and the shear. The non-dimensional solution derived above is further converted into the dimensional form to obtain the physiological values of pore-pressure which is defined as:

$$p_{\text{steady}}(y, t) = \frac{MB(1 + \nu_{\mathrm{u}})}{6b^2} \left\{ (1 + \Lambda(y/b)) \sin\omega t - \operatorname{Im}\left\{ \left(\frac{H\sqrt{i\Omega}\cosh\sqrt{i\Omega}(y/b) - \sinh\sqrt{i\Omega}}{H\sqrt{i\Omega}\cosh\sqrt{i\Omega} - \sinh\sqrt{i\Omega}} + \Lambda \frac{i\Omega H \sinh\sqrt{i\Omega}(y/b) + 3(y/b)(\sinh\sqrt{i\Omega} - \cosh\sqrt{i\Omega})}{i\Omega H \sinh\sqrt{i\Omega} + 3(y/b)(\sinh\sqrt{i\Omega} - \cosh\sqrt{i\Omega})} \right) \mathrm{e}^{\mathrm{i}\omega t} \right\} \right\} \tag{20}$$

Similarly, steady-state pore-pressure can be derived for cantilever, four-point, and three-point bending models as indicated by Tiwari and Prasad [16] and Tiwari et al. [28]:

$$p_{\text{steady}}(y,t) = \frac{MB(1+\nu_u)}{6b^2}\left[\begin{array}{l}(1+3(y/b))\sin\omega t \\ -\mathrm{Im}\left\{\left(\begin{array}{l}\frac{H\sqrt{i\Omega}\cosh(\sqrt{i\Omega}(y/b))-\sinh\sqrt{i\Omega}}{H\sqrt{i\Omega}\cosh\sqrt{i\Omega}-\sinh\sqrt{i\Omega}} \\ +\frac{3i\Omega H\sinh\sqrt{i\Omega}(y/b)+3(y/b)(\sinh\sqrt{i\Omega}-\sqrt{i\Omega}\cosh\sqrt{i\Omega})}{i\Omega H\sinh\sqrt{i\Omega}+3(\sinh\sqrt{i\Omega}-\sqrt{i\Omega}\cosh\sqrt{i\Omega})}\end{array}\right)e^{i\omega t}\right\}\end{array}\right] \tag{21}$$

5.5 Calculation of Fluid Velocity and Fluid Shear

Pressure gradients developed between compression–tension regions guide canalicular fluid motion. The pressure gradient of steady pore-pressure $\left(\frac{\partial p}{\partial y}\right)$ is calculated in the y-direction at the cortex. Governing equation is formulated to calculate the fluid motion as:

$$\frac{1}{\mu}\frac{\partial p}{\partial y} = \left(\frac{1}{r}\frac{\partial}{\partial r}\left(r\frac{\partial u}{\partial r}\right)\right) + \frac{u}{k} \tag{22}$$

Boundary conditions used to obtain the solution for fluid motion are defined as follows:

$$u|_{r=R_m} = 0;\ u|_{r=R_w} = 0; \tag{23}$$

where r is the radial distance from the center of the canaliculus, R_m and R_w are the radii of cell process and canaliculus, respectively.

The solution of Eq. (22) defines fluid velocity (u) as [28]:

$$u = C_1 I_o\left(\frac{r}{\sqrt{k}}\right) + I_o\left(\frac{r}{\sqrt{k}}\right) - \frac{k}{\mu}\frac{\partial p}{\partial y} \tag{24}$$

where I_0 and K_0 are zero-order-modified Bessel functions of the first kind and second kind, respectively, and constants C_1 and C_2 are defined as:

$$C_1 = \frac{k}{\mu}\frac{\partial p}{\partial y}\frac{A_1}{A_2} \text{ and } C_2 = \frac{k}{\mu}\frac{\partial p}{\partial y}\frac{1}{A_2} \tag{25}$$

in which

$$A_1 = \frac{K_0\left(\frac{R_w}{\sqrt{k}}\right) - K_0\left(\frac{R_m}{\sqrt{k}}\right)}{I_0\left(\frac{R_w}{\sqrt{k}}\right) - I_0\left(\frac{R_m}{\sqrt{k}}\right)} \text{ and } A_2 = K_0\left(\frac{R_m}{\sqrt{k}}\right) + A_1 I_0\left(\frac{R_m}{\sqrt{k}}\right) \tag{26}$$

Fluid motion-induced shear stress (S) at osteocyte cell process can be calculated as:

$$S = \frac{\partial u}{\partial r}|r = R_m \tag{27}$$

which can be further expressed as:

$$S_i = \left(\frac{\sqrt{k}}{\mu}\frac{\partial p}{\partial y_i}\frac{A_1}{A_2}\right) I_1\left(\frac{R_m}{\sqrt{k}}\right) - \left(\frac{\sqrt{k}}{\mu}\frac{\partial p}{\partial y_i}\frac{1}{A_2}\right) K_1\left(\frac{R_m}{\sqrt{k}}\right) \tag{28}$$

6 The Model

6.1 Mechanobiological Stimulus

Strain energy density (SED) is a prime stimulus of osteogenesis. In contrast, fluid motion in lacunar space has also been reported osteogenic [50]. It is noticed that canalicular fluid motion induces shear stress on osteocyte cell body and process which is sensed by the osteocyte. These mechanical signals are then transformed into biochemical signals which are assumed responsible to initiate osteogenic activities at cortical envelopes. A numerical model is proposed to simulate experimental new bone formation as a function of strain energy density (SED) or fluid shear. Based on these assumptions, the mechanobiological stimuli (ψ) are defined as:

$$\psi_{\text{SED}} = U_i^{\varepsilon} \tag{29}$$

$$\psi_{\text{fluid}} = S_i \tag{30}$$

6.2 Bone Adaptation Law and Assumptions

Mechanobiological stimulus is calculated at cortical envelopes (periosteal and endocortical) of mid-diaphyseal cross sections. The model assumes that loading encourages new bone formation at those sites where the stimulus surpasses the reference stimulus. The difference between estimated stimulus (ψ) and the reference stimulus (ψ_{ref}) decides the site and the amount of new bone formation. The model utilizes the bone adaptation law to predict osteogenesis which is defined as [16, 28, 51]:

$$\frac{\partial c_i}{\partial t} = \left\{ \begin{array}{ll} C(\psi_i - \psi_{\text{ref}}), & \text{if } \psi_i > \psi_{\text{ref}} \\ 0, & \text{otherwise} \end{array} \right\} \tag{31}$$

where $\frac{\partial c_i}{\partial t}$ is rate of bone modeling and C is the rate constant.

This work uses similar formulations and assumptions made in the previously developed model of Tiwari et al. [28]. The assumptions are listed as:

i. The nature (e.g., compressive or tensile) of strain guides the new bone response [52].
ii. Endocortical surface experiences greater new bone formation as compared to periosteal surface [53].
iii. The model incorporates natural bone growth based on the differences in mineral apposition rate (MAR, μm/day) of loaded and intact groups [16].
iv. Osteocytes near the cortical surfaces sense the fluid motion/shear which activates osteogenic activity at the cortical envelopes [16].

Under the aforementioned assumptions, the growth law is modified as follows:

(i) *SED as the stimulus at the periosteal surface*:

$$\frac{\partial c_i^{\psi_U^{\text{p}}}}{\partial t} = \left\{ \begin{array}{ll} C_U^{\text{p}}(\psi_i^{\text{p}}(U_i) - \psi_{\text{ref}}^{\text{p}}(U_{\text{ref}})) + \Delta_i^{\text{p}}, & \text{if } \psi_i^{\text{p}}(U_i) > \psi_{\text{ref}}^{\text{p}}(U_{\text{ref}}) \\ \Delta_i^{\text{p}}, & \text{otherwise} \end{array} \right\} \tag{32}$$

and endocortical surface:

$$\frac{\partial c_i^{\psi_U^{\text{e}}}}{\partial t} = \left\{ \begin{array}{ll} C_U^{\text{e}}(\psi_i^{\text{e}}(U_i) - \psi_{\text{ref}}^{\text{e}}(U_{\text{ref}})) + \Delta_i^{\text{e}}, & \text{if } \psi_i^{\text{e}}(U_i) > \psi_{\text{ref}}^{\text{e}}(U_{\text{ref}}) \\ \Delta_i^{\text{e}}, & \text{otherwise} \end{array} \right\} \tag{33}$$

(ii) *Fluid shear as stimulus at periosteal surface*:

$$\frac{\partial c_i^{\psi_S^{\text{p}}}}{\partial t} = \left\{ \begin{array}{ll} C_S^{\text{p}}(\psi_i^{\text{p}}(S_i) - \psi_{\text{ref}}^{\text{p}}(U_{\text{ref}})) + \Delta_i^{\text{p}}, & \text{if } \psi_i^{\text{p}}(S_i) > \psi_{\text{ref}}^{\text{p}}(S_{\text{ref}}) \\ \Delta_i^{\text{p}}, & \text{otherwise} \end{array} \right\} \tag{34}$$

and endocortical surface:

$$\frac{\partial c_i^{\psi_S^{\text{e}}}}{\partial t} = \left\{ \begin{array}{ll} C_S^{\text{e}}(\psi_i^{\text{e}}(S_i) - \psi_{\text{ref}}^{\text{e}}(S_{\text{ref}})) + \Delta_i^{\text{e}}, & \text{if } \psi_i^{\text{e}}(S_i) > \psi_{\text{ref}}^{\text{e}}(S_{\text{ref}}) \\ \Delta_i^{\text{e}}, & \text{otherwise} \end{array} \right\} \tag{35}$$

where $\psi_i^{\text{p}}(U_i)$ and $\psi_i^{\text{e}}(U_i)$ are stimuli calculated as a function of SED (U_i) at any surface coordinate "*i*" of periosteal and endocortical surfaces, respectively; whereas, $\psi_i^{\text{p}}(s_i)$ and $\psi_i^{\text{e}}(s_i)$ represent the stimuli as a function fluid shear (S_i) at periosteal and endocortical surfaces, respectively. Δ_i is the parameter which takes care of the natural bone growth in the model. Δ_i^{p} and Δ_i^{e} are natural bone growth rates at any surface coordinate "*i*" of periosteal and endosteal surfaces, respectively. These

Table 3 Values of parameters used in the simulation

Parameters	Symbol	Units	Value
Young's modulus	E	GPa	17–20
Undrained Poisson's ratio	ν_u		0.33
Diffusion coefficient	c	(m^2/s)	5.1×10^{-7}
Skempton coefficient	B	–	0.367
Radius of canaliculus	R_w	nm	300
Radius of cell process	R_m	nm	130
Stress coefficient	α	–	0.14
Bulk modulus	K_f	MPa	2.3×10^3
Dynamic viscosity	μ	(Pa s)	8.9×10^{-4}
Reference stimulus	(i) $\psi^{p}_{ref}(U_{ref}) = \psi^{e}_{ref}(U_{ref})$ (ii) $\psi^{p}_{ref}(S_{ref}) = \psi^{e}_{ref}(S_{ref})$	J/m^3 dyne/cm^2	6400 6

Table 4 Rate coefficients calculated for each in vivo study during simulation

Case	Rate coefficients, C (unit^4/time step-Joule)			
	C^{p}_{U}	C^{e}_{U}	C^{p}_{S}	C^{e}_{S}
Cantilever	1×10^{-8}	1×10^{-8}	1×10^{-6}	1×10^{-6}
Cantilever	6.2×10^{-9}	1.2×10^{-9}	3×10^{-6}	1×10^{-6}
Four-point	1.48×10^{-9}	9.07×10^{-9}	2.9×10^{-7}	1.23×10^{-7}
Three-point	7.53×10^{-8}	5.22×10^{-8}	1.14×10^{-6}	9.79×10^{-7}
Axial	3.14×10^{-8}	1.33×10^{-8}	5.9×10^{-6}	2.3×10^{-6}

parameters are kept zero for those in vivo studies in which natural growth was absent at cortical surfaces in the control group. For the steps involved in the calculation of Δ_i, please refer to [28]. Literature indicates that normal strain of 800 $\mu\varepsilon$ and fluid shear of 6 dyne/cm^2 (0.6 Pa) are osteogenic at the cellular level. Based on these values, a reference stimulus (ψ_{ref}) is calculated. C^p and C^e represent the rate coefficients at periosteal and endosteal surfaces, respectively, which are calculated using the approach reported by Tiwari et al. [19]. An average mineral apposition rate (MAR) is calculated based on MAR data reported in the modelled in vivo studies. The averaged MAR is then divided by the difference between the peak stimulus and the reference stimulus to obtain rate coefficients (C^p and C^e). Material properties and other parameters used in the simulation are listed in Tables 3 and 4.

The model shifts surface coordinate "i" in the direction normal to the cortical surfaces at those locations where the stimulus surpasses the reference stimulus. The updated coordinate positions are calculated as:

$$X_i(t + dt) = X_i(t) + \frac{\partial c_i}{\partial t} dt \cos\theta \tag{36}$$

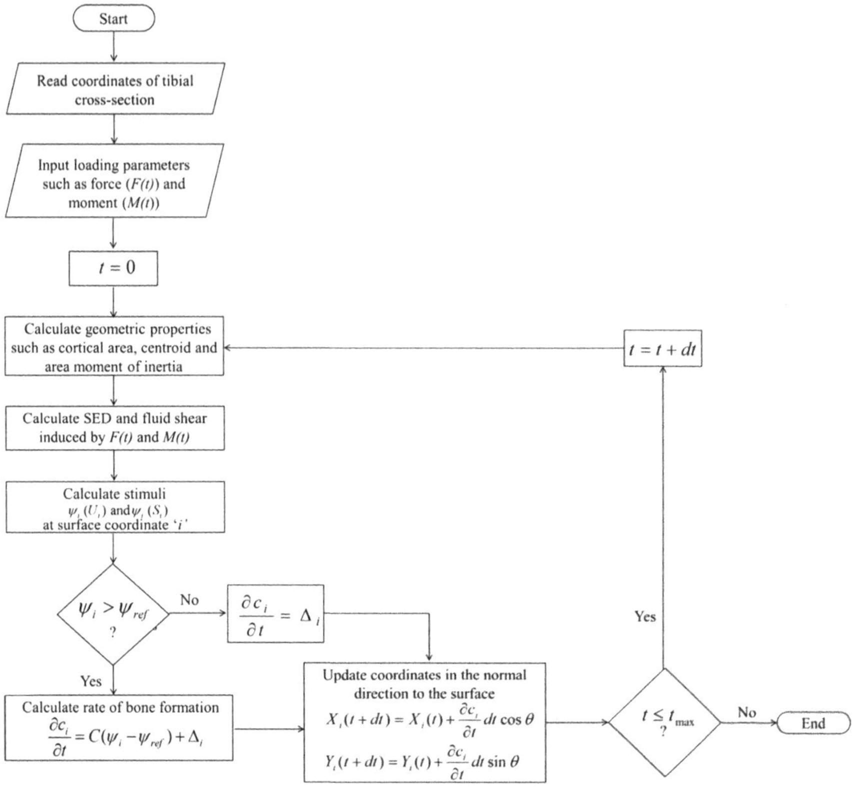

Fig. 3 Algorithm designed to simulate osteogenesis

$$Y_i(t + \mathrm{d}t) = Y_i(t) + \frac{\partial c_i}{\partial t}\mathrm{d}t \, \sin\theta \tag{37}$$

where "θ" is the angle made by the surface normal at any location i with the X-direction; $X_i(t)$ and $Y_i(t)$ are surface co-ordinates at any time t, whereas $X_i(t + \mathrm{d}t)$ and $Y_i(t + \mathrm{d}t)$ are the updated coordinates at time $t + \mathrm{d}t$. dt is time step size. The algorithm used to simulate the osteogenesis is presented in Fig. 3. Loading cycle time of simulated in vivo loading case is used to decide the iteration time and step size.

7 Results (In Silico Versus In Vivo)

The model computes strain energy density and fluid shear induced by loading. Strain magnitude and distributions calculated at mid-diaphyseal cross section closely align with experimental results. Fluid shear is noticed in a range of 6–20 dyne/cm^2 which is reported osteogenic [26]. The model simulates in vivo new bone formation noticed around mid-diaphyseal cross sections of long bones. SED and fluid-flow-induced shear stress are considered as stimuli of osteogenesis individually, as well as in their combination. In silico new bone formation is site-specifically compared with experimental new bone formation. Accordingly, the circumference of the cortex is divided using 36 radial lines (each 10° apart from the neighboring line). The intersection points made by radial lines with periosteal and endocortical surfaces are numbered in counter-clockwise direction, namely 1–36 at periosteal surface and $1'$–$36'$ at endocortical surface. New bone formation is finally compared at these points or sites.

Based on the loading configurations, simulation results can be described as follows:

7.1 Cantilever Bending

7.1.1 Case 1 (Srinivasan et al. [45])

The model predicts osteogenesis for an in vivo study of Srinivasan et al. [45], where, a tibia of a 10-week-old female C57BL/6J mouse was subjected to cyclic cantilever bending (0.5 N at 1 Hz for 500 cycles). The load is applied at the distal end in the medial-lateral plane. The proximal end of tibia is fixed (Fig. 4a). The new bone formation is noticed at mid-diaphyseal cross section located approximately at 1.8 mm proximal to tibia–fibula junction (Fig. 5a). Normal strain distribution in the cortex is estimated using beam theory as shown in Fig. 4b, whereas the fluid shear is estimated using Eq. (28).

In silico results explained in Fig. 5 indicate that *SED* as stimulus predicts the new bone formation away from the neutral axis at periosteal points 1–10 and 18–26, and endocortical points $3'$–$9'$ and $16'$–$23'$ which align with experimental results (Fig. 5a, b). The model does not fit new bone formation near the sites of minimal strain, namely at endocortical points $26'$–$35'$ and $10'$–$14'$ (Fig. 5b). When *fluid shear* is considered osteogenic, new bone is modeled near the neutral axis of bending, namely at periosteal points 12–17 and 29–35, and endocortical points $10'$–$17'$, $20'$–$23'$ and $27'$–$35'$ (Fig. 5c). However, the model fails to explain the experimental new bone formation present at the periosteal points 1–10 (Fig. 5a, c). In silico results improve when the osteogenic potentials of *fluid shear* and *SED* are added together (Fig. 5d). These findings substantiate that (i) fluid shear acts as a secondary stimulus of osteogenesis, and (ii) different components of mechanical environment contribute collectively to bone adaptation.

7.1.2 Case 2 (Srinivasan et al. [47])

The model predicts new bone formation noticed at the mid-diaphyseal cross section of a tibia of C57BL/6J female mouse (16-week-old) (Fig. 4c). Normal strain distribution computed can be observed in Fig. 4d. Computational results indicate that *SED* as a sole stimulus predicts osteogenesis at the sites away from the neutral axis, namely at periosteal points 1–9 and 16–25, and endocortical points 1′–9′ and 18′–25′ (Fig. 6b). The model predictions match within vivo observations except at endocortical points 1′–9′. In addition, the model fails to fit the experimental new bone formation at endocortical points 25′–36′ (Fig. 6b). The model with *fluid shear* as stimulus predicts osteogenesis near the neutral axis at periosteal and endocortical surfaces, namely at points 10–16 and 28–33, and 9′–18′ and 25′–36′, respectively (Fig. 6c). Out of these points, the new bone formation predicted at points 28–33 are found in accordance with experimental results. The model, however, does not predict experimental new bone distribution present at periosteal points 1–8 and 16–23 (Fig. 6c). When the osteogenic potentials of *SED and fluid shear* are added together, the results improve (Fig. 6d). The model overestimates the new bone formation at endocortical points 1′–17′ (out of 72 points).

7.1.3 Four-Point Bending (Akhter et al. [9])

The model simulates new bone formation for an in vivo four-point bending experiment. Akhter et al. [9] exposed tibia of C57BL/6J female mouse (16-week-old) to cyclic four-point bending (5 N at 2 Hz for 324 cycles) (Fig. 7a). This loading pro-

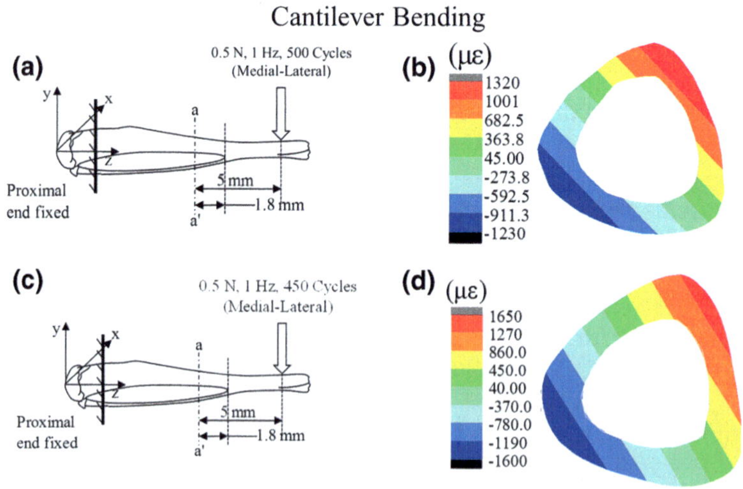

Fig. 4 Cantilever bending configurations and normal strain distributions calculated at mid-shaft section a–a′ for: (**a** and **b**) Srinivasan et al. [45]; and (**c** and **d**) Srinivasan et al. [47]

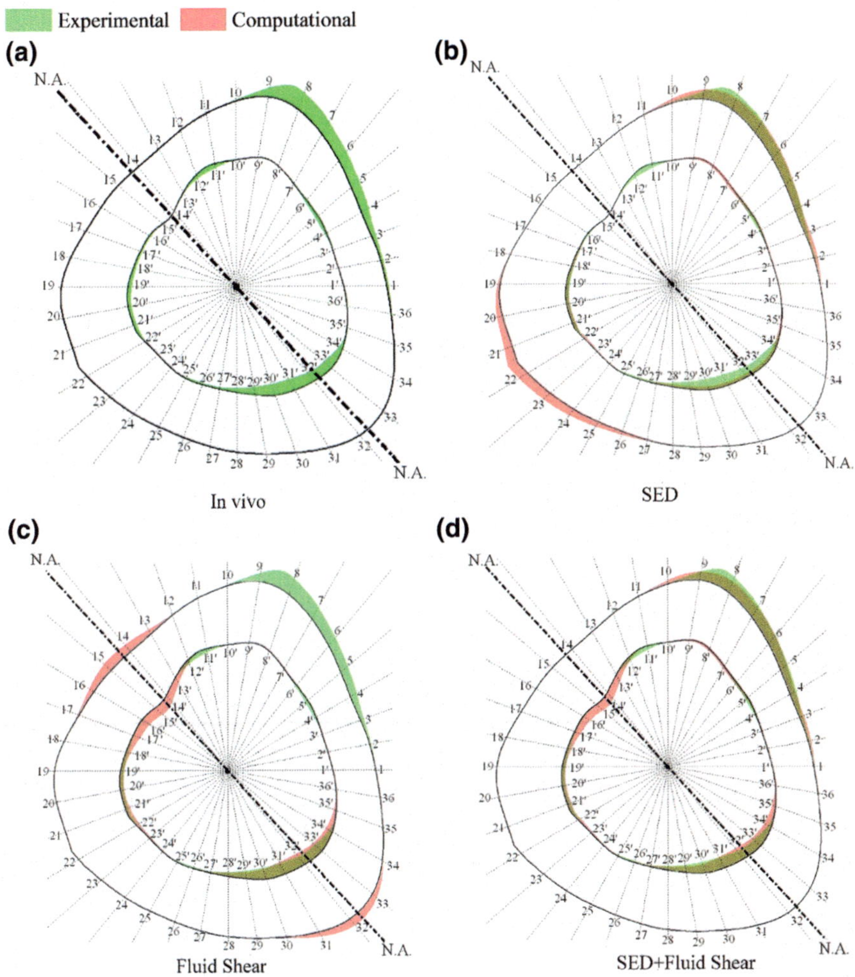

Fig. 5 Simulation versus experimental results: **a** in vivo new bone formation (green) in murine tibia (adapted from Fig. 3b of Srinivasan et al. [45]); and computer model predictions of osteogenesis (red) due to: **b** *SED*, **c** *fluid shear*, and **d** the combination of *SED and fluid shear* [16]. The overlap of experimental and computational osteogenesis is shown in red-green color mixture

duced a constant bending moment of 9.38 N-mm at mid-diaphyseal sections lying between the two loading points. Normal strain distribution induced due to loading is shown in Fig. 7b. The new bone distribution was noticed at mid-diaphyseal cross section, approximately 4 mm from tibia–fibula junction (Fig. 8a). *SED* and *fluid shear* distributions at mid-diaphyseal cross section are estimated to predict osteogenesis.

When *SED* is considered as stimulus, the model predicts new bone formation at periosteal points 5–14 and slightly at points 24–32, and endocortical points 6′–13 and 24′–32′ (Fig. 8b). Computational predictions align with experimental observa-

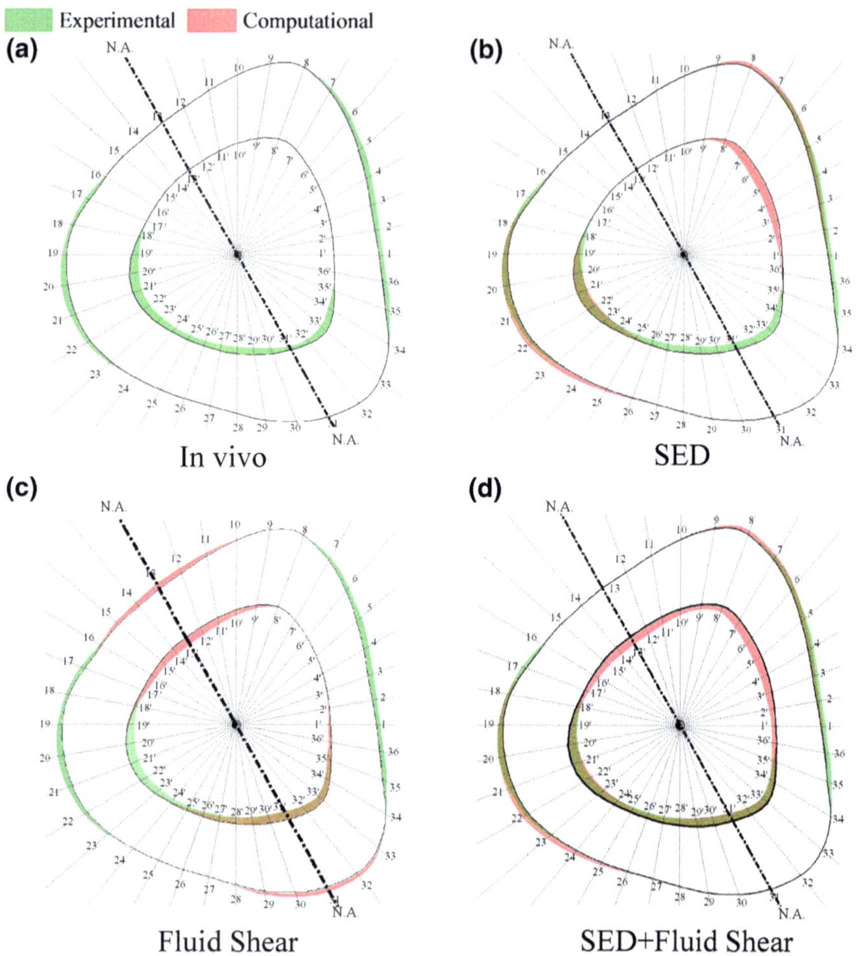

Fig. 6 Computational versus experimental results: **a** in vivo new bone formation (green) in murine tibia (adapted from Fig. 2g of Srinivasan et al. [47]); and in silico predictions of osteogenesis (red) due to: **b** *SED*, **c** *fluid shear*, and **d** when osteogenic potentials of *SED* and *fluid shear* are added

tions at periosteal points 5–14 and endocortical points 6′–13′. The model fails to fit experimental new bone distribution around the neutral axis at periosteal points 14–23, and endocortical points 1′–4′, 13′–28′ and 32′–36′ (Fig. 8b). *Fluid shear* as a sole osteogenic stimulus in the model predicts new bone formation at the sites of minimal normal strain or around the neutral axis, namely at periosteal points 1–4, 13–22, and 34–36, and endocortical points 1′–4′, 13′–24′, and 33′–36′. These locations align with in vivo observations (Fig. 8c). Fluid shear fails to fit in vivo new bone formation at periosteal points 5–14 and 24–32, and endocortical points 6′–13′ and 24′–32′, which occurred at the sites of peak normal strain magnitude (Fig. 8c). Computational results, however, improves when *SED* and *fluid shear* are considered

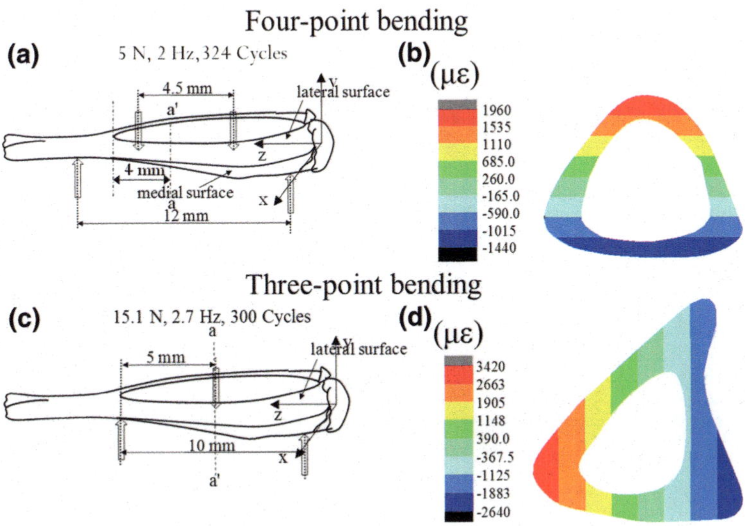

Fig. 7 Four-point and three-point bending, and normal strain distributions calculated at mid-shaft section a–a′ for: (**a** and **b**) Akhter et al. [9]; and (**c** and **d**) Silva and Brodt [10]

together (Fig. 8d). The model closely fits the in vivo new bone distribution with a slight overestimation at periosteal points 1–4 and 24–32 (Fig. 8d).

7.1.4 Three-Point Bending (Silva and Brodt [10])

The new bone distribution noticed in an in vivo three-point bending study of Silva and Brodt [10] is simulated. In the in vivo study, tibiae of SAMR1 female mice (16-week-old) were exposed to three-point bending (15.1 N at 2.7 Hz for 300 cycles) (Fig. 7c). The new bone formation is observed at mid-diaphyseal cross section near the loading point (Fig. 9a). *SED* is derived from normal strain distribution (Fig. 7d) and *fluid shear* is estimated using Eq. (28). When *SED* is considered as a sole osteogenic stimulus, the model predicts new bone formation at locations away from the neutral axis, namely at periosteal points 1–4, 15–24, and 32–36, and endocortical points 18′–25′ (Fig. 9b). A correlation can be observed with the sites of experimental new bone formation. *SED* again fails to explain the new bone formation present around neutral axis, namely at periosteal points 9–14, and endocortical points 1′–18′ and 26′–31′ (Fig. 9a). *Fluid shear* as the stimulus predicts newly formed bone near the minimal strain sites, namely at periosteal points 9–14 and 25–31, and at endocortical points 25′–36′ and 1′–16′ (Fig. 9c). Same as the previous cases, the model fails to fit experimental osteogenesis at the sites away from neutral axis. The model's predictability improves when *SED plus fluid shear* is assumed osteogenic (Fig. 9d). Out of 72 points, the model slightly overestimates new bone formation at periosteal points 25–31 and endocortical points 31′–34′ (Fig. 9d).

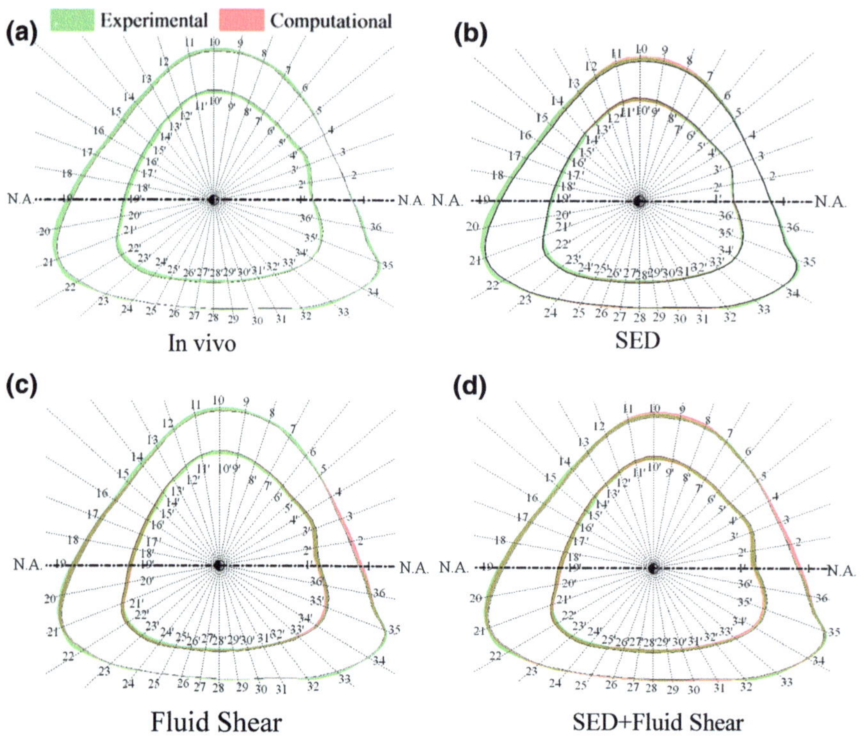

Fig. 8 In silico versus in vivo results: **a** new bone formation (green) at mid-diaphyseal cross section of mouse tibia (adapted from Fig. 5.2c of Akhter et al. [9]); and computational predictions of osteogenesis (red) due to: **b** *SED*, **c** *fluid shear*, and **d** when osteogenic effects of *SED* and *fluid shear* are added together

7.1.5 Axial Loading (Weatherholt et al. [38])

The model also attempts to predict osteogenesis for an axial loading experiment, where Weatherholt et al. [38] exposed tibia of a C57BL/6 female mouse (16-week-old) to an axial load of 9 N at 2 Hz for 4320 cycles (Fig. 10a, b). The new bone formation is reported at mid-diaphyseal cross section located at 7 mm from the proximal end of tibia (Fig. 11a). The model simulates osteogenesis at the cortex as a function of *SED* and fluid shear distribution. Figure 11b explains that *SED* as a stimulus forms new bone at periosteal points 1–3 and 27–36, and endocortical points 1–3 and 27′–36′, whereas the same stimulus fails to predict in vivo new bone formation near periosteal points 7–12 and 16–22, and at endocortical points 8′–14′ specifically near the neutral axis of bending. When *fluid shear* is considered osteogenic, the model predicts new bone formation at periosteal points 7–12 and 17–21, and endocortical points 10′–18′ (Fig. 11c). The model does not fit osteogenesis around the neutral axis, namely at periosteal points 1–3 and 27–36, and at endocortical points 1–3 and 27′–36′ (Fig. 11c). The combined stimulus, i.e., *SED plus fluid shear*,

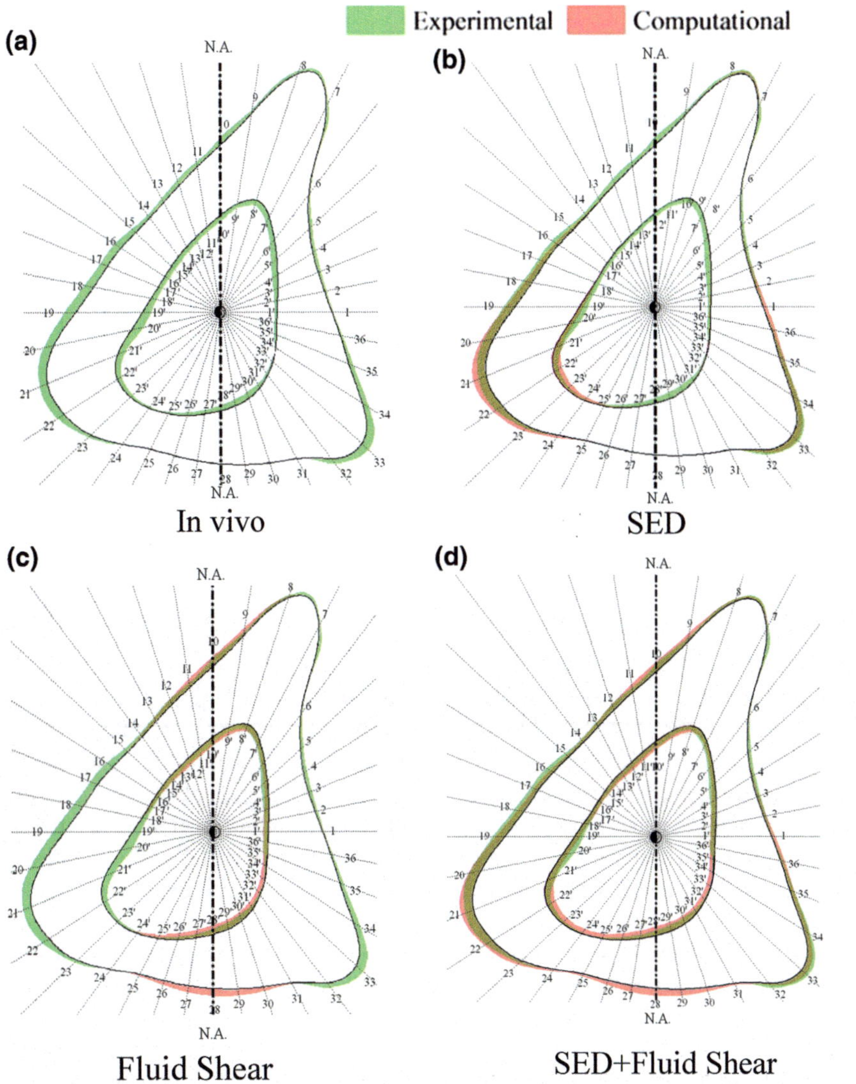

Fig. 9 In silico versus in vivo results: **a** new bone formation (green) at mid-diaphyseal cross section of mouse tibia (adapted from Fig. 2 of Silva and Brodt [10]); and computational predictions of osteogenesis (red) due to: **b** *SED*, **c** *fluid shear*, and **d** *SED plus fluid shear*

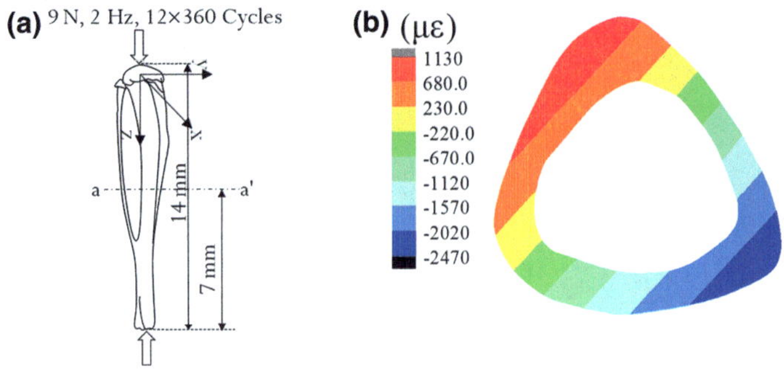

Fig. 10 Axial loading configurations and normal strain distributions calculated at mid-shaft section a–a′ for: (**a** and **b**) Weatherholt et al. [38]

closely predicts experimental new bone distribution with a slight overestimation of osteogenesis at periosteal points 18–21, and endocortical points 18′–21′ (Fig. 11d).

8 Major Findings and Advantages

This chapter presents an in silico investigation on osteogenic potentials of normal strain energy density (SED) and fluid motion-induced shear stress. The findings suggest that *SED* is partially successful as a stimulus in explaining the site-specific new bone formation. For example, *SED* fails to fit experimental new bone formation present near the neutral axis of bending. In contrast, fluid motion induced *shear* as a stimulus which predicts osteogenesis at the sites of minimal strain magnitude. Gross et al. [17] and Judex et al. [18] observed that circumferential strain gradient distribution (an indirect measurement of fluid flow) positively correlates with the sites of the new bone formation. Srinivasan et al. [54] also observed that the distribution of fluid motion in circumferential direction correlates with the sites of new bone formation near the neutral axis of bending of an avian ulna cross section. Thus, experimental results also align with the computational results presented in the chapter. In silico results presented in this chapter, however, affirm that neither *SED* nor fluid shear solely guides the new bone formation. In fact, both the stimuli collectively regulate osteogenesis. The predictability improves in most of the cases when osteogenic potentials of *SED* and *fluid shear* are combined. Thus, two or more stimuli, e.g., normal strain or strain energy density (SED) with fluid motion, fluid shear, or strain gradients, may be required as the stimulus to precisely fit the mechanical loading guided site-specific new bone formation. Finite element (FE) models have been widely used to study the cortical bone adaptation and remodeling. A finite element bone remodeling simulation involves large computation time since it calculates the stimulus at a large number of nodes repeatedly with the iterations. The method presented in this chapter is computationally easy and requires less time as compared

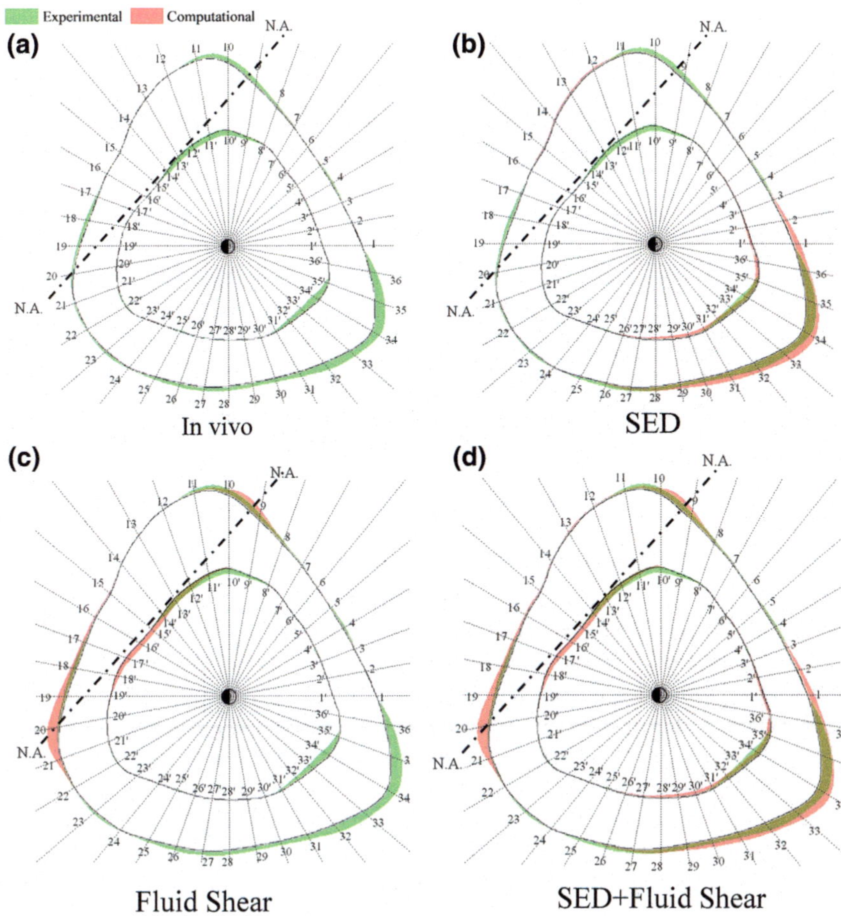

Fig. 11 In silico versus in vivo results: **a** in vivo new bone formation (green) at mid-diaphyseal cross section of mouse tibia (adapted from Fig. 4 of Weatherholt et al. [38]); and computational predictions of osteogenesis (red) due to: **b** *SED*, **c** *fluid shear*, and **d** when osteogenic effects of *SED* and *fluid shear* are combined

to FE approaches; nevertheless, it can be used only to simulate the new bone formation only at bone cross sections. Thus, there is a scope to develop a simple and robust model to predict new bone formation along the length of the bone. SED does not involve any time-dependent parameters such as frequency, number of cycles, and time period which significantly affect cortical bone adaptation. The computation of fluid motion, however, includes these parameters as input. Thus, the incorporation of fluid motion or fluid shear in the computer models may be useful in simulating the effect of loading parameters on the new bone formation. Poroelastic formulations presented in the chapter may also be useful in the calculation of fluid motion to establish a relationship between the fluid flow and cortical bone adaptation. Each stimulus opts a unique mechanism to pursue mechanotransduction. For example, osteocyte, a mechanosensory cell, and osteoblast directly sense the bone matrix strain. This amplifies the calcium signaling and the new bone formation. Cytoskeleton of bone cells restructures in response to mechanical strain to down- or up-regulate the bone remodeling activities. In addition, bone tissue deformations generate pressure gradients which govern the interstitial fluid flow through the osteocyte's canalicular network. This canalicular flow imparts shear stress on the cell process and the cell membrane of osteocytes. Signaling molecules such as prostaglandin (PGE2) and NO are released in response to fluid shear. These molecules establish intercellular communication between osteocytes and osteoblasts to initiate the new bone formation. This indicates that both normal strain-derived SED and fluid motion-induced shear are important stimuli of osteogenesis which must be incorporated in the in silico models to capture the characteristics of cellular modeling activities.

Hellmich and co-workers [55] recently reported that hydrostatic pressure also stimulates the new bone formation. Hydrostatic pressure, however, has not been widely accepted as the stimulus in orthopedic research community and requires more investigations on its osteogenic potential. Nullend et al. [56] pointed out that fluid shear is more effective than the hydrostatic pressure in eliciting a bone cell response. This could be one of the reasons that the researchers overlooked the hydrostatic pressure as the mechanobiological stimulus. Wang et al. [57] also explained that the loading-induced fluid motion in a murine cortical bone correlates with the sites of new bone formation. L Fan et al. [58] suggested a strong correlation between loading-induced fluid distribution and the sites of new bone formation. In a recent study, Carriero et al. [27] mentioned that strain energy density distribution correlates only with the periosteal new bone formation, whereas fluid velocity distribution closely associates with the osteogenesis at both the cortical surfaces. These findings also substantiate that the fluid motion may be incorporated as mechanobiological stimulus in the in silico models of bone adaptation.

9 Limitations and Future Scope

The model has certain limitations which are presented here. For example, the adaptation law used in the model framework considers bone growth and neglects bone resorption process since the overall effect noticed in the in vivo studies is the new bone formation. In addition, the poroelastic formulation assumes intrinsic permeability of the bone constant at the macroscopic level; however, it may vary at the microscopic level with the locations. Bone cell sensitization or desensitization, which affects bone adaptation, is not considered in the adaptation law. The bone's adaptability to saturate after a certain increase in the strain magnitude, frequency, or load cycles is not effectively modeled. The model may fail to explain those cases where the new bone formation is enhanced when rest is inserted between loading cycles. In a recent study, Tiwari et al. [28], however, incorporated the bone mechanosensitivity as a function of fluid motion and pore-pressure relaxation time to model osteogenesis. The model does not include the effect of distribution of lacunar porosity and vascular porosity which may affect the fluid flow within the bone. The model assumes that osteocytes located near the cortical surfaces sense the mechanical signals. These osteocytes receive signals from other osteocytes, integrate them, and then communicate these signals to lining cells or surface osteoblasts through the canalicular network to initiate bone-modeling activities. The model considers the local effect of the fluid flow which is to be estimated across the whole osteocyte network. Kerschnitzki et al. [59] presented a strong correlation between osteocyte network distribution and mineralized bone which may be useful in improving the model. However, the in vivo studies modeled did not report any data of mineralized bone, osteocyte network, micro-CT data, or density distribution (required to characterize the mineralized bone and osteocyte network) for murine bone cortex. This may be taken as a constraint in incorporating the osteocyte network in the model. Thus, the osteocyte network is overlooked in the model to keep the computation easy.

10 Concluding Remarks

This chapter concludes that strain energy density and fluid shear both are the important stimuli to model osteogenesis. This consideration may be useful in establishing a unifying principle to predict loading-induced osteogenesis. A robust computer model can also be developed in future which can predict the site-specific new bone formation. Based on such findings better biomechanical strategies such as prophylactic exercises may be designed to stimulate or guide the new bone formation at the sites of bone loss. In addition, improved computational strategies may be designed to understand the osteogenesis around bone-implant interface which ultimately may be useful in improving the design and the performance of orthopedic implants

References

1. Lau RY, Guo X (2011) A review on current osteoporosis research: with special focus on disuse bone loss. J Osteoporos 2011:1–6
2. Kennel KA, Drake MT (2009) Adverse effects of bisphosphonates: implications for osteoporosis management. Mayo Clin Proc 84(7):632–638
3. Rubin C, Recker R, Cullen D, Ryaby J, McCabe J, McLeod K (2003) Prevention of postmenopausal bone loss by a low-magnitude, high-frequency mechanical stimuli: a clinical trial assessing compliance, efficacy, and safety. J Bone Miner Res 19:343–351
4. Rubin CT, Lanyon LE (1985) Regulation of bone mass by mechanical strain magnitude. Calcif Tissue Int 37:411–417
5. Kotha SP, Hsieh Y-F, Strigel RM, Müller R, Silva MJ (2004) Experimental and finite element analysis of the rat ulnar loading model—correlations between strain and bone formation following fatigue loading. J Biomech 37:541–548
6. De Souza RL, Matsuura M, Eckstein F, Rawlinson SC, Lanyon LE, Pitsillides AA (2005) Non-invasive axial loading of mouse tibiae increases cortical bone formation and modifies

trabecular organization: a new model to study cortical and cancellous compartments in a single loaded element. Bone 37:810–818
7. Sample SJ, Collins RJ, Wilson AP, Racette MA, Behan M, Markel MD, Kalscheur VL, Hao Z, Muir P (2010) Systemic effects of ulna loading in male rats during functional adaptation. J Bone Miner Res 25:2016–2028
8. Gross TS, Srinivasan S, Liu CC, Clemens TL, Bain SD (2002) Noninvasive loading of the murine tibia: an in vivo model for the study of mechanotransduction. J Bone Miner Res 17:493–501
9. Akhter MP, Cullen DM, Recker RR (2002) Bone adaptation response to sham and bending stimuli in mice. J Clin Densitom 5:207–216
10. Silva MJ, Brodt MD (2008) Mechanical stimulation of bone formation is normal in the samp6 mouse. Calcif Tissue Int 82:489–497
11. Zhang P, Tanaka SM, Jiang H, Su M, Yokota H (2006) Diaphyseal bone formation in murine tibiae in response to knee loading. J Appl Physiol 100:1452–1459
12. Meakin LB, Price JS, Lanyon LE (2014) The contribution of experimental in vivo models to understanding the mechanisms of adaptation to mechanical loading in bone. Front Endocrinol. https://doi.org/10.3389/fendo.2014.00154
13. Sugiyama T, Price JS, Lanyon LE (2010) Functional adaptation to mechanical loading in both cortical and cancellous bone is controlled locally and is confined to the loaded bones. Bone 46:314–321
14. Chennimalai Kumar N, Dantzig JA, Jasiuk IM (2012) Modeling of cortical bone adaptation in a rat ulna: effect of frequency. Bone 50:792–797
15. Hambli R, Katerchi H, Benhamou C-L (2011) Multiscale methodology for bone remodelling simulation using coupled finite element and neural network computation. Biomech Model Mechanobiol 10:133–145
16. Tiwari AK, Prasad J (2017) Computer modelling of bone's adaptation: the role of normal strain, shear strain and fluid flow. Biomech Model Mechanobiol 16:395–410
17. Gross TS, Edwards JL, Mcleod KJ, Rubin CT (1997) Strain gradients correlate with sites of periosteal bone formation. J Bone Miner Res 12:982–988
18. Judex S, Gross TS, Zernicke RF (1997) Strain gradients correlate with sites of exercise-induced bone-forming surfaces in the adult skeleton. J Bone Miner Res 12:1737–1745
19. Srinivasan S, Gross T (1999) Canalicular fluid flow in bone: a basis for bone formation at sites of minimal strain. In: 45th annual meeting, Orthopaedic Research Society, 1999
20. Qin YX, Lin W, Rubin C (2001) The relationship between bone fluid flow and adaptation as stimulated by intramedullary hydraulic loading. Trans Orth Res Soc 26:319
21. Qin Y-X, Kaplan T, Saldanha A, Rubin C (2003) Fluid pressure gradients, arising from oscillations in intramedullary pressure, is correlated with the formation of bone and inhibition of intracortical porosity. J Biomech 36:1427–1437
22. Hu M, Cheng J, Bethel N, Serra-Hsu F, Ferreri S, Lin L, Qin Y-X (2014) Interrelation between external oscillatory muscle coupling amplitude and in vivo intramedullary pressure related bone adaptation. Bone 66:178–181
23. Steck R, Niederer P, Tate MK (2000) A finite difference model of load-induced fluid displacements within bone under mechanical loading. Med Eng Phys 22:117–125
24. Tate MLK, Knothe U, Niederer P (1998) Experimental elucidation of mechanical load-induced fluid flow and its potential role in bone metabolism and functional adaptation. Am J Med Sci 316:189–195
25. Tate MK, Steck R, Forwood M, Niederer P (2000) In vivo demonstration of load-induced fluid flow in the rat tibia and its potential implications for processes associated with functional adaptation. J Exp Biol 203:2737–2745
26. Weinbaum S, Cowin S, Zeng Y (1994) A model for the excitation of osteocytes by mechanical loading-induced bone fluid shear stresses. J Biomech 27:339–360
27. Carriero A, Pereira AF, Wilson AJ, Castagno S, Javaheri B, Pitsillides AA, Marenzana M, Shefelbine SJ (2018) Spatial relationship between bone formation and mechanical stimulus within cortical bone: combining 3D fluorochrome mapping and poroelastic finite element modelling. Bone Rep 8:72–80

28. Tiwari AK, Kumar R, Tripathi D, Badhyal S (2018) In silico modeling of bone adaptation to rest-inserted loading: strain energy density versus fluid flow as stimulus. J Theor Biol 446:110–127
29. Kuruvilla SJ, Fox SD, Cullen DM, Akhter MP (2008) Site specific bone adaptation response to mechanical loading. J Musculoskelet Neuronal Interact 8:71–78
30. Wagner DW, Chan S, Castillo AB, Beaupre GS (2013) Geometric mouse variation: implications to the axial ulnar loading protocol and animal specific calibration. J Biomech 46:2271–2276
31. LaMothe JM (2004) Rest insertion combined with high-frequency loading enhances osteogenesis. J Appl Physiol 96:1788–1793
32. Srinivasan S, Gross T (2013) Concurrent optimization of Cyclosporin A and mechanical loading identifies multiple optima to rescue senescentbone adaptation. In: Orthopaedic Research Society, 2013
33. Lynch ME, Main RP, Xu Q, Schmicker TL, Schaffler MB, Wright TM, van der Meulen MCH (2011) Tibial compression is anabolic in the adult mouse skeleton despite reduced responsiveness with aging. Bone 49:439–446
34. Yang H, Embry RE, Main RP (2017) Effects of loading duration and short rest insertion on cancellous and cortical bone adaptation in the mouse tibia. PLoS ONE 12:e0169519
35. Matsumoto HN, Koyama Y, Takakuda K (2008) Effect of mechanical loading timeline on periosteal bone formation. J Biomech Sci Eng 3:176–187
36. Roberts MD, Santner TJ, Hart RT (2009) Local bone formation due to combined mechanical loading and intermittent hPTH-(1-34) treatment and its correlation to mechanical signal distributions. J Biomech 42:2431–2438
37. Sakai D, Kii I, Nakagawa K, Matsumoto HN, Takahashi M, Yoshida S, Hosoya T, Takakuda K, Kudo A (2011) Remodeling of actin cytoskeleton in mouse periosteal cells under mechanical loading induces periosteal cell proliferation during bone formation. PLoS ONE 6:e24847
38. Weatherholt AM, Fuchs RK, Warden SJ (2013) Cortical and trabecular bone adaptation to incremental load magnitudes using the mouse tibial axial compression loading model. Bone 52:372–379
39. Patel TK, Brodt MD, Silva MJ (2014) Experimental and finite element analysis of strains induced by axial tibial compression in young-adult and old female C57Bl/6 mice. J Biomech 47:451–457
40. Mosley J, March B, Lynch J, Lanyon L (1997) Strain magnitude related changes in whole bone architecture in growing rats. Bone 20:191–198
41. Wallace I, Demes B, Mongle C, Pearson O, Polk J, Carrier D et al (2014) Exercise-induced bone formation is poorly linked to local strain magnitude in the sheep tibia. PLoS ONE 9(6):e99108
42. Birkhold AI, Razi H, Duda GN, Weinkamer R, Checa S, Willie BM (2014) Mineralizing surface is the main target of mechanical stimulation independent of age: 3D dynamic in vivo morphometry. Bone 66:15–25
43. Yang H, Butz KD, Duffy D, Niebur GL, Nauman EA, Main RP (2014) Characterization of cancellous and cortical bone strain in the in vivo mouse tibial loading model using microCT-based finite element analysis. Bone 66:131–139
44. Razi H, Birkhold AI, Zaslansky P, Weinkamer R, Duda GN, Willie BM, Checa S (2015) Skeletal maturity leads to a reduction in the strain magnitudes induced within the bone: a murine tibia study. Acta Biomater 13:301–310
45. Srinivasan S, Weimer DA, Agans SC, Bain SD, Gross TS (2002) Low-magnitude mechanical loading becomes osteogenic when rest is inserted between each load cycle. J Bone Miner Res 17:1613–1620
46. Srinivasan S, Agans SC, King KA, Moy NY, Poliachik SL, Gross TS (2003) Enabling bone formation in the aged skeleton via rest-inserted mechanical loading. Bone 33:946–955
47. Srinivasan S, Ausk BJ, Prasad J, Threet D, Bain SD, Richardson TS, Gross TS (2010) Rescuing loading induced bone formation at senescence. PLoS Comput Biol 6:e1000924
48. Kameo Y, Adachi T, Hojo M (2009) Fluid pressure response in poroelastic materials subjected to cyclic loading. J Mech Phys Solids 57:1815–1827
49. Biot MA (1955) Theory of elasticity and consolidation for a porous anisotropic solid. J Appl Phys 26:182–185

50. Pereira AF, Javaheri B, Pitsillides A, Shefelbine S (2015) Predicting cortical bone adaptation to axial loading in the mouse tibia. J R Soc Interface 12:20150590
51. Kumar NC, Dantzig JA, Jasiuk IM, Robling AG, Turner CH (2010) Numerical modeling of long bone adaptation due to mechanical loading: correlation with experiments. Ann Biomed Eng 38:594–604
52. Burr DB, Allen MR (eds) (2013) Basic and applied bone biology. Elsevier/Academic Press, Amsterdam
53. Birkhold AI, Razi H, Duda GN, Weinkamer R, Checa S, Willie BM (2016) The periosteal bone surface is less mechano-responsive than the endocortical. Sci Rep 6:23480
54. Srinivasan S, Gross T (2000) Canalicular fluid flow induced by bending of a long bone. Med Eng Phys 22:127–133
55. Liu J, Shao XJ, Liu YS, Yue ZF (2008) Effect of cold expansion on fatigue performance of open holes. Mater Sci Eng, A 477:271–276
56. Klein-Nulend J, Van der Plas A, Semeins C, Ajubi N, Frangos J, Nijweide P, Burger E (1995) Sensitivity of osteocytes to biomechanical stress in vitro. FASEB J 9:441–445
57. Wang L, Wang Y, Han Y, Henderson SC, Majeska RJ, Weinbaum S, Schaffler MB (2005) In situ measurement of solute transport in the bone lacunar-canalicular system. Proc Natl Acad Sci USA 102:11911–11916
58. Fan L, Pei S, Lucas LuX, Wang L (2016) A multiscale 3D finite element analysis of fluid/solute transport in mechanically loaded bone. Bone Res 4:16032
59. Kerschnitzki M, Kollmannsberger P, Burghammer M, Duda GN, Weinkamer R, Wagermaier W, Fratzl P (2013) Architecture of the osteocyte network correlates with bone material quality. J Bone Miner Res 28:1837–1845

Dr. Abhishek Kumar Tiwari is an Assistant Professor with Department of Applied Mechanics at Motilal Nehru National Institute of Technology (MNNIT) Allahabad, Prayagraj, Uttar Pradesh. He completed his degree of Bachelor of Technology (Honours) in Mechanical Engineering from Uttar Pradesh Technical University, Lucknow, Uttar Pradesh in 2011. He then joined Department of Mechanical Engineering at Indian Institute of Technology (IIT) Ropar for doctoral studies. The topic of his doctoral thesis was "Computer modeling of bone's adaptation: The role of normal and shear strains, fluid shear and strain gradients". He graduated with the degree of Doctor of Philosophy in 2017. Later on, his doctoral work also appeared in one of the leading newspaper "The Telegraph" with title "Bone Again". He then worked as Assistant Professor with Department of Mechanical Engineering at Manipal University Jaipur till April 2018 and then moved to MNNIT Allahabad. His research group at MNNIT Allahabad is interested in exploring the mechanobiology involved in bone's adaptation, canalicular fluid dynamics within the bone, and design of orthopedic implants and instruments. His research interests also lie bone fracture healing, poromechanics, finite element simulations, and computational fluid dynamics.

Dr. Jitendra Prasad is an Assistant Professor at Indian Institute of Technology (IIT) Ropar. He received his B.Tech. (Honours) in Ocean Engineering & Naval Architecture from IIT Kharagpur in 1997. His undergraduate research project was entitled "Analysis of ship structures by using three dimensional finite element method" and was supervised by Prof. Madhujit Mukhopadhyay. He then worked as Assistant Ship Surveyor at Research & Training Center of Indian Register of Shipping at Powai, Mumbai for four years. His work there was focused on ship structural analysis including computer programming, structural plan approval and software approval etc. He then did his Master of Engineering (by Research) at Nanyang Technological University (NTU), Singapore in the Department of Mechanical & Aerospace Engineering. His Masters thesis was entitled "Design of compliant mechanisms by evolutionary optimization of topology and shape" and was guided by Prof. Tai Kang. In 2003, he joined Michigan State University to do his Ph.D. in Mechanical Engineering under guidance of Prof. Alejandro R. Diaz and graduated in 2007. His

Ph.D. thesis was entitled “Design of materials with special dynamic properties using negative stiffness components”. He then joined Ortheopaedic Science Lab supervised by Prof. Ted S. Gross in the department of Orthopaedics & Sports Medicine at University of Washington, Seattle. His postdoctoral research was focused on bone fracture healing and bone’s adaptation to mechanical loading. Since Jitendra Prasad joined IIT Ropar in April 2011, he has taught numerous courses such as Material Science & Engineering, Biomechanics, Theory of Machines, Finite Element Analysis, Solid Mechanics, Medical Devices & Equipment etc. He has also developed and taught Biomechanics Lab, Medical Devices & Equipment Lab, and Technology Museum Lab. He has been especially interested in curriculum development and pedagogy, and was the chairman of the committee that developed new curriculum, which has been implemented at IIT Ropar starting 2017. This curriculum was also one of the subjects of his research publication. Jitendra Prasad’s research group at IIT Ropar is focused on development of novel theoretical and computer models of bone fracture healing and bone’s adaptation to exogenous mechanical loading. He has also collaborated with Prof. Nirmal Raj Gopinathan from Post Graduate Institute of Medical Education and Research (PGIMER) Chandigarh to work on implants for Osteogenesis Imperfecta (OI) patients. This collaboration has led to three patents (that have been applied) for the novel design of implants for OI patients. Jitendra Prasad was the founding member of the Biomedical Engineering (BME) Program at IIT Ropar and officially served as the Coordinator of the BME Program, which was the precursor to the Center of Biomedical Engineering at IIT Ropar. As the Coordinator, he built a molecular biology research lab and also started collaboration of IIT Ropar with PGIMER Chandigarh.

Printed by Printforce, the Netherlands